Gamma-Hydroxybutyrate

Gamma-Hydroxybutyrate

Molecular, functional and clinical aspects

Edited by
Godfrey Tunnicliff
and Christopher D. Cash

Routledge
Taylor & Francis Group

LONDON AND NEW YORK

First published 2002
by Taylor & Francis

2 Park Square, Milton Park, Abingdon, Oxfordshire OX14 4RN
52 Vanderbilt Avenue, New York, NY 10017

Routledge is an imprint of the Taylor & Francis Group, an informa business

First issued in paperback 2019

Typeset in 10/12 Baskerville by
Newgen Imaging Systems (P) Ltd, Chennai, India

British Library Cataloguing in Publication Data
A catalogue record for this book is available
from the British Library

Library of Congress Cataloging in Publication Data
A catalog record for this book has been requested

ISBN 978-0-415-28499-8 (hbk)
ISBN 978-0-367-39596-4 (pbk)

Contents

Figures

Tables

Contributors

R. Agabio, 'Bernard B. Brodie' Department of Neuroscience, University of Cagliari, I-09124 Cagliari, Italy.

C. Andriamampandry, Institut de Chimie Biologique and INSERM U 338, 11 rue Humann, 67085 Strasbourg, France.

D. Aunis, Institut de Chimie Biologique and INSERM U 338, 11 rue Humann, 67085 Strasbourg, France.

P. K. Banerjee, Department of Molecular and Medical Pharmacology, Mental Retardation Research Unit and Brain Research Institute, University of California, 760 Westwood Plaza, Los Angeles, CA 90024, USA.

R. Bernasconi, INSERM Unit 398, Neurobiologie et Neuropharmacologie des Epilepsies Généralisées, F-67091 Strasbourg, France.

B. Bettler, Department of Physiology, Biozentrum/Pharmazentrum, Medical Faculty, University of Basel, CH-4056 Basel, Switzerland.

S. Bischoff, Novartis Pharma AG, TA Nervous System, CH-4002 Basel, Switzerland.

C. D. Cash, U416 du CNRS, 11 rue Humann 67085 Strasbourg, France.

G. Colombo, CNR Institute of Neurogenetics and Neuropharmacology, 'Bernard B. Brodie' Department of Neuroscience, University of Cagliari, I-09042 Monserrato (CA), Italy.

V. Crunelli, School of Biosciences, Cardiff University, Museum Avenue, Cardiff CF10 3US, UK.

G. L. Gessa, CNR Center for Neuropharmacology, Cagliari, Italy; 'Bernard B. Brodie' Department of Neuroscience, University of Cagliari, I-09124 Cagliari, Italy.

K. M. Gibson, Departments of Molecular and Medical Genetics and Pediatrics, Oregon Health Sciences University, 2525 SW 3rd Avenue, Portland, OR 79201, USA.

S. Gobaille, Institut de Chimie Biologique and INSERM U 338, 11 rue Humann, 67085 Strasbourg, France.

R. Godbout, Centre de Biomédicine, Hôpital du Sacré-Coeur, Université de Montréal, Montreal, Quebec H4J 1C5, Canada.

S. G. Howard, Department of Molecular and Medical Pharmacology, Mental Retardation Research Unit and Brain Research Institute, University of California, 760 Westwood Plaza, Los Angeles, CA 90024, USA.

D. Hyndman, Department of Biochemistry, Queens University, Kingston, Ontario K7L 3N6, Canada.

E. E. Kaufman, Laboratory of Cerebral Metabolism, NIMH, NIH, Building 36, Room 1A-20, 9000 Rockville Pike, Bethesda, MD 20892, USA.

V. Kemmel, Institut de Chimie Biologique and INSERM U 338, 11 rue Humann, 67085 Strasbourg, France.

N. Leresche, UMR CNRS Neurobiologie des Processus Adaptalifs, Université Pierre et Marie Curie, 9 quai Saint-Bernard, F-75005 Paris, France.

C. Lobina, 'Bernard B. Brodie' Department of Neuroscience, University of Cagliari, I-09124 Cagliari, Italy.

M. Maitre, Institut de Chimie Biologique and INSERM U 338, 11 rue Humann, 67085, Strasbourg, France.

M. Mamelak, Department of Psychiatry, University of Toronto, Baycrest Centre for Geriatric Care, 3560 Bathurst Street, Toronto, Ontario M6A 2E1, Canada.

C. Marescaux, INSERM Unit 398, Neurobiologie et Neuropharmacologie des Epilepsies Généralisées, F-67091 Strasbourg, France.

P. Mathivet, INSERM Unit 398, Neurobiologie et Neuropharmacologie des Epilepsies Gènèralisèes, F-67091 Strasbourg, France.

J. Montplaisir, Department of Psychiatry, Centre de Recherche de Montréal, Hôpital du Sacré-Coeur, Université de Montréal, Montreal, Quebec H4J 1C5, Canada.

U. Otten, Department of Physiology, University of Basel, Vesalanium, Vesalgasse 1, CH-4051 Basel, Switzerland.

B. U. Raess, Department of Pharmacology, Indiana University School of Medicine, 8600 University, Boulevard, Evansville, IN 47712, USA.

O. C. Snead, Departments of Pediatrics, Medicine and Pharmacology, University of Toronto, Hospital for Sick Children, 555 University Avenue, Toronto, Ontario M5G 1X8, Canada.

G. Tunnicliff, Department of Biochemistry & Molecular Biology, Indiana University School of Medicine, 8600 University Boulevard, Evansville, IN 47712, USA.

Preface

Some of us who had been following the scientific progress of γ-hydroxybutyrate (GHB) for several decades were surprised to encounter this drug on the national news and in popular news magazines a few years ago. Its 'recreational' properties had been discovered. Thus the focus of GHB seemed to have moved from the research laboratory to the streets. Not long ago the US Food and Drug Administration designated GHB a Schedule I drug. It was all very engaging for neuroscientists studying this intriguing compound.

GHB has come a long way from those early experiments in animals where it was found to induce a sleep-like state, and from its use in general anesthesia in human subjects. Soon it was found to be a naturally occurring compound in the brain, a metabolite of GABA, the emerging ubiquitous inhibitory neurotransmitter. This opened up a completely new line of research in an effort to establish a function for GHB. One idea that followed was the possible regulation of central dopaminergic activity. At the same time, details of the enzymes involved in the metabolism of GHB began to be uncovered. Further biochemical studies revealed a sodium-dependent transport system and the presence of a specific, high-affinity binding site which now seems to represent a plasma membrane receptor. Results from other experiments suggested that the GABA$_B$ receptor might be an additional mediator of GHB action. As a drug, GHB has certain features of a CNS depressant and, physiologically, it has many characteristics of an inhibitory neurotransmitter. One noteworthy pharmacological property of GHB is that it can effect absence seizure-type activity in animals. This has led to an important model of petit mal epilepsy. Recently it has gained a reputation as a popular drug of abuse with the attendant toxicological risks. Nevertheless, it might have some redeeming features as a drug since it is currently being investigated for clinical utility in sleep disorders, drug addiction and fibromyalgia. Indeed, it is anticipated that GHB will soon be marketed in the United States as a treatment for narcolepsy.

This book has brought together the combined expertise of many of the leading authorities on the biochemistry, physiology and pharmacology of GHB in the CNS. Each chapter is an in-depth review of the field. In the final chapter, Professor Michel Maitre and colleagues digest and integrate the key features of our knowledge of GHB and attempt to lay out a distinct role for this molecule in brain function. We expect that this volume will appeal to all neurobiologists interested in neurotransmitter mechanisms, as well as to clinicians and other health-care workers who see the results of the misuse of this fascinating chemical but who also anticipate its valuable therapeutic use in certain sleep disorders.

G. Tunnicliff
C. D. Cash

1 Metabolism and distribution of γ-hydroxybutyrate in the brain

E. E. Kaufman

Introduction

γ-Hydroxybutyrate (GHB) is a naturally occurring compound found in micromolar concentration in both brain (Roth 1970) and peripheral tissues (Nelson *et al.* 1981). This compound was first synthesized (Laborit 1964) in an attempt to find an analogue of the inhibitory neurotransmitter, γ-aminobutyrate (GABA) that would cross the blood–brain barrier. When GHB is administered to rats in pharmacological doses, it can enter the brain where it produces a sleep-like state, a flattening of the EEG (Winters and Spooner 1965; Marcus *et al.* 1967), a profound depression of cerebral glucose utilization (Wolfson *et al.* 1977), changes in striatal dopamine and decreased impulse flow in dopaminergic neurons (Roth and Suhr 1970).

Clinically, GHB has been used as an adjuvant to anesthesia (Laborit *et al.* 1960; Vickers 1969) and in the treatment of narcolepsy (Mamelak *et al.* 1986; Scrima *et al.* 1990). More recently it has been suggested that GHB may be useful in the treatment of opiate withdrawal syndrome (Gallimberti *et al.* 1993) and alcohol dependence (Gallimberti *et al.* 1992).

Since any physiological or therapeutic effect of GHB will depend on the tissue concentration of the compound, it is important to understand both the metabolism of this compound and the factors that regulate it. *In vivo* studies have demonstrated that a cytosolic $NADP^+$-dependent oxidoreductase plays a key role in the metabolism of GHB and that either inhibitors or activating co-substrates of this enzyme can significantly alter the tissue concentration of GHB both brain and peripheral tissue (Kaufman and Nelson 1987).

Biosynthesis

Roth and Giarman (1969) demonstrated by gas chromatographic analysis that when [³H] GABA is administered intracisternally, rat brain is capable of converting it to [³H]GHB. Roth (1965) had previously identified a compound in the brains of both cats and rats that was chromatographically identical with authentic GHB. This research group proposed a pathway for GHB synthesis in which (1) GABA and α-ketoglutarate are transaminated by GABA α-ketoglutarate transaminase to form succinic semialdehyde (SSA) and glutamate, and (2) SSA is reduced to GHB. The second step requires either NADPH or NADH. The finding that the *in vivo* conversion of GABA to GHB was markedly inhibited by amino-oxyacetic acid supported the proposal that the first step in the conversion is carried out by GABA transaminase (E.C.2.6.1.19). The overall reaction was not, however, inhibited by pyrazole, a compound known to inhibit alcohol dehydrogenase. This result suggested that reaction (2) was not catalyzed by the brain alcohol dehydrogenase as had been proposed by Taberner (1974).

Santaniello *et al.* (1978) demonstrated that a mouse brain homogenate could catalyze the formation of [^{14}C]GHB from [^{14}C]glutamate. The identity of the [^{14}C]-labeled product was definitively established by the isolation of the phenylhydrazide derivative of GHB. This work not only confirmed the presence of GHB in brain at a concentration of approximately 10^{-6} M but also gave support to a pathway of biosynthesis via the GABA shunt (glutamate $\rightarrow$ GABA $\rightarrow$ SSA $\rightarrow$ GHB), a pathway proposed earlier by Roth and Giarman (1969).

Whereas the early work on the biosynthesis of GHB established the overall pathway, it did not identify the specific enzyme responsible for the biosynthesis of this compound. Tabakoff and von Wartburg (1975) succeeded in separating fractions from a rat brain homogenate which catalyzed the reduction of SSA to GHB and were distinct from the fractions which contained either alcohol dehydrogenase or lactic dehydrogenase.

The isolation from human brain of two NADPH-linked aldehyde reductases capable of reducing SSA to GHB was reported by Cash *et al.* (1979). Both enzymes were purified to electrophoretic homogeneity: one of these proteins has a molecular weight of about 45,000, is strongly inhibited by hypnotics and anticonvulsants, and is not specific for SSA; the other is a dimer with a molecular weight of approximately 90,000, is specific for SSA, and is not inhibited by hypnotics and anticonvulsants. The K_m of this enzyme for SSA was found to be 2.4×10^{-5} M. Matsuda and Hoshung (1977) have reported a concentration of SSA in mouse brain of 0.5–1.0×10^{-10} M. Both these reductases were found in the cytosolic fraction.

Hoffman *et al.* (1980) also isolated two aldehyde reducing enzymes from human brain. They found that one of these enzymes was a low molecular weight, NADPH-dependent aldehyde reductase that was relatively non-specific and was inhibited by barbiturates, chlorpromazine and diphenylhydantoin. They suggested that this enzyme is probably identical to aldehyde reductase (E.C.1.1.1.2), L-hexanoate reductase (E.C.1.1.1.9) (Mano *et al.* 1961) and GHB dehydrogenase (Kaufman *et al.* 1979). The second enzyme which has a high specificity for SSA was designated as SSA reductase. They also found that SSA reductase had approximately twice the molecular weight of the monomeric aldehyde reductases and, as suggested by Cash *et al.* (1979), is probably a dimer. In addition, it differs from the large family of aldo-keto reductases in the stereospecificity of the hydrogen transfer from NADPH to carbonyl-containing compounds; the aldo-keto reductases in the large family were specific for the A(pro-4*R*) hydrogen whereas SSA reductase is specific for the B(pro-4*S*) hydrogen of NADPH.

Rumigny *et al.* (1981) carried out experiments to determine the relative contribution of the non-specific aldehyde reductase and of the specific SSA reductase to the biosynthesis of GHB from labeled GABA in rat brain cerebellar slices. The two inhibitors that were used in these experiments were 2,3-di-*n*-propylhexan 3-doic acid (GT 16), a compound which at 10^{-3} M produced an 85% inhibition of the non-specific reductase with no detectable inhibition of the specific SSA reductase, and 4-*n*-propylheptanoic acid (GT 43), which produced a 74% inhibition of the non-specific reductase and a 40% inhibition of the specific SSA reductase. It is important to note that both compounds inhibit SSA dehydrogenase, and would tend to increase the normally very low tissue concentration of SSA and therefore the rate of synthesis of GHB. The addition of GT 43, the compound that inhibited both enzymes, produced a small (10%) but significant decrease in the conversion of GABA to GHB. By contrast, the addition of GT 16, the compound that inhibited only the non-specific reductase, produced a concentration-dependent increase in the conversion of GABA to GHB (a 42.5% increase at 1×10^{-3} M and a 73.6% increase at 2×10^{-3} M). The fact that a large concentration-dependent increase in GHB formation is only seen in the presence of GT 16, the compound which inhibits the non-specific reductase and SSA dehydrogenase, strongly supports the idea that the specific SSA reductase is the enzyme that functions *in vivo* to convert SSA to GHB.

The question of the subcellular localization of SSA reductase has been raised by the work of Hearl and Churchich (1985), who have purified to homogeneity an NADPH-dependent SSA reductase from both pig brain homogenate and pig brain mitochondria. These two enzymes were found to be identical: both are dimers with a molecular weight of 110,000 and have identical kinetic constants for SSA, NADPH and NADH. This raises the question of whether the SSA reductase is cytosolic (as reported by several other groups including Cash *et al.* (1979) and Hoffman *et al.* (1980)) or mitochondrial (as proposed by Hearl and Churchich (1985)).

Andriamampandry *et al.* (1998) have reported on the cloning and expression of a c-DNA sequence from rat hippocampus which encodes a brain-specific enzyme whose activity is identical with that of SSA reductase purified from rat brain. Human brain SSA reductase was also cloned and expressed in *Escherichia coli*. A data bank search indicated that this enzyme is identical with aflatoxin B1-aldehyde reductase from human liver, an enzyme involved in the detoxification of zenobiotic carbonyl compounds. The recombinant protein is identical to native brain SSA reductase on analysis by SDS/PAGE. The enzyme catalyzes the reduction of 9,10-phenanthrenequinone, phenylglyoxal, 4-nitrobenzaldehyde as well as SSA. This suggests that this enzyme functions both in the biosynthesis of GHB and in the detoxification of xenobiotic compounds. Although this enzyme belongs to the aldo-keto reductase superfamily, it has been proposed that aflatoxin aldehyde reductase constitutes a separate subfamily. It is distinguished by the stereospecificity of the hydrogen transfer from NADPH to carbonyl-containing compounds. This reductase, like the SSA reductase isolated from human brain by Hoffman *et al.* (1980), transfers the $4S$ hydrogen atom of NADPH. It is therefore a B-specific enzyme, in contrast to the enzymes of the aldo-keto reductase subfamily which use the $4R$ hydrogen atom and are therefore A-specific enzymes. Unlike the classic aldo-keto reductases which have a broad range of substrates, SSA is the only known naturally occurring substrate for this enzyme. Ellis *et al.* (1993) also demonstrated that this enzyme, which is ethoxyquin-inducible in rat liver, metabolizes aflatoxin B-1.

That GABA is not the sole source of GHB has been demonstrated by the work of Gibson *et al.* (1995), who found that therapy with vigabatrin, an irreversible inhibitor of GABA transaminase, is beneficial to varying degrees to only 35% of their patients with 4-hydroxy-butyric aciduria, a genetic disease caused by a deficiency in SSA dehydrogenase. Failure to block GHB formation by blocking the GABA shunt pathway suggests an alternative pathway for the biosynthesis of GHB. Snead and Morley (1981) have suggested that polyamines may act as precursors of GHB. Seiler and Al-Therib (1974) proposed that, in the brain, the conversion of putrescine to GABA takes place via the following series of reactions: putrescine → monoacetyl putrescine → *N*-acetyl-γ-aminobutyraldehyde → *N*-acetyl GABA → GABA. Snead *et al.* (1980) found that though the GABA transaminase inhibitor, amino-oxyacetic acid, blocked the increase in GHB produced by the administration of putrescine, it did not block the increase produced by the administration of 1,4-butanediol. This latter compound has been proposed as a second alternative source of GHB. Poldrugo and Snead (1984) reported that 1,4-butanediol produced behavioral and EEG effects identical to those of GHB. Finally, Mamelak and Hyndman in Chapter 13 of this volume have proposed that GHB may be produced by lipid peroxidation. All these sources deserve further consideration.

Catabolism of GHB

Much of the early knowledge concerning the metabolism of GHB in brain was learned from studies carried out *in vivo*. Möhler *et al.* (1976) and Doherty *et al.* (1975) demonstrated that, *in vivo*, GHB enters the citric acid cycle predominantly as succinate rather than as acetyl CoA, as had previously been proposed (Walkenstein *et al.* 1964). These findings indicate that the degradative pathway for GHB proceeds through the following three steps: (1) GHB → SSA; (2) SSA → succinate; (3) succinate → → → → CO_2 and H_2O. The reactions in step (3) are catalyzed by the enzymes of the citric acid cycle; the reaction in step (2) is catalyzed by SSA dehydrogenase. Two enzymes that catalyze the reaction in step (1) have now been identified. One of them is a cytosolic $NADP^+$-dependent oxidoreductase, 'GHB dehydrogenase', that can catalyze the oxidation of GHB to SSA *in vitro* under conditions that approximate those that are found in the cytosol of brain and some peripheral tissues. This enzyme was isolated from hamster liver and brain (Kaufman *et al.* 1979). Previously, alcohol dehydrogenase (E.C.1.1.1.1) (Taberner 1974) and lactic dehydrogenase (E.C.1.1.1.27) (Fishbein and Bessman 1964), both NAD^+-dependent oxidoreductases, were proposed as the enzymes that catalyzed the oxidation of GHB to SSA in brain. By contrast, the $NADP^+$-dependent oxidoreductase isolated from brain was not inhibited by pyrazole; this result is consistent with the work of Doherty *et al.* (1975), who showed that pyrazole, an inhibitor of alcohol dehydrogenase, had no effect on the *in vivo* metabolism of GHB. In addition, with the soluble fraction prepared from either hamster or rat brain, the cofactor requirement for the reaction in which GHB is oxidized to SSA was found to be specific for $NADP^+$ at low concentrations of GHB (10 mM or less); activity with NAD^+ is seen only with very high (100 mM) concentrations of GHB. These findings indicated that an $NADP^+$-dependent oxidoreductase, not alcohol dehydrogenase or lactic dehydrogenase, catalyzes the oxidation of GHB to SSA in brain. GHB dehydrogenase has been purified to electrophoretic homogeneity and found to have a molecular weight of ~31,000. The enzymes isolated from brain and liver appear to be very similar with regard to their molecular weights and their kinetic constants for GHB and SSA. In both tissues, the enzyme is predominantly in the cytosol. The naturally occurring aldehydes that are substrates for this enzyme include SSA, D-glucuronate and D,L-glyceraldehyde; synthetic substrates include *p*-nitrobenzaldehyde and *p*-carboxybenzaldehyde. The enzyme is inhibited by amobarbital, diphenylhydantion, valproate, octanoate, diethyldithiocarbamate, cyanide, and the biological intermediates *p*-hydroxyphenylacetate and phenyllactate. Salicylate, α-ketoisocaproate, phenylacetate and valproate were also reported to be good inhibitors of GHB dehydrogenase with K_i values in the range of 10^{-4} M (Kaufman and Nelson 1987).

Like lysozyme and ribonucleolase, GHB dehydrogenase may contain disulfide bridges that are essential for its activity. It is inhibited by compounds such as β-mercaptoethanol and dithiothreitol (DTT) that can reduce disulfide bonds. The 85% inhibition produced by the addition of 2.5 mM DTT can be completely reversed by the addition of oxidized glutathione (Kaufman *et al.* 1983).

When tested *in vivo*, phenylacetate and salicylate increased the concentration of GHB in brain to ~200% of the saline control; sodium valproate produced an increase of 142% of the control (Kaufman and Nelson 1987). The *in vivo* studies with known inhibitors of GHB dehydrogenase support a significant role for this enzyme in a degradative pathway that leads to the complete oxidation of GHB. Since the activity of SSA dehydrogenase (step 2) is approximately 1000 times greater (Pitts and Quick 1967) than that of GHB dehydrogenase (Kaufman *et al.* 1979), step (1) is probably the rate-limiting step in the catabolic pathway outlined above.

Although the oxidation of GHB to SSA catalyzed by GHB dehydrogenase proceeds at an easily measurable rate when the assay was carried out *in vitro* under optimal conditions, it can be calculated that when it is assayed at concentrations of $NADP^+$ and NADPH found *in vivo*, the reaction would proceed at only 1% of V_{max}; the low tissue concentrations of GHB would reduce the rate even more.

An investigation of the substrate specificity of GHB dehydrogenase had revealed that D-glucuronate and L-gulonate are good substrates for the purified enzyme. A comparison of the physical characteristics and the substrate and inhibitor specificity of GHB dehydrogenase suggested that it represented a new activity for the NADP$^+$-dependent oxidoreductase commonly known as D-glucuronate reductase (E.C.1.1.1.19) (York *et al.* 1961). The enzyme may also be identical to the group of enzymes known as ALR-1, the high K_m aldehyde reductases, or L-hexanoate dehydrogenase (Cromlish and Flynn 1985). Kaufman and Nelson (1981) demonstrated that these two activities (the oxidation of GHB and the reduction of D-glucuronate) could be coupled. The coupled reaction is depicted below as the sum of the two partial reactions:

$$GHB + NADP^+ \rightarrow SSA + NADPH \qquad \text{(uncoupled reaction)}$$
$$\text{D-glucuronate} + NADPH + H^+ \rightarrow \text{L-gulonate} + NADP^+$$

$$GHB + \text{D-glucuronate} \rightarrow SSA + \text{L-gulonate} \qquad \text{(coupled reaction)}$$

In the uncoupled reaction, SSA and NADPH are formed in stoichiometric amounts (Kaufman *et al.* 1979); in the coupled reaction, NADPH is being used for the reduction of D-glucuronate to L-gulonate so that there is a rapid formation of SSA with little or no formation of NADPH (Kaufman and Nelson 1981).

Both D-glucuronate and L-gulonate had profound effects on the oxidation of GHB to SSA. For example, there is a marked change in the kinetic constants in the coupled reaction compared to the uncoupled reaction: K_m for GHB is 2.3×10^{-3} M in the uncoupled reaction and 4.5×10^{-4} M in the coupled reaction, and K_m for NADP$^+$ is 2×10^{-5} M in the uncoupled reaction and 1.4×10^{-6} M in the coupled reaction. Furthermore, NADPH, which in the uncoupled reaction is inhibitory with a K_i of 7×10^{-6} M, has no inhibitory effect in the coupled reaction (Kaufman and Nelson 1981). The kinetic constants found in the coupled reaction were all much closer to the actual tissue concentrations of these substrates. K_m for NADP$^+$ is within the reported tissue concentration, and K_m for GHB is an order of magnitude closer to the tissue concentration. Perhaps the most important change is that GHB can be oxidized in the presence of an otherwise inhibitory concentration of NADPH. The effect of the coupling of the two reactions was demonstrated in both *in vitro* and *in vivo* experiments. *In vitro*, when the oxidation of GHB was coupled to the reduction of D-glucuronate in the presence of a limiting amount of NADP$^+$ and an inhibitory amount of NADPH, an increase in the concentration of D-glucuronate from 0 to 2 mM produced an eightfold increase in the rate of SSA formation. *In vivo*, the effect of D-glucuronate was seen when $t_{1/2}$ for the disappearance of [^{14}C]GHB from the plasma was measured following a dose of D-glucuronate; a 33% decrease in $t_{1/2}$ was observed as compared to the saline control (40 min as compared to 60 min for the control). On the other hand, when L-gulonate was administered, $t_{1/2}$ increased from 60 to 80 min, a result consistent with an inhibition by L-gulonate of the reaction catalyzed by GHB dehydrogenase (Kaufman and Nelson 1987). These results provide at least a partial explanation of how GHB dehydrogenase, an enzyme with such unfavorable kinetic constants in the uncoupled reaction, can function under physiological conditions. The interaction of the pathway leading to GHB oxidation with the pathway leading to D-glucuronate reduction is shown in Fig. 1.1.

Reddy *et al.* (1981) found that in hog kidney, myoinositol oxygenase (the enzyme which catalyzes the synthesis of D-glucuronate) exists as a complex with D-glucuronate reductase (GHB dehydrogenase) and at least two other proteins. The authors propose that this complex

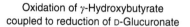

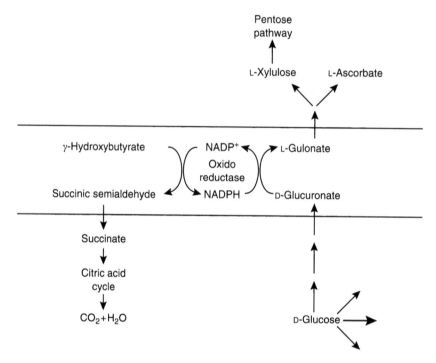

Figure 1.1 Coupled metabolism of γ-hydroxybutyrate and D-glucuronate.

facilitates the transfer of D-glucuronate from myoinositol oxygenase to D-glucuronate reductase/GHB dehydrogenase. A similar system has not yet been found in brain or other tissues.

With the discovery of the cytosolic GHB dehydrogenase, the question remained as to whether there were other oxidoreductases, either in the cytosol or other subcellular compartments, which could oxidize GHB to SSA. The development of an antibody to the cytosolic GHB dehydrogenase (Kaufman *et al.* 1988) led to the discovery of a second enzyme capable of catalyzing this reaction. Although this antibody was found to inhibit over 95% of the activity of the cytosolic GHB dehydrogenase in rat brain and kidney, it inhibited only 50% of the activity in a rat kidney homogenate. This suggested the presence of an additional enzyme or enzymes. The mitochondrial fractions from rat brain and kidney were found to catalyze the conversion of $1-[^{14}C]GHB$ to $^{14}CO_2$. The oxidation of GHB to SSA by a dialyzed mitochondrial fraction did not require the addition of either NAD^+ or $NADP^+$, and the reaction was not inhibited by sodium valproate or by the antibodies to the cytosolic oxidoreductase. This mitochondrial enzyme was found to be a hydroxyacid–oxoacid transhydrogenase (Kaufman *et al.* 1988). In the reaction catalyzed by the mitochondrial enzyme, the oxidation of GHB to SSA is both dependent on, and stoichiometric with, the metabolism of α-ketoglutarate. The possibility that this enzyme might be an α-ketoglutarate dioxygenase was eliminated when it was shown (1) that the reaction did not require oxygen, it could proceed under anaerobic conditions, and (2) that no $^{14}CO_2$ was formed when the reaction was carried out with $1-[^{14}C]α$-ketoglutarate. The ^{14}C label was retained. It was also demonstrated chromatographically that a new ^{14}C-labeled product was formed. A combination of gas chromatography and mass spectroscopy demonstrated that the product of the reaction is 2-hydroxyglutarate. It was also shown that this enzyme is a transhydrogenase which catalyzes a reaction in which the hydrogen attached to the hydroxyl-bearing carbon of GHB is transferred to the ketone-bearing carbon of α-ketoglutarate. The reaction is reversible; the enzyme catalyzes the reaction in which SSA and D-α-hydroxyglutarate react to form stoichiometric amounts of GHB and α-ketoglutarate. It is also specific for the D-isomer; no activity is seen with the L-isomer of α-hydroxyglutarate. It is of interest that two different metabolic diseases have been described: in one of these L-α-hydroxyglutarate accumulates (Barth *et al.* 1992), and in the other the D-isomer accumulates (Gibson *et al.* 1993). L-β-Hydroxybutyrate is also a substrate for this enzyme; no activity is seen with the D-isomer. Of the oxoacids tested as substrates, only α-ketoadipate and oxaloacetate were found to have some activity, ~20–25% of that seen with α-ketoglutarate. The mitochondrial fraction of kidney, liver and brain were all found to contain this enzyme with the highest amounts in kidney and liver.

A transhydrogenase that catalyzes a similar reaction (i.e. L-lactate + oxaloacetate → pyruvate + L-malate) has been isolated and purified from *Micrococcus lactilyticus* (Allen and Patil 1972). This enzyme contains a tightly bound NAD$^+$/NADH prosthetic group that is essential for the enzymatic activity. The enzyme also contains a tightly bound pyruvate whose role is not known. The mitochondrial enzyme has not yet been sufficiently purified to determine whether or not it also contains any tightly bound prosthetic groups.

It is interesting to compare the role of an oxoacid in the reaction catalyzed by the mitochondrial enzyme with that catalyzed by the cytosolic enzyme. In the case of the mitochondrial hydroxyacid–oxoacid transhydrogenase, there is an absolute requirement for an oxoacid in order for the enzyme to catalyze oxidation of the hydroxyacid, GHB. In the case of the cytosolic oxidoreductase, although the reaction can proceed without the oxoacid (D-glucuronate), the oxidation of GHB is greatly facilitated in the coupled reaction. The pathways involved in both the synthesis and the degradation of GHB are shown in Fig. 1.2.

It has been proposed that GHB can be metabolized to GABA (Vayer *et al.* 1985). These workers have demonstrated the formation of [^{3}H]GABA in an *in vitro* system containing [^{3}H]GHB, glutamate, NADP$^+$, purified GHB dehydrogenase and GABA transaminase. This work does not provide convincing evidence that, *in vivo*, there is any net flux of the carbon chain of GHB to GABA. *In vivo*, the SSA formed either as a product of the oxidation of GHB by the GHB dehydrogenase, or the mitochondrial transhydrogenase would encounter not only GABA transaminase but also SSA dehydrogenase. The latter enzyme is extremely active with a K_m for SSA of 10^{-5} M or less, and catalyzes a reaction (SSA + NAD$^+$ → succinate + NADH) which is essentially irreversible (Jakoby and Scott 1959). Because of this, when GABA transaminase and SSA dehydrogenase are coupled, either *in vitro* or *in vivo*, the reaction catalyzed by SSA dehydrogenase 'pulls' the GABA transaminase reaction in the direction of SSA formation (Baxter 1970). Hearl and Churchich (1985) have demonstrated that the oxidation of GHB to SSA by GHB dehydrogenase is also 'pulled' by SSA dehydrogenase.

When Möhler *et al.* (1976) followed the fate of intravenously administered [^{14}C]GHB in mouse brain, they found that the conversion of this compound to [^{14}C]GABA, before oxidation via the citric acid cycle, was negligible. Moreover, the incorporation of the [^{14}C] label into amino acids associated with the citric acid cycle was consistent with the oxidation of GHB via succinate before entering the cycle. Doherty *et al.* (1975) demonstrated that, *in vivo*, the administration of 1-[^{14}C]succinate and of 1-[^{14}C]GHB resulted in a similar profile of

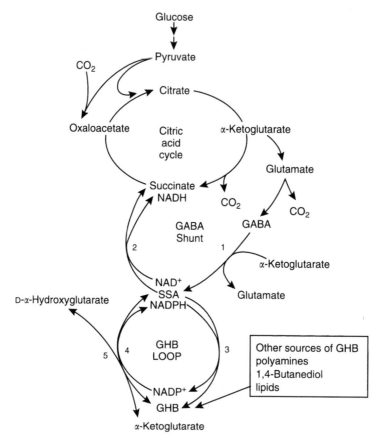

Enzyme	Cofactor/Co-substrate	Location
1 GABA transaminase	α-Ketoglutarate	Mitochondrion
2 SSA dehydrogenase	$NAD^+/NADH$	Mitochondrion
3 SSA reductase	$NADP^+/NADPH$	Cytosol/mitochondrion
4 GHB dehydrogenase	$NADP^+/NADPH$	Cytosol
5 GHB transhydrogenase	α-Ketoglutarate	Mitochondrion

Figure 1.2 Relationship of γ-hydroxybutyrate metabolism and glucose metabolism.
(SSA: succinic semialdehyde)

amino acid formation and that the production of GABA from GHB occurred primarily after GHB had been metabolized to succinate and not by the transamination of SSA.

Collier and De Feudis (1970) followed the appearance of radioactivity in GABA, glutamate and alanine in rat brain after an i.p. injection of $1-[^{14}C]GHB$. They found a sevenfold increase in the d.p.m./g in the fraction containing GABA between 15 min postinjection and 120 min postinjection. When the actual concentration of GABA in these fractions was measured over the same time period, it was found to be 4.0 μmol/g at 15 min and 3.7 μmol/g at 120 min. Thus, there was no increase in the tissue concentration of GABA with time, in spite of a sevenfold increase in the radioactivity in the GABA fraction during this same period. Margolis (1969) also examined the effect of an i.p. injection of GHB (1 g/kg) on the GABA concentration in rat brain; no change was observed.

An explanation for the seemingly contradictory results of Collier and De Feudis (1970) may be found in the work of Albers and Jakoby (1960), who demonstrated that GABA transaminase can catalyze not only the overall reaction in which GABA and α-ketoglutarate $\rightarrow$ SSA and glutamate, but also exchange reactions between [^{14}C]GABA and SSA and between [^{14}C]glutamate and α-ketoglutarate. The initial rate of the GABA/SSA exchange reaction was found to be approximately the same as that of the complete reaction; it is also inhibited by the same compounds that inhibit the complete reaction. Such an exchange reaction would explain the results obtained by Collier and De Feudis (1970) in which there is a partition of the radioactive label with no change in the concentration of GABA. These experiments, especially those of Collier and de Feudis, should serve as a caution that any attempt to demonstrate the enzymatic conversion of GHB to GABA, either *in vivo* or *in vitro*, must rely on enzymatic or chemical assays of the GABA concentration. The measurement of changes in radioactivity in the GABA fraction following incubation with radioactive GHB cannot be interpreted as the conversion of GHB to GABA without such assays.

The interaction of GHB with the GABA$_B$ receptor has been a subject of interest and one which will be covered in Chapter 3 of this volume. The conversion of GHB to GABA by (1) oxidation of GHB to SSA and (2) transamination back to GABA has been proposed by Cash (1994) as the mechanism by which GHB affects the GABA$_B$ receptor. Mathivet *et al.* (1997) tested this hypothesis by carrying out binding experiments in the presence of concentrations of valproate and ethosuximide (30–50 $\times$ K_i) sufficient to completely inhibit the oxidation of GHB to SSA by GHB dehydrogenase, the first step in this reversal. Neither valproate nor ethosuximide had any effect on the displacement of the GABA agonist [^{3}H]CGP 27494 by GHB from cortical membranes. The results of these experiments appear to demonstrate that the affinity of GHB for the GABA$_B$ receptor is not due to GHB-derived GABA, but rather that GHB is a direct agonist of the GABA$_B$ receptor. Similar binding experiments have been carried out by Hechler *et al.* (1997). They found, however, that both valproate and ethosuximide markedly inhibited the 'GABA-like' effect of GHB at the GABA$_B$ receptor. Since both these groups are attempting to determine whether or not GHB has an effect at the GABA$_B$ receptor by first undergoing conversion to GABA, both these experiments need to include controls in which the ability of the membrane preparation to convert GHB first to SSA and then to GABA (as opposed to [^{3}H]GABA) is determined. It should be noted that GHB dehydrogenase, the enzyme which catalyzes the valproate-sensitive oxidation of GHB to SSA, is a soluble cytosolic enzyme, not a membrane-bound enzyme (Kaufman *et al.* 1979).

Distribution of GHB and of the enzymes that synthesize and degrade GHB

This discussion will include the findings on both the subcellular and the tissue distribution of these enzymes. GHB is a product of GABA metabolism (Roth and Giarman 1969) through the GABA shunt pathway. The two main enzymes of this pathway, GABA transaminase (E.C.2.6.1.19) and SSA dehydrogenase (E.C.1.2.1.24), are both mitochondrial (Baxter 1970; Pitts and Quick 1967).

Two aldehyde reductases, one of which is specific for SSA, have been isolated and purified from brain homogenate; both are believed to be in the cytosolic fraction (Cash *et al.* 1979; Rumigny *et al.* 1982). These findings raise the question of whether SSA, which is formed in the mitochondrion, is transported to the cytosol where it serves as a substrate for SSA reductase. Hearl and Churchich (1985) have raised another possibility. They have purified a specific SSA reductase from pig brain mitochondria; this enzyme is identical to the reductase purified from whole brain homogenate. These results raise the possibility that SSA may be reduced to GHB within the mitochondrion and then transported to the cytosol. Any GHB which was either transported to the cytosol or formed in the cytosol would be a substrate for the non-specific cytosolic aldehyde reductase, GHB dehydrogenase, which would oxidize GHB and $NADP^+$ to SSA and NADPH. The GHB in the mitochondrion would be a substrate for the hydroxyacid–oxoacid transhydrogenase which oxidizes GHB + α-ketoglutarate to SSA + α-hydroxyglutarate. The SSA formed by either reaction would be rapidly oxidized to succinate by SSA dehydrogenase, an enzyme with a high affinity for SSA ($\sim 10^{-5}$ M) and a specific activity that is at least sixteen times that of SSA reductase (Hearl and Churchich 1985). These authors also demonstrated, in an *in vitro* assay, that SSA dehydrogenase can 'pull' the GHB dehydrogenase-catalyzed oxidation of GHB to the formation of succinate (Fig. 1.2).

Thus, a considerable portion of GHB metabolism appears to be located in the mitochondrion. Lee *et al.* (1997) used an anti-SSA reductase antibody to investigate the distribution of this enzyme. They found positive structures mainly around microtubules in the axoplasm and within the mitochondrial matrix. Immunoreactive cells were distributed widely throughout the brain; high reactivity was seen in areas associated with the limbic system and reticular formation.

The distribution of the enzymes involved in the biosynthesis and catabolism of GHB varies with the age of the animal, as does the tissue concentration of the metabolite. GHB concentrations are relatively high in the fetal brain and fall until postnatal day 20, when they reach a fairly constant level (Snead and Morley 1981; Nelson and Kaufman 1994). Rumigny *et al.* (1982) found that the activity of SSA reductase, the enzyme responsible for the synthesis of GHB, reached its peak a few days after birth and then remained stable until 63 days of age. This enzyme probably provides a rather constant capacity for GHB synthesis during a period when the concentration of GHB in the brain is falling. The developmental time course of the biosynthetic enzyme is quite different from that of the two enzymes which catalyze the initial, rate-limiting step in the catabolism of GHB (Nelson and Kaufman 1994); these enzymes, by contrast, show a large variation with age. The activity of GHB dehydrogenase is low in both fetal and newborn brain; however, between birth and 20 days the activity of this enzyme increases by almost 50%, with a decrease to lower adult levels by 70 days of age. The activity of mitochondrial GHB transhydrogenase was determined over the same time course. This activity was essentially absent at gestational day 18 but appeared soon after birth. By postnatal day 10, it had risen to 5–10% of the activity found in the adult brain. Whereas GHB dehydrogenase reached a maximum level at approximately 20 days after birth, the mitochondrial transhydrogenase activity continued to rise between 20 and 70 days of age.

It is interesting to note that the decrease in the concentration of GHB in rat brain from 4 nmol/g at birth to less than 2 nmol/g at 20 days of age occurs during the same time period in which there is a rapid increase in GHB dehydrogenase. These results are in agreement with those of Snead and Morley (1981).

The developmental time course of the mitochondrial GHB transhydrogenase shows a 100-fold increase in the activity of this enzyme between birth and adulthood. Other mitochondrial enzymes such as SSA dehydrogenase and cytochrome C oxidase also increase, but only by three- to fivefold. What effect this remarkable increase may have on neural function and metabolism is not known. It is likely, however, that GHB concentration in the mitochondrion is much lower in the adult than in the fetus. GHB transhydrogenase has been shown to contribute approximately 65% of the oxidizing capacity for GHB in the adult brain; cytosolic GHB dehydrogenase, however, is the predominant, if not sole, degradative enzyme for GHB during the fetal and early neonatal period.

Snead and Morley (1981) have examined the regional concentrations of GHB in developing rat, monkey and human brain. They have reported levels in fetal brain (15 days gestational age) that are 400% of the adult levels. This concentration decreased to adult levels by 20 days of age. During this time period, there is a slow increase in GABA levels and in the activity of both GABA transaminase and SSA dehydrogenase. The authors point out that the activity of SSA dehydrogenase is increasing more rapidly than that of GABA transaminase; this, in turn, might affect tissue levels of SSA.

Snead and Morley have also carried out extensive regional studies on GHB concentration in both human and rat brain. In human brain, the highest concentrations are found in the cerebellum, cortex, brain stem and striatum. At a gestational age of 12–19 weeks, the concentration in the cerebellum of the human brain is ~90 nmol/g; it decreases to ~20 nmol/g at birth. The pattern is similar to that found in rat brain (Snead and Morley 1981; Nelson and Kaufman 1994). Age was found to be an important factor in the distribution of GHB in the human brain. In children, the concentration of GHB is higher in the hippocampus and in all the subcortical structures than it is in adults; in adults, however, the level of GHB in the hypothalamus is twice that in the younger brain (Snead and Morley 1981). Clearly, both the tissue distribution and the concentration of GHB vary greatly with age.

It has been shown that a number of biological intermediates such as α-ketoisocaproate and phenylacetate, drugs such as salicylate and sodium valproate, as well as physiological changes can each affect the tissue levels of GHB (Snead *et al.* 1980; Kaufman and Nelson 1987). These changes in tissue level of GHB reflect the effect of these compounds on the various enzymes that synthesize and degrade GHB. Of the enzymes involved in the synthesis and degradation of this compound (SSA reductase, GHB dehydrogenase and GHB transhydrogenase), GHB dehydrogenase has the kinetic characteristics which make it the most sensitive to these regulatory influences. For example, NADPH is an endogenous inhibitor of GHB dehydrogenase with a K_i value of 7×10^{-6} M. In the absence of maximal coupling of GHB dehydrogenase and D-glucuronate reductase, a change in the ratio of NADPH/ NADP$^+$, such as would occur with low ambient oxygen, would inhibit GHB dehydrogenase and lead to an increase in tissue levels of GHB. This effect would be augmented by the inhibition of SSA dehydrogenase by the parallel change in the ratio of NADH and NAD$^+$. Since the tissue concentration of SSA ($\sim 10^{-9}$ M) (Matsuda and Hoshung 1977) is so far below the K_m of SSA reductase (10^{-5} M), even a small increase in the concentration of SSA could lead to an increase in the synthesis of GHB. In experiments in which rats were exposed to low oxygen (5.6%) for a period of 2 h, there was a 3.6-fold increase in the concentration of GHB in brain and a 2.3-fold increase in kidney when compared to the concentrations found in the tissues of rats exposed to room air for the same period.

Biological intermediates such as phenylacetate and α-ketoisocaproate, as well as the drugs valproate and salicylate, all have K_i values in the range of 10^{-4}–10^{-5} M when assayed with GHB dehydrogenase. *In vivo*, all these compounds were capable of increasing the concentration of GHB by ~200% in one or more of the tissues assayed (Kaufman and Nelson 1991). These results are in agreement with those of Snead *et al.* (1980), who found elevated levels of GHB in brain following the administration of sodium valproate. A small inhibition of

SSA dehydrogenase may also occur with the administration of valproate; the K_i of SSA dehydrogenase for valproate is 4×10^{-3} M as compared to 5.7×10^{-5} M with GHB dehydrogenase. As in the case of inhibition by low ambient oxygen, this could lead to a small increase in the tissue concentration of SSA and therefore to an increase in the synthesis of GHB. For comparison, it has been calculated (Kaufman and Nelson 1987) that sodium valproate at a tissue level of 0.3 mM, a reasonable tissue concentration expected in therapeutic use, would produce an 85% inhibition of GHB dehydrogenase but only a 5% inhibition of SSA dehydrogenase and a negligible inhibition of GABA transaminase. Although the contribution of inhibition of SSA dehydrogenase to increases in tissue levels of GHB is probably quite small in the examples cited above, in the case of 4-hydroxybutyric aciduria where SSA dehydrogenase is missing (Gibson *et al.* 1983), the increase in GHB in cerebrospinal fluid, for example, can be substantial. In a study of six patients with this disorder, the concentration of GHB in the cerebrospinal fluid ranged from 116 to 1,110 µmol/l compared to control levels of 2.6 µmol/l (Gibson *et al.* 1995). This will be covered in Chapter 12 of this volume. We predict that a genetic disease in which either GHB dehydrogenase or GHB transhydrogenase are missing would give rise to another form of 4-hydroxybutyric aciduria.

Inhibitors of the mitochondrial GHB transhydrogenase have not yet been reported. However, the absolute requirement of this enzyme for an oxoacid suggests that the metabolism of GHB by the mitochondrial enzyme would depend on the steady-state levels of citric acid intermediates and of α-ketoglutarate in particular. Factors which stimulate anaplerotic reactions such as pyruvate carboxylase, phosphoenolpyruvate carboxykinase and the malic enzyme, and thus increase the level of these intermediates, may indirectly affect the activity of the mitochondrial transhydrogenase (Fig. 1.2).

Summary

The metabolism of GHB can be regarded as a small loop in which SSA derived from the transamination of GABA is first reduced to GHB (Fig. 1.2). This reduction is catalyzed by a specific aldehyde reductase for which SSA is the only known naturally occurring substrate. Although this is the most well documented source of GHB, as has been discussed above, other sources of GHB have been identified, for example, polyamines, 1,4-butanediol (Poldrugo and Snead 1984) and lipids (Chapter 13 of this volume). The relative contribution of each of these sources has yet to be determined.

The loop is completed by the oxidation of GHB back to SSA (Fig. 1.2). Two enzymes which can catalyze oxidation of GHB back to SSA have been identified; one of these is the cytosolic GHB dehydrogenase, also an oxidoreductase but one that belongs to the large family of aldo-keto reductases rather than to the small subfamily to which SSA reductase has been shown to belong. In contrast to SSA reductase, this enzyme has several naturally occurring substrates and is inhibited by a number of drugs and biological intermediates as described above. These properties make it the site in the loop where regulation, leading to changes in the tissue concentration of GHB, is most likely to occur; such changes have been discussed earlier (Snead *et al.* 1980; Kaufman and Nelson 1987). The second enzyme involved in the oxidation of GHB back to SSA is the mitochondrial GHB transhydrogenase. This enzyme catalyzes a reaction in which the oxidation of GHB to SSA is coupled to the reduction of α-ketoglutarate to D-α-hydroxyglutarate. Regulation of GHB metabolism by either inhibition or by activation of enzymes in the 'GHB loop' has already been described.

Since the GABA shunt pathway and therefore the 'GHB loop' are derived from the citric acid cycle intermediate α-ketoglutarate, an intermediate which also functions as the co-substrate for GABA transaminase and for GHB transhydrogenase (Fig. 1.2), drugs or physiological changes which affect carbohydrate metabolism should also be considered as potential regulators of tissue levels of both GABA and GHB.

Once GHB has been metabolized to SSA, either by GHB dehydrogenase or by GHB transhydrogenase, it is oxidized to succinate by SSA dehydrogenase; the evidence indicates that none is converted back to GABA. Thus, the important role of SSA dehydrogenase in facilitating the oxidation of SSA derived from either GABA or GHB to CO_2 and H_2O should be emphasized.

Acknowledgements

I would like to thank Mona Pedersen for her help in the preparation of the manuscript and Dr Thomas Nelson for a critical evaluation of the manuscript.

References

Albers R. W. and Jakoby W. B. (1960) Exchange reactions catalyzed by GABA transaminase. *Biochem. Biophys. Acta* 40, 457–61.

Allen S. H. and Patil J. R. (1972) Studies on the structure and mechanism of action of the malate–lactate transhydrogenase. *J. Biol. Chem.* 247, 909–16.

Andriamampandry C., Shiffert J. C., Schmitt M., Garnier J., Staub A., Muller C., Gobaille S., Mark J. and Maitre M. (1998) Cloning of rat brain succinic semialdehyde reductase involved in the synthesis of the neuromodulator γ-hydroxybutyrate. *Biochem J.* 334, 43–50.

Barth P. G., Hoffman G. F. and Jacken J. (1992) L-2-Hydroxyglutaric acidemia: a novel inherited neurometabolic disease. *Ann. Neurol.* 32, 66–71.

Baxter C. F. (1970) The nature of γ-aminobutyric acid. In *Handbook of Neurochemistry*, Vol. 3 (Lajtha A., ed.), pp. 289–353. Plenum, New York.

Cash C. D. (1994) Gamma-hydroxybutyrate: an overview of the pros and cons for it being a neurotransmitter and/or a useful therapeutic agent. *Neurosci. Behav. Rev.* 18, 4375–79.

Cash C. D., Maitre M. and Mandel P. (1979) Purification from human brain and some properties of two NADPH-linked aldehyde reductases which reduce succinic semialdehyde to 4-hydroxybutyrate. *J. Neurochem.* 33, 1169–75.

Collier B. and De Feudis F. V. (1970) Conversion of γ-hydroxybutyrate to γ-aminobutyrate by mouse brain *in vivo*. *Experientia* 26, 1072–73.

Cromlish J. A. and Flynn T. G. (1985) Identification of pig brain aldehyde reductases with the high K_m aldehyde reductase, the low-K_m aldehyde reductase and aldose reductase, carbonyl reductase and succinic semialdehyde reductase. *J. Neurochem.* 44, 1485–93.

Doherty J. D., Stout R. W. and Roth R. H. (1975) Metabolism of [1-^{14}C]γ-hydroxybutyric acid by rat brain after intraventricular injection. *Biochem. Pharmacol.* 24, 469–74.

Ellis E. M., Judah D. J., Neal G. E. and Hayes J. D. (1993) An ethoxyquin-inducible aldehyde reductase from rat liver that metabolizes aflatoxin B 1 defines a subfamily of aldoketo-reductases. *Proc. Natl Acad. Sci. USA* 90, 10350–54.

Fishbein W. N. and Bessman S. P. (1964) γ-Hydroxybutyrate in mammalian brain. Reversible oxidation by lactic dehydrogenase. *J. Biol. Chem.* 239, 357–61.

Gallimberti L., Ferri M., Ferrara S. D., Fadda F. and Gessa G. L. (1992) Gamma-hydroxybutyric acid in the treatment of alcohol dependence: a double-blind study. *Alcohol Clin. Exp. Res.* 16, 673–76.

Gallimberti L., Cibib M., Pagnin P., Sabbion R., Pani P.P., Pirastu R., Ferrara S.D. and Gessa G.L. (1993) Gamma-hydroxybutyric acid for treatment of opiate withdrawal syndrome. *Neuropsychopharmacology* 9, 77–81.

Gibson K. M., Sweetman L., Nyhan W.L., Jakobs C., Rating D., Siemes H. and Hanefeld F. (1983) Succinic semialdehyde dehydrogenase deficiency: an inborn error of gamma-aminobutyric acid metabolism. *Clin. Chim. Acta* 133, 33–42.

Gibson K. M., Craigen W., Herman G. E. and Jakobs C. (1993) D-2-Hydroxyglutaric aciduria in a newborn with neurological abnormalities: a new neurological disorder? *J. Inher. Metab. Dis.* 16, 497–500.

Gibson K. M., Jacobs C. and Ogier H. (1995) Vigabratin therapy in six patients with succinic semialdehyde dehydrogenase deficiency. *J. Inher. Metab. Dis.* 18, 143–46.

Hearl W. G. and Churchich J. E. (1985) A mitochondrial $NADP^+$-dependent reductase related to the 4-aminobutyrate shunt. *J. Biol. Chem.* 260, 16361–66.

Hechler V., Ratomponirina C. and Maitre M. (1997) γ-Hydroxybutyrate conversion into GABA induces displacement of $GABA_B$ binding that is blocked by valproate and ethosuximide. *J. Pharm. Exp. Ther.* 281, 753–60.

Hoffman P. L., Wermuth B. and Wartburg J. P. (1980) Human brain aldehyde reductases: relationship to succinic semialdehyde reductase and aldose reductase. *J. Neurochem.* 35, 354–66.

Jakoby W. B. and Scott E. M. (1959) Aldehyde oxidation: succinic semialdehyde dehydrogenase. *J. Biol. Chem.* 234, 937–40.

Kaufman E. E. and Nelson T. (1981) Kinetics of coupled γ-hydroxybutyrate oxidation and D-glucuronate reduction by an $NADP^+$-dependent oxidoreductase. *J. Biol. Chem.* 265, 6890–94.

Kaufman E. E. and Nelson T. (1987) Evidence for the participation of a cytosolic $NADP^+$-dependent oxidoreductase in the catabolism of γ-hydroxybutyrate *in vivo. J. Neurochem.* 48, 1935–38.

Kaufman E. E. and Nelson T. (1991) An overview of γ-hydroxybutyrate catabolism: the role of the cytosolic $NADP^+$-dependent oxidoreductase E.C.1.1.1.19 and of a mitochondrial hydroxyacid– oxoacid transhydrogenase in the initial, rate-limiting step in this pathway. *Neurochem. Res.* 16, 965–74.

Kaufman E. E., Nelson T., Goochee C. and Sokoloff L. (1979) Purification and characterization of an $NADP^+$-linked alcohol oxido-reductase which catalyzes the interconversion of γ-hydroxybutyrate and succinic semialdehyde. *J. Neurochem.* 32, 699–712.

Kaufman E. E., Relkin N. and Nelson T. (1983) Regulation and properties of an $NADP^+$ oxidoreductase which functions as a γ-hydroxybutyrate dehydrogenase. *J. Neurochem.* 40, 1639–46.

Kaufman E. E., Nelson T., Fales H. M. and Levin D. M. (1988) Isolation and characterization of a hydroxyacid–oxoacid transhydrogenase from rat kidney mitochondria. *J. Biol. Chem.* 263, 16872–79.

Laborit H. (1964) Sodium 4-hydroxybutyrate. *Int. J. Neuropharmacol.* 3, 433–52.

Laborit H., Jouany J. M., Gerard J. and Fabiani F. (1960) Résumé d'une etude expérimentale et clinique sur un substrat métabolique a action centrale inhibitrice le 4-hydroxybutyrate de Na. *Presse Medicale* 68, 1867–69.

Lee J. E., Choi S. Y. and Suk J. W. (1997) Distribution of succinic semialdehyde reductase in rat brain. *Mol. Cells* 7, 13–20.

Mamelak M., Scharf M. and Woods M. (1986) Treatment of narcolepsy with gamma-hydroxybutyrate: a review of the clinical and sleep laboratory findings. *Sleep* 9, 285–89.

Mano Y., Suzuki K., Yamada K. and Shimzono N. (1961) Enzymatic studies on TPN L-hexonate dehydrogenase from rat liver. *J. Biochem. Tokyo* 49, 618–34.

Marcus R. J., Winters W. D., Mori K. and Spooner C. E. (1967) EEG and behavioral comparison of the effects of gamma-hydroxybutyrate, gamma-butyrolactone and short chain fatty acids in the rat. *Int. J. Neuropharmacol.* 6, 175–85.

Margolis R. K. (1969) The effect of γ-hydroxybutyric acid on amino acid levels in brain. *Biochem. Pharmacol.* 18, 1243–46.

Mathivet P., Bernasconi R. and De Barry J. (1997) Binding characteristics of γ-hydroxybutyric acid as a weak but selective $GABA_B$ receptor agonist. *Eur. J. Pharmacol.* 321, 67–75.

Matsuda M. and Hoshung M. (1977) Natural occurrence of succinic semialdehyde in mouse brain. *Jikeikai Med. J.* 24, 33–36.

Möhler H., Patel A. J. and Balázs R. (1976) Gamma-hydroxybutyrate degradation in brain *in vivo*: negligible direct conversion to GABA. *J. Neurochem.* 27, 253–58.

Nelson T. and Kaufman E. E. (1994) Developmental time course in the brain and kidney of two enzymes that oxidize gamma-hydroxybutyrate. *Dev. Neurochem.* 16, 352–58.

Nelson T., Kaufman E. E., Kline J. E. and Sokoloff L. (1981) The extraneural distribution of γ-hydroxybutyrate. *J. Neurochem.* 37, 1345–48.

Pitts F. N. Jr and Quick C. (1967) Brain succinic semialdehyde dehydrogenase. II. Changes in developing rat brain. *J. Neurochem.* 14, 561–70.

Poldrugo F. and Snead O. C. (1984) 1, 4-Butanediol, γ-hydroxybutyric acid and ethanol: relationships and interactions. *Neuropharmacology* 23, 109–13.

Reddy C. C., Swan J. S. and Hamilton G. (1981) Myo-inositol oxygenase from hog kidney. *J. Biol. Chem.* 256, 8510–18.

Roth R. H. (1965) A pharmacological investigation of γ-butyrolactone and γ-hydroxybutyric acid general depressants of the central nervous system. Ph.D. Thesis, Yale University.

Roth R. H. (1970) Formation and regional distribution of γ-hydroxybutyric acid in mammalian brain. *Biochem. Pharmacol.* 19, 3013–19.

Roth R. H. and Giarman N. J. (1969) *In vivo* conversion of γ-aminobutyric acid to γ-hydroxybutyric acid in the rat. *Biochem. Pharmacol.* 18, 247–50.

Roth R. H. and Suhr Y. (1970) The mechanism of the γ-hydroxybutyrate-induced increase in brain dopamine and its relationship to 'sleep'. *Biochem. Pharmacol.* 19, 3001–12.

Rumigny J. F., Cash C., Mandel P., Vincendon G. and Maitre M. (1981) Evidence that a specific succinic semialdehyde reductase is responsible for γ-hydroxybutyrate synthesis in brain tissue slices. *FEB Lett.* S134, 96–98.

Rumigny J. E., Cash C., Mandel P. and Maitre M. (1982) Ontogeny and distribution of specific succinic semialdehyde reductase apoenzyme in the rat brain. *Neurochem. Res.* 7, 555–61.

Santaniello E., Manzocchi A. and Tosi L. (1978) Evaluation of gamma-hydroxybutyrate formed from L-glutamate by mouse brain homogenate. *J. Neurochem.* 31, 1117–18.

Scrima L., Hartman P. G., Johnson F. H., Thomas E. E. and Hiller F. C. (1990) The effects of γ-hydroxybutyrate on the sleep of narcolepsy patients: a double-blind study. *Sleep* 13, 479–90.

Seiler N. and Al-Therib M. J. (1974) Putrescine catabolism in mammalian brain. *Biochem. J.* 144, 29–35.

Snead O. C. and Morley B. J. (1981) Ontogeny of γ-hydroxybutyric acid. 1. Regional concentration in developing rat, monkey and human brain. *Dev. Brain Res.* 1, 579–89.

Snead O. C., Bearden L. J. and Pegram V. (1980) Effect of acute and chronic anticonvulsant administration on endogenous γ-hydroxybutyrate in rat brain. *Neuropharmacology* 19, 47–52.

Snead O. C., Liu C. C. and Bearden L. J. (1982) Studies on the relation of gamma-hydroxybutyric acid (GHB) to gamma-aminobutyric acid (GABA). Evidence that GABA is not the sole source for GHB in rat brain. *Biochem. Pharmacol.* 31, 3917–23.

Tabakoff B. and von Wartburg J. P. (1975) Separation of aldehyde reductases and alcohol dehydrogenase from brain by affinity chromatography: metabolism of succinic semialdehyde and ethanol. *Biochim. Biophys. Res. Commun.* 63, 956–66.

Taberner P. V. (1974) Alcohol dehydrogenase activity in rat brain: evidence for the metabolism of succinic semialdehyde to gamma-hydroxybutyrate. *Biochem. Pharmacol.* 23, 1219–20.

Vayer P., Mandel P. and Maitre M. (1985) Conversion of γ-hydroxybutyrate to γ-aminobutyrate *in vivo*. *J. Neurochem.* 45, 810–14.

Vickers M. D. (1969) Gammahydroxybutyric acid. *Int. Anesth. Clinics* 7, 75–89.

Walkenstein S., Wiser R., Gudmundsen C. and Kimel H. (1964) Metabolism of γ-hydroxybutyric acid. *Biochim. Biophys. Acta* 86, 640–42.

Winters W. D. and Spooner C. F. (1965) A neurophysiological comparison of gamma-hydroxybutyrate with pentobarbital in cats. *EEG Clin. Neurophysiol.* 18, 287–96.

Wolfson L. I., Sakurada O. and Sokoloff L. (1977) Effects of γ-butyrolactone on local cerebral glucose utilization in the rat. *J. Neurochem.* 29, 777–83.

York J. L., Grollman A. P. and Bublitz C. (1961) TPN-L-gulonate dehydrogenase. *Biochim. Biophys. Acta* 47, 298–306.

2 The γ-hydroxybutyrate receptor in the brain

C. D. Cash and G. Tunnicliff

Introduction

In the central nervous system (CNS) γ-hydroxybutyrate (GHB) is a metabolite of γ-aminobutyric acid (GABA), the principal inhibitory neurotransmitter (Bessman and Fishbein 1963; Roth and Giarman 1968). Although GABA is present in brain at millimolar concentrations, GHB is found only in the lower micromolar range (Ehrhardt *et al.* 1988). It has been well documented that administration of relatively high doses of GHB to animals produces a hypnotic state (White and Samson 1956; Laborit 1964). Indeed, in humans GHB has been used as an induction agent in general anesthesia (Vickers 1969). GHB has been reported to exert a euphoric effect, and probably as a result of this it has developed a reputation as a drug of abuse (Ropero-Miller and Goldberger 1998). Many of the actions of GHB are consistent with it being a CNS depressant, although it also possesses seizure-type properties, especially those resembling petit mal epilepsy (Winters and Spooner 1965; Godschalk *et al.* 1977).

In addition to its pharmacological actions on the brain at relatively high concentrations, GHB exhibits several actions that might be described as physiological in nature. These properties are consistent with GHB being an inhibitory substance (Maitre 1997). Enough data is now accumulating to place GHB in the category of a putative neurotransmitter or modulator of CNS activity. This leads to the concept of a receptor that can mediate the observed actions of GHB. Receptors that mediate neurotransmitter action typically are located in neuronal plasma membranes and are either associated with ligand-gated ion channels (e.g. GABA$_A$ receptors) or linked to intracellular G protein activation (e.g. GABA$_B$ receptors).

Much has been written about the biochemistry and pharmacology of GHB but comparatively little about the GHB receptor, even though considerable evidence supports the existence of this receptor. Most neuropharmacologists who study the effects of GHB, both *in vivo* and *in vitro*, employ concentrations of the drug which are orders of magnitude higher than the affinity constant of this ligand for its receptor. The administration of pharmacological doses of GHB gives rise to far higher brain levels (Guidotti and Ballotti 1970) than those measured postmortem under normal conditions (Roth and Giarman 1970). These data, though interesting in themselves, are often difficult to extrapolate in order to define physiological actions of GHB at physiological concentrations.

This chapter will concentrate on the evidence for a role of the GHB receptor in the physiological actions of GHB, document its properties, and discuss, where possible, the plausible functions of the receptor at physiological GHB concentrations. These lower concentrations probably are more pertinent to its role as a receptor agonist. Much of the available data, though, stems from the pharmacological action rather than the physiological action of GHB.

Activation of GABA$_B$ receptors by GHB

Accumulated evidence suggests that GHB can act as an agonist at two distinct types of receptors in the brain. The first of these is the GABA$_B$ receptor. This is a G protein-linked site that mediates a portion of the inhibitory action of GABA (Bowery 2000). Mathivet *et al.* (1997) have demonstrated that GHB can bind selectively, but weakly, to the GABA$_B$ receptor ($K_i = 0.1$ mM). It has further been shown that GHB can act as an agonist at this receptor (Xie and Smart 1992; Erhardt *et al.* 1998; Madden and Johnson 1998) although, again, it has a much lower affinity for the site than does GABA. Because of this, it is conceivable that certain pharmacological effects of GHB – its hypnotic properties, for instance – are a result of an interaction with this receptor. It is clear that the activation of GABA$_B$ receptors by GHB leads to a hyperpolarization, the basis of which seems to be the efflux of K^+ ions from the neuron (Lorente *et al.* 2000). A detailed account of this interaction is the subject of Chapter 3 of this volume (Bernasconi *et al.* 2002). The physiological actions of GHB might not be mediated by the GABA$_B$ receptor, though. The existence of another type of GHB receptor site strongly suggests its function is to mediate the influences of endogenous GHB in the brain.

General properties of the GHB receptor

The second receptor that GHB can activate is known as the GHB receptor. High-affinity binding sites for [^{3}H]GHB were first described in membranes from rat cerebral cortex nearly twenty years ago (Benavides *et al.* 1982a) (Fig. 2.1). Ligand binding displayed saturation

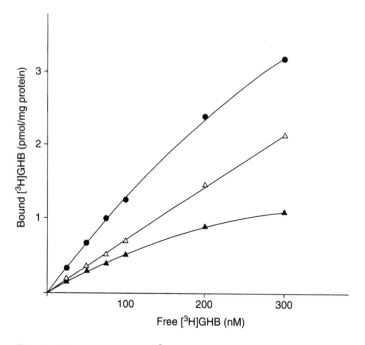

Figure 2.1 Saturation curve of [^{3}H]GHB binding to a rat brain membrane fraction. Each point is the mean of three separate experiments performed in quadruplicate at each concentration. ●, total binding; △, non-specific binding; ▲, binding displacable by 5 mM GHB. Redrawn from Benavides *et al.* (1982a).

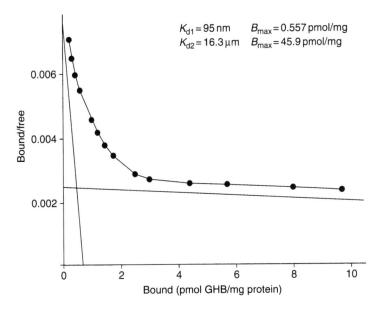

Figure 2.2 Scatchard plot of [³H]GHB binding to rat brain membranes. Data derived from two separate experiments performed in quadruplicate at each concentration (variation <5%). Redrawn from Benavides *et al.* (1982a).

kinetics with a dissociation constant (K_d) of 95 nM and a B_{max} of 0.56 pmol/mg protein. When a Scatchard plot was drawn to display the binding data, a second saturable binding site became apparent (Fig. 2.2), but this had a much lower affinity for GHB ($K_d = 16$ μM). When a follow-up study was carried out on a similar crude membrane preparation from rat brain (Maitre *et al.* 1983b), after a more extensive washing of the membranes and at a range of [³H]GHB concentrations below 20 nM, the existence of an even higher affinity binding site was noted ($K_d = 19$ nM). The kinetics of this binding suggested that positive cooperativity was occurring.

The existence of two binding sites for [³H]GHB was also observed by Snead and Liu (1984). In rat brain homogenates they observed a relatively high-affinity site ($K_d = 0.6$ μM) as well as a low-affinity site (2.3 μM). Further, these investigators reported that similar specific [³H]GHB binding occurred in membranes from human postmortem brain.

For each of these studies, none of the binding sites involved could have represented part of the GABA_B receptor because the presence of GABA failed to displace the radiolabelled ligand. Furthermore, it is unlikely that the low-affinity component represents binding to a GHB receptor, as the K_d is much higher than any known neurotransmitter receptor–ligand interaction. As [³H]GHB binding did not depend on the presence of Na⁺ (Benavides *et al.* 1982b), the specific binding site in the membrane preparations was not a component of the GHB transport system known to be present in mammalian brain since GHB transport is clearly Na⁺-dependent (Benavides *et al.* 1982b; McCormick and Tunnicliff 1998).

Benavides *et al.* (1982a) observed that [³H]GHB binding to the high-affinity site was extremely sensitive to pH. Optimum binding occurred at pH 5.5, but at each side of this value (pH 5.2 or 6.0), relatively little binding was detected. The initial experiments were done using rat cerebral cortex, but greater binding was seen in the olfactory bulbs and hippocampus (Benavides *et al.* 1982a). The high-affinity binding sites are thought to be located in neuronal membranes since [^{3}H]GHB binding was seen in chick embryo cultured neurons but not in cultured glial cells (Maitre 1997).

Distribution of GHB binding sites in the brain

These saturable binding sites fulfill many of the requirements of a receptor for GHB as a neurotransmitter, although certain doubt remains as to whether GHB is the primary endogenous ligand. A number of reports have established that GHB binding sites are heterogeneously distributed in the brain (Benavides *et al.* 1982a; Hechler *et al.* 1992; Castelli *et al.* 2000). In rat CNS, the highest amount of [^{3}H]GHB binding was reported in the hippocampus, particularly the CA$_1$ field. The CA$_2$ and CA$_3$ fields, as well as the dentate gyrus, were also well endowed with receptor sites (Maitre 1997). Several areas of the cerebral cortex exhibited moderate binding, as did the amygdala and thalamus. Brain regions associated with dopaminergic pathways – for example, caudate putamen, substantia nigra and ventral tegmental area – also contained a significant density of receptors. Hypothalamus, cerebellum, pons and medulla, and spinal cord did not exhibit any [^{3}H]GHB binding (Maitre 1997). The autoradiographic study of Castelli *et al.* (2000) involving squirrel monkey and human brain revealed a similar distribution to that of the rat, including an absence of receptors in the hindbrain and cerebellum. It is noteworthy, however, that Snead and Liu (1984) found significant binding of GHB to hypothalamus and pons/medulla in neuronal membranes from postmortem human brain. The regional distribution of the GHB receptor does not correspond to that of either the GABA$_A$ receptor or the GABA$_B$ receptor (Maitre 1997; Snead 1994).

Snead (1994) measured [^{3}H]GHB binding in the developing rat brain with the aid of autoradiography. Binding was first observed at postnatal day 17 in the hippocampus, followed the next day by binding in the frontal cortex. By postnatal day 40, an adult distribution pattern was seen. Both the ontogeny and distribution profile of GABA$_B$ binding was vastly different from GHB receptor binding (Snead 1994).

GHB receptor function linked to G protein activation

The first indication that the GHB receptor might be metabotropic in nature was provided by Snead (1992), who measured an inhibition of GHB receptor binding by 100 μM GTP in rat brain membranes. Later, Ratomponirina *et al.* (1995) extended these findings by confirming that GTP produced a concentration-dependent reduction in GHB binding. The mechanism of this inhibition was based on a decrease in the affinity between GHB and its binding site. In the same study, pertussis toxin was shown to inhibit GHB binding, also by increasing the K_d. The obvious conclusion was that activation of the GHB receptor leads to intracellular G protein modification which, presumably, is the basis of GHB's action on neurons via GHB receptors.

Recently, Snead (2000) has provided evidence that these G protein-linked GHB receptors are presynaptically situated and that they are distinct from GABA$_B$ receptors. Neither GHB nor the GABA$_B$ agonist, baclofen, had an appreciable effect on basal cAMP production in rat brain homogenates, but both drugs inhibited forskolin-induced cAMP formation. In both cases, these effects were concentration-dependent (ED$_{50}$ = 4 and 2 μM, respectively). Pertussis toxin prevented the effects of GHB on forskolin-induced cAMP formation. Whereas GHB had no effect on isoproterenol-evoked cAMP production, baclofen was able to enhance this production. Interestingly, the effects of GHB on forskolin-induced cAMP formation were observed only in cerebral cortex and hippocampus but not in thalamus or cerebellum. On the other hand, baclofen was effective in all brain regions examined. Snead reported that the GHB receptor antagonist NCS-382 blocked the action of GHB on forskolin-induced cAMP formation, whereas the GABA$_B$ receptor antagonist CGP 35348 had no such effect. In the case of baclofen, however, the reverse was true. That is, the GHB receptor antagonist had no effect, while the GABA$_B$ antagonist blocked the action of baclofen on forskolin-induced cAMP production.

The influence of GHB and baclofen on forskolin-induced cAMP formation is, of course, a measure of their effects on adenylyl cyclase activity. Snead (2000) prepared both synaptosomes (a model of presynaptic events) and synaptoneurosomes (a model of postsynaptic events). He found that GHB inhibited forskolin-activated adenylyl cyclase in synaptosomes but not in synaptoneurosomes. Baclofen, however, inhibited the enzyme from both sources. In addition, both GHB and baclofen stimulated [^{35}S]GTPγS binding and low K_m GTPase activity. The GHB effect was inhibited by NCS-382 but not by CGP 35348. Conversely, the baclofen effect was inhibited by CGP 35348 but not by NCS-382. Consequently, these data strongly support the idea that the GHB receptors are located presynaptically and that they are unrelated to GABA$_B$ receptors. If any GHB receptors are located in postsynaptic membranes, they do not appear to be linked to cAMP formation in the same way that they are in presynaptic membranes.

The above results are a strong indication that the GHB receptor is a member of the family of G protein-associated protein binding sites and that the G proteins involved belong to the G$_i$ or G$_o$ family.

Molecular structure of the GHB receptor

The subunit composition and the amino acid sequences of several neurotransmitter receptors have been established. For instance, the GABA$_A$ receptor is a heteromeric pentamer that can be assembled in the plasma membrane from at least fifteen different subunit types. Four transmembrane domains have been deduced. The amino acid sequence of the subunits and various isoforms has been well studied (Mehta and Ticku 1999). The GABA$_B$ receptor, on the other hand, consists of just two isoforms and has seven transmembrane domains, as do other metabotropic receptors (Kuner *et al.* 1999; Bowery 2000).

Studies of the structure of the GHB receptor are underway. Initially the receptor was solubilized from rat brain using high concentrations of NaCl and Triton X-100 (Cash *et al.* 1996). Under these extraction conditions, only one class of binding sites was evident, exhibiting a K_d close to the low-affinity binding component. Purification employing affinity chromatography and electrophoresis was later achieved. A band corresponding to 54 kDa was reported. This led to the production of a cDNA which coded for a protein of 50 kDa (Andriamampandry *et al.* 1999). Surprisingly, the amino acid sequence is not similar to other G protein-linked receptors. Instead there is a homology with tetraspanin proteins (Andriamampandry and Maitre 2000). Proteins belonging to the tetraspanin superfamily (TM4SF) are transmembrane glycoproteins (Maecker *et al.* 1997). Tetraspanin genes are expressed in several types of tissue, including nervous tissue (Garcia-Frigola *et al.* 2000). They might have a role in synaptogenesis (Kopczynski *et al.* 1996), and recently evidence has become available implicating a specific tetraspanin in human non-specific X-linked mental retardation (Zemni *et al.* 2000). Further light will be shed on the molecular structure of the GHB receptor, and on its relationship to the TM4SF family, with the full publication of the sequence data.

Not known at present is the degree of GHB receptor heterogeneity, if any. Results from indirect studies, however, have hinted at the existence of GHB receptor isoforms. As mentioned in an earlier paragraph, Benavides *et al.* (1982a) discovered that two distinct [^{3}H]GHB binding sites could be seen, each with a different affinity for GHB. Further, Hechler *et al.* (1990) found that the receptor agonist *trans*-hydroxycrotonic acid (THCA) had a greater affinity for GHB receptor sites than did GHB in striatum, but the reverse situation was true in the hippocampus. Another regional difference in the properties of the GHB receptor appears in the work of Snead (2000). Adenylyl cyclase is inhibited by GHB in the cerebral cortex but not in the thalamus. Obviously, results from such indirect studies have to be clarified by techniques utilizing molecular biology and protein biochemistry. Only then can we unequivocally say that there are different forms of the GHB receptor.

Ion channels linked to the GHB receptor

Both receptors for which GHB has an affinity are of the metabotropic type. As discussed above, occupation of the GABA$_B$ receptor by GHB produces an activation of G proteins. This in turn leads to an activation of inwardly rectifying K^+ currents, which forms the basis of the hyperpolarization induced by the agonist (Lorente *et al.* 2000). In contrast, the events underlying the neuronal action of GHB mediated by GHB receptors are not well understood. It is apparent that G protein activation is involved (Snead 2000), but subsequent events have not been established. Studying membranes from guinea-pig pars compacta neurons of the substantia nigra, Harris *et al.* (1989) concluded that their observed hyperpolarizations induced by GHB were related to a facilitation of Ca^{2+} conductances. They also pointed out that the efflux of K^+ would be consistent with their findings, yet, to date, no data exist to support this notion.

Cultured NCB-20 neurons have also been used to study the electrophysiological actions of GHB. Kemmel *et al.* (1998) reported that in patch-clamp experiments using these cells, GHB reduced voltage-dependent Ca^{2+} conductances, an effect blocked by NCS-382 but not by a GABA$_B$ antagonist. Thus, the best evidence implicates Ca^{2+} fluxes in the genesis of hyperpolarizations produced by GHB receptor stimulation and subsequent G protein activation.

Pharmacology of the GHB receptor

Though inhibitors of [^{3}H]GHB receptor binding are known, little information is available on possible agonists or antagonists at the GHB receptor. Bourguignon *et al.* (1988) investigated the effects of a number of GHB structural analogues on [^{3}H]GHB binding to rat brain membranes. The most effective inhibitors were THCA and 4-phenyl-THCA, each of which was more potent than non-radioactive GHB at displacing the ligand. Also more effective than GHB itself were 4-methyl GHB and 3-methyl GHB. Equally effective as GHB were 4-phenyl GHB and 3-methyl-THCA. Slightly less effective than GHB itself were 4-methyl-THCA and 5-hydroxyvaleric acid. Unfortunately, there is no evidence that any of these compounds, apart from THCA, have agonist properties at the GHB receptor.

THCA, like GHB, has been shown to possess both antidopaminergic and neuroleptic properties, both of which can be blocked by a specific GHB antagonist (Hechler *et al.* 1993). It is noteworthy that low levels of THCA have been found to occur naturally in the CNS (Ehrhardt *et al.* 1988). Although Hechler *et al.* (1990) measured the specific binding of [^{3}H]THCA to rat brain homogenates and found that there were two kinetically distinct binding sites, the authors concluded that the ligand had an affinity for a subpopulation of GHB receptors. No evidence of a functional role for THCA has yet emerged.

Other structural analogues of GHB have been synthesized and tested as inhibitors of [^{3}H]GHB receptor binding (Hechler *et al.* 1993). These compounds were γ-*p*-fluoro-phenyl-THCA, γ-*p*-trifluoromethyl-phenyl-THCA, γ-*p*-chloro-phenyl-THCA, γ-*σ*-fluoro-phenyl-THCA and γ-*p*-nitro-phenyl-THCA. Each analogue was more effective at inhibiting [^{3}H]GHB binding than GHB itself. Moreover, each inhibitor was more potent than GHB at inhibiting dopamine release from striatal slices. The compounds also exhibited neuroleptic activity in a series of intact animal tests which was comparable to the effects of GHB. The GHB antagonist NCS-382 antagonized these effects. This is evidence for the GHB agonist properties of these analogues.

A group of currently used clinical drugs exhibit a high affinity for the GHB receptor. These are some substituted benzamides which include (−)sulpiride, amisulpride and sulto-pride (Maitre *et al.* 1994). Though these compounds bear no structural resemblance to GHB, they readily displace [^{3}H]GHB from its receptor recognition site. These drugs possess antipsychotic properties and are antagonists for dopamine receptors. These observations lead to the intriguing possibility that the beneficial effects of these drugs in schizophrenia might, at least in part, be due to a modulation of GHBergic activity (and thus to an inhibition of dopamine function) as well as to a direct reduction in dopaminergic activity. Later experiments revealed that (−)sulpiride, after repeated administration, can up-regulate GHB receptors in rat brain. This is in contrast to GHB administration which led to receptor down-regulation (Ratomponirina *et al.* 1998). A similar result was obtained when NCB-20 cells in culture were exposed to the two drugs.

The most important GHB antagonist discovered is the GHB analogue NCS-382. This compound was first reported to inhibit [^{3}H]GHB receptor binding by Maitre *et al.* (1990). These investigators also demonstrated that NCS-382 possessed several properties that were consistent with it being an antagonist. First, it abolished cGMP increases in rat brain induced by GHB treatment. Second, pretreatment of rats with NCS-382 completely prevented the GHB-induced release of dopamine in the striatum. Third, the drug blocked the abnormal EEG epileptoform spike profile evoked by GHB administration. In addition to its effects on GHB-induced events, NCS-382 was shown to possess anticonvulsant properties. The drug protected mice against audiogenic seizures in the Swiss Rb strain. Finally, NCS-382 diminished generalized non-convulsive seizures in a rat model of petit mal epilepsy (Maitre *et al.* 1990). Further studies support the idea that NCS-382 is a GHB antagonist. After a series of sensorimotor tests in Wistar rats, Schmidt *et al.* (1991) reported that the compound reduced the sedative and cataleptic effects of GHB treatment. At low doses, GHB increased the spontaneous firing rate of rat cortical neurons (Godbout *et al.* 1995). Administration of NCS-382 blocked the excitatory effects of GHB. At higher doses GHB had an inhibitory effect on cell firing rate, but in this case NCS-382 was unable to antagonize the effect. NCS-382 has also been used in behavioral studies. Colombo *et al.* (1995) found that the discriminative properties of GHB in rat could be blocked by treatment with the antagonist. This effect was shown at both low and high doses of GHB. Therefore, it provides evidence of a functional role for GHB in certain types of behavior.

Role of the GHB receptor in CNS function

Undoubtedly, when GHB was first shown to be present in the CNS (Bessman and Fishbein 1963), it evoked the question: why is it there? We previously attempted to address this question in an earlier commentary (Cash 1996). Almost forty years after the original discovery of GHB in the brain, we are closer to an answer. The distribution has been found to be uneven in the brain (Doherty *et al.* 1978). The existence of enzymes necessary for its biosynthesis and metabolic degradation have been characterized (Kaufman 2002). Unique pharmacological properties have been described, some of which can be blocked with NCS-382 (Maitre *et al.* 1990; Tunnicliff 1992; Colombo *et al.* 1995; Godbout *et al.* 1995). Depolarization of neurons stimulates GHB release (Maitre *et al.* 1983a; Kemmel *et al.* 1998). Neurons exposed to GHB consistently undergo a hyperpolarization (Olpe and Koella 1979; Kozhechkin 1980). A sodium-dependent active transport system has been documented (Benavides *et al.* 1982b; McCormick and Tunnicliff 1998). This transport process could be responsible for rapidly removing GHB from the synaptic gap after hyperpolarizing adjacent nerve cells. All these observations are consistent with GHB being a neurotransmitter. The studies on [^{3}H]GHB binding, and more recently those on the molecular structure of the GHB receptor, lend further robust support to this idea. Early studies suggested an intimate link between GHB action and dopaminergic activity (Gessa *et al.* 1966, 1968; Walters and Roth 1972). In the light of accumulated evidence since then, it appears that the effect of GHB on dopamine neurons is to decrease the release of this neurotransmitter. Many of these effects are undoubtedly mediated through GHB receptors, but an indirect activation of GABAergic mechanisms also seems to be part of the overall effects on dopamine function (Maitre 1997).

Evidence has been presented that GHB acts as an agonist at least at two different plasma membrane located receptors – the GABA$_B$ receptor and the GHB receptor. Since it is accepted that GABA is the native ligand for GABA$_B$ receptors, and the affinity of GHB for this receptor is relatively low, it is tempting to suggest that exogenously applied GHB acts through both GABA$_B$ receptors and GHB receptors whereas the physiological action of GHB is mediated by GHB receptors only.

References

Andriamampandry C. and Maitre M. (2000) Cloning, expression and characterization of a cDNA coding for a brain gamma-hydroxybutyrate (GHB) receptor. WO 00/78948 A2.

Andriamampandry C., Kemmel V., Hechler V., *et al.* (1999) Characterization of the receptor for the neurotransmitter and addictive drug γ-hydroxybutyrate (GHB). *Biochimie* 81 (Suppl. 6), 107.

Benavides J., Rumigny J. F., Bourguignon J. J., Cash C. D., Wermuth C. G., Mandel P., Vincendon G. and Maitre M. (1982a) High affinity binding site for γ-hydroxybutyric acid in rat brain. *Life Sci.* 30, 953–96.

Benavides J., Rumigny J. F., Bourguignon J. J., Wermuth C. J., Mandel P. and Maitre M. (1982b) A high affinity Na$^+$-dependent uptake system for γ-hydroxybutyrate in membrane vesicles prepared from rat brain. *J. Neurochem.* 38, 1570–75.

Bernasconi R., Mathivet P., Otten U., Bettler B., Bischoff S. and Marescaux C. (2002) Part of γ-hydroxybutyrate pharmacological actions are mediated by GABA$_B$ receptors. In *Gamma-Hydroxybutyrate: Molecular, Functional and Clinical Aspects* (Tunnicliff G. and Cash C.D., eds), pp. 28–63. Taylor and Francis, London.

Bessman S. P. and Fishbein W. M. (1963) Gamma-hydroxybutyrate, a normal brain metabolite. *Nature* 200, 1207–08.

Bourguignon J. J., Schoenfelder A., Schmitt M., Wermuth C.-G., Hechler V., Charlier B. and Maitre M. (1988) Analogues of γ-hydroxybutyric acid. Synthesis and binding studies. *J. Med. Chem.* 31, 893–97.

Bowery N. G. (2000) GABA$_B$ receptors: structure and function. In *GABA in the Nervous System: The View at Fifty Years* (Martin D.L. and Olsen R.W., eds), pp. 233–44. Lippincott Williams and Wilkins, Philadelphia.

Cash C. D. (1996) What is the role of the gamma-hydroxybutyrate receptor? *Med. Hypotheses* 47, 455–59.

Cash C. D., Hechler V., Mersel M. and Maitre M. (1996) Kinetic characterisation and solubilisation of γ-hydroxybutyrate receptors from rat brain. *Neurosci. Lett.* 209, 25–28.

Castelli M. P., Mocci I., Langlois X., Gommerendagger W., Luyten W. H., Leysen J. E. and Gessa G. L. (2000) Quantitative autoradiographic distribution of gamma-hydroxybutyric acid binding sites in human and monkey brain. *Brain Res. Mol. Brain Res.* 78, 91–99.

Colombo G., Agabio R., Bourguignon J., Fadda F., Lobina C., Maitre M., Reali R., Schmitt M. and Gessa G. L. (1995) Blockade of the discriminative stimulus effects of γ-hydroxybutyric acid (GHB) by the GHB receptor antagonist NCS-382. *Physiol. Behav.* 58, 587–90.

Doherty J. D., Hattox S. E., Snead O. C. and Roth R. H. (1978) Identification of endogenous γ-hydroxybutyrate in human and bovine brain and its regional distribution in human, guinea pig and rhesus monkey brain. *J. Pharmacol. Exp. Ther.* 207, 130–39.

Ehrhardt J. D., Vayer P. and Maitre M. (1988) A rapid and sensitive method for the determination of gamma-hydroxybutyric acid and *trans*-gamma-hydroxycrotonic acid in rat brain tissue by gas chromatography/mass spectrometry with negative ion detection. *Biomed. Env. Mass Spectrom.* 15, 521–24.

Erhardt S., Andersson B., Nissbrandt H. and Engeberg G. (1998) Inhibition of firing rate and changes in the firing pattern of nigral dopamine neurons by gamma-hydroxybutyric acid (GHBA) are specifically induced by activation of GABA$_B$ receptors. *Naunyn Schmiedebergs Arch. Pharmacol.* 357, 611–19.

Garcia-Frigola C., Burgaya F., Calbet M., de Lecea L. and Soriano F. (2000) Mouse Tspan-5, a member of the tetraspanin superfamily, is highly expressed in brain cortical structures. *NeuroReport* 11, 3181–85.

Gessa G. L., Vargiou L., Crabai F., Boero G. C., Caboni F. and Camba R. (1966) Selective increase of brain dopamine induced by gamma-hydroxybutyrate. *Life Sci.* 5, 1921–31.

Gessa G. L., Crabai L., Vargiu L. and Spano P. F. (1968) Selective increase of brain dopamine induced by γ-hydroxybutyrate: study of the mechanism of action. *J. Neurochem.* 15, 377–81.

Godbout R., Jelenic P., Labrie C., Schmitt M. and Bourguignon J. J. (1995) Effects of gamma-hydroxybutyrate and its antagonist NCS-382 on spontaneous cell firing in the prefrontal cortex of the rat. *Brain Res.* 673, 157–60.

Godschalk M., Dzoljic M. R. and Bonta I. L. (1977) Slow wave sleep and a state resembling absence epilepsy induced in the rat by γ-hydroxybutyrate. *Eur. J. Pharmacol.* 44, 105–11.

Guidotti A. and Balloti P. L. (1970) Relationship between pharmacological effects and blood and brain levels of gamma-butyrolactone and gamma-hydroxybutyrate. *Biochem. Pharmacol.* 19, 883–94.

Harris N. C., Webb C. and Greenfield S. A. (1989) The effects of gamma-hydroxybutyrate on the membrane properties of guinea-pig pars compacta neurons in the substantia nigra *in vitro*. *Neuroscience* 31, 363–70.

Hechler V., Schmitt M., Bourguignon J. J. and Maitre M. (1990) *trans*-γ-Hydroxycrotonic acid binding sites in brain: evidence for a subpopulation of γ-hydroxybutyrate sites. *Neurosci. Lett.* 110, 204–09.

Hechler V., Gobaille S. and Maitre M. (1992) Selective distribution pattern of γ-hydroxybutyrate receptors in the rat forebrain and midbrain as revealed by quantitative autoradiography. *Brain Res.* 572, 345–48.

Hechler V., Peter P., Gobaille S., Bourguignon J. J., Schmitt M., Ehrhardt J. D., Mark J. and Maitre M. (1993) γ-Hydroxybutyrate ligands possess antidopaminergic and neuroleptic-like activities. *J. Pharmacol. Exp. Ther.* 264, 1406–14.

Kaufman E. E. (2002) Metabolism and distribution of γ-hydroxybutyrate in the brain: In *Gamma-Hydroxybutyrate: Molecular, Functional and Clinical Aspects.* (Tunnicliff G. and Cash C.D., eds), pp. 1–16. Taylor and Francis, London.

Kemmel V., Taleb O., Perard A., Andriamampandry C., Siffert J. C., Mark J. and Maitre M. (1998) Neurochemical and electrophysiological evidence for the existence of a functional γ-hydroxybutyrate system in NCB-20 neurons. *Neuroscience* 86, 989–1000.

Kozhechkin S. X. (1980) Microiontophoretic study of the mechanism of action of gamma-hydroxybutyric acid. *Bull. Exp. Biol. Med.* 88, 1293–96.

Kopczynski C. C., Davis G. W. and Goodman C. S. (1996) A neural tetraspanin encoded by *late bloomer* that facilitates synaptic formation. *Science* 271, 1867–70.

Kuner R., Köhr G., Grünewald S., Eisenhardt G., Bach A. and Kornau H.-C. (1999) Role of heteromeric formation in GABA$_B$ receptor function. *Science* 283, 74–77.

Laborit H. (1964) Sodium 4-hydroxybutyrate. *Int. J. Neuropharmacol.* 3, 433–52.

Lorente P., Lacampagne A., Pouzeratte Y., Richards S., Malitschek B., Kuhn R., Bettler B. and Vassort G. (2000) γ-Aminobutyric acid type B receptors are expressed and functional in mammalian cardiomyocytes. *Proc. Natl. Acad. Sci. USA* 97, 8664–69.

McCormick S. J. and Tunnicliff G. (1998) Inhibitors of synaptosomal γ-hydroxybutyrate transport. *Pharmacology* 57, 124–31.

Madden T. E. and Johnson S. W. (1998) Gamma-hydroxybutyrate is a GABA$_B$ receptor agonist that increases a potassium conductance in rat ventral tegmental dopamine neurons. *J. Pharmacol. Exp. Ther.* 287, 261–65.

Maecker H. T., Todd S. C. and Levy S. (1997) The tetraspanin superfamily: molecular facilitators. *FASEB J.* 11, 428–42.

Maitre M. (1997) The γ-hydroxybutyrate signalling system in brain: organization and functional implications. *Prog. Neurobiol.* 51, 337–61.

Maitre M., Cash C. D., Weissmann-Nanopoulos D. and Mandel M. (1983a) Depolarization-evoked release of γ-hydroxybutyrate from rat brain slices. *J. Neurochem.* 41, 287–90.

Maitre M., Rumigny J. F. and Mandel P. (1983b) Positive cooperativity in high affinity binding sites for γ-hydroxybutyric acid in rat brain. *Neurochem. Res.* 8, 113–20.

Maitre M., Hechler V., Vayer P., Gobaille S., Cash C. D., Schmitt M. and Bourguignon J. J. (1990) A specific γ-hydroxybutyrate receptor ligand possesses both antagonistic and anticonvulsant properties. *J. Pharmacol. Exp. Ther.* 225, 657–63.

Maitre M., Ratomponirina C., Gobaille S., Hodé Y. and Hechler V. (1994) Displacement of [^{3}H]γ-hydroxybutyrate binding by benzamide neuroleptics and prochlorperazine but not by other antipsychotics. *Eur. J. Pharmacol.* 256, 211–14.

Mathivet P., Bernasconi R., De Barry J., Marescaux C. and Bittiger H. (1997) Binding characteristics of γ-hydroxybutyric acid as a weak but selective GABA$_B$ receptor agonist. *Eur. J. Pharmacol.* 321, 65–75.

Mehta A. K. and Ticku M. K. (1999) An update on GABA$_A$ receptors. *Brain Res. Rev.* 29, 196–217.

Olpe H. R. and Koella W. P. (1979) Inhibition of nigral and neocortical cells by γ-hydroxybutyrate: a microiontophoretic effect. *Eur. J. Pharmacol.* 53, 359–64.

Ratomponirina C., Yodé Y., Hechler V. and Maitre M. (1995) γ-Hydroxybutyrate receptor binding in rat brain is inhibited by guanylyl nucleotides and pertussis toxin. *Neurosci. Lett.* 189, 51–53.

Ratomponirina C., Gobaille S., Yodé Y., Kemmel V. and Maitre M. (1998) Sulpiride, but not haloperidol, up-regulates γ-hydroxybutyrate receptors *in vivo* and in cultured cells. *Eur. J. Pharmacol.* 346, 331–37.

Ropero-Miller J. D. and Goldberger B. A. (1998) Recreational drugs: current trends in the 90s. *Clin. Lab. Med. Tox.* 18, 727–46.

Roth R. H. and Giarman N. J. (1968) Conversion *in vivo* of γ-aminobutyric acid to γ-hydroxybutyric acid in the rat. *Biochem. Pharmacol.* 18, 247–50.

Roth R. H. and Giarman N. J. (1970) Natural occurrence of gamma-hydroxybutyrate in mammalian brain. *Biochem. Pharmacol.* 19, 1087–93.

Schmidt C., Gobaille S., Hechler V., Schmitt M., Bourguignon J. J. and Maitre M. (1991) Anti-sedative and anti-cataleptic properties of NCS-382, a γ-hydroxybutyrate receptor antagonist. *Eur. J. Pharmacol.* 203, 393–97.

Snead O. C. (1992) Evidence for G protein modulation of experimental-generalized absence seizures in rat. *Neurosci. Lett.* 148, 15–18.

Snead O. C. (1994) The ontogeny of [^{3}H]γ-hydroxybutyrate and [^{3}H]GABA$_B$ binding and binding sites: relation to the development of experimental absence seizures. *Brain Res.* 659, 147–56.

Snead O. C. (2000) Evidence for a G protein-coupled γ-hydroxybutyric acid receptor. *J. Neurochem.* 75, 1986–96.

Snead O. C. and Liu C.-C. (1984) Gamma-hydroxybutyric acid binding sites in rat and human brain synaptosomal membranes. *Biochem. Pharmacol.* 33, 2587–90.

Tunnicliff G. (1992) Significance of γ-hydroxybutyric acid in the brain. *Gen. Pharmacol.* 23, 1027–34.

Vickers M. D. (1969) Gammahydroxybutyric acid. *Int. Anaesth. Clin.* 7, 75–89.

Walters J. R. and Roth R. H. (1972) Effect of gamma-hydroxybutyrate on dopamine and dopamine metabolites in the rat striatum. *Biochem. Pharmacol.* 21, 2111–21.

White R. P. and Samson F. E. (1956) Effects of fatty acid anions on electroencephalogram of unanesthetized rabbits. *Am. J. Physiol.* 186, 271–78.

Winters W. D. and Spooner C. E. (1965) Various seizure activities following γ-hydroxybutyrate. *Int. J. Neuropharmacol.* 4, 197–200.

Xie X. and Smart T. G. (1992) γ-Hydroxybutyrate hyperpolarizes hippocampal neurones by activating GABA$_B$ receptors. *Eur. J. Pharmacol.* 212, 291–94.

Zemni R., Bienvenu T., Vinet M. C., Sefiani A., Carrie A., Billuart P., McDonell N., Couvert P., Francis F., Chafey P., Fauchereau F., Friocourt G., Portes V. D., Cardona A., Frints S., Meindl A., Brandau O., Ronce N., Moraine C., Bokhoven H. V., Ropers H. H., Sudbrak R., Kahn A., *et al.* (2000) A new gene involved in X-linked mental retardation identified by analysis of an X;2 balanced translocation. *Nature Genet.* 24, 167–70.

3 Part of the pharmacological actions of γ-hydroxybutyrate are mediated by GABA$_B$ receptors

R. Bernasconi, P. Mathivet, U. Otten, B. Bettler,
S. Bischoff and C. Marescaux

Introduction

γ-Hydroxybutyric acid (GHB) is a natural constituent of mammalian brain, which is formed primarily following γ-aminobutyric acid (GABA) metabolism through GABA-transaminase (E.C.2.6.1.19) and specific succinic semialdehyde reductase (E.C.1.1.1.1.2) (Fig. 3.1) (Maitre 1997). Concentrations of GHB in rat brain are in the range 2–5 nmol/g (Bernasconi *et al.* 1992). Numerous experimental data have provided evidence that endogenous GHB is not just an incidental by-product of GABA metabolism. Indeed, it fulfills some of the criteria generally accepted for defining a neurotransmitter such as conditions of synthesis, release, turnover, degradation, presence of active transport and uptake mechanisms and binding to specific sites (Benavides *et al.* 1982a,b; Maitre 1997) ($K_{d1} = 30$–580 nM; $K_{d2} = 2.3$–16 μM, Table 3.1). However, the proposal that GHB is a neurotransmitter has been questioned for a number of reasons: (1) It has not been shown that GHB is stored in presynaptic vesicles, which are ready for fusion with the cellular membrane and subsequent exocytosis. (2) No specific neuronal tracts that use GHB have been identified. (3) The effects of endogenous GHB on signal transmission and its physiological role are unknown. (4) GHB has been found at higher concentrations in non-neuronal tissue such as the heart, kidney, liver, muscle and brown fat than in the central nervous system (CNS). Additionally, peripheral GHB is able to cross the blood–brain barrier (Nelson *et al.* 1981), unprecedented for a neurotransmitter.

While the physiological role of GHB is unclear, it is known that a large variety of pharmacological responses are produced by the administration of high doses (100–400 mg/kg) of exogenous GHB (Maitre 1997; Table 3.2a). Oral, intraperitoneal or intravenous application of GHB or of its metabolic precursor γ-butyrolactone (GBL, Fig. 3.1) dose-dependently (1) depresses CNS function and induces sedation, sleep and coma (Maitre 1997), (2) induces EEG hypersynchronization sometimes associated with absence-like seizures (Snead 1991), (3) hyperpolarizes neuronal membranes (Olpe and Koella 1979; Harris *et al.* 1989; Xie and Smart 1992; Williams *et al.* 1995), and (4) alters synthesis and/or release of various neurotransmitters (GABA, glutamate or dopamine) (Waldmeier 1991; Engberg and Nissbrandt 1993; Nissbrandt *et al.* 1994; Banerjee and Snead 1995; Bernasconi *et al.* 1995; Emri *et al.* 1996; Feigenbaum and Howard 1997; Hu *et al.* 2000). These behavioral, EEG, cellular and biochemical responses occur only when GHB levels are increased 50–100-fold above their endogenous concentrations (Cash 1994). This discrepancy suggests that the effects of exogenous GHB are not only mediated through stimulation of GHB binding sites, but also

On the occasion of his eightieth birthday, the authors would like to dedicate this work to Dr H. Keberle, who synthesized baclofen about forty years ago.

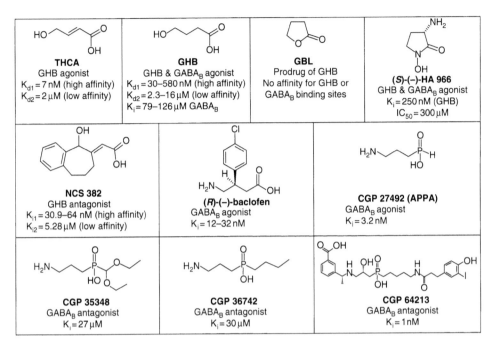

Figure 3.1 Structures of some agonists and antagonists of GHB binding sites and GABA$_B$ receptors, respectively. GBL does not bind to GHB binding sites or to GABA$_B$ receptors and is inactive *per se*. However, as GBL is rapidly and irreversibly hydrolyzed to GHB by peripheral γ-lactonase and is more rapidly and reproducibly absorbed, it is often used as a GHB precursor in animal experiments. (GHB: γ-hydroxybutyric acid; HA-966: S-(−)-3-amino-1-hydroxy-pyrrolidin-2-one; GBL: γ-butyrolactone; TCHA: *trans*-γ-hydroxycrotonic acid; NCS-382: 6,7,8,9-tetrahydro-5-hydroxy-5H-benzocyclohept-6-ylideneacetic acid; *R*-(−)-baclofen: 4-amino-3-(*R*)-3-(4-*p*-chlorophenyl)-butyric acid; CGP 27492: (3-amino-propyl)-phosphinic acid; CGP 35348: (3-amino-propyl)-diethoxymethyl-phosphinic acid; CGP 36742: (3-amino-propyl)-*n*-butyl-phosphinic acid; CGP 64213: 3-{1-(*R*)-[2-(*S*)-hydroxy-3-(hydroxy-{5-[3-(4-hydroxy-3-iodo-phenyl)-propionylamino]-pentyl}-phosphinoyl)-propyl-amino]-ethyl}-benzoic acid.)

through other mechanisms. Indeed, many biological effects secondary to systemic administration of high doses (100–400 mg/kg) can be blocked by GABA$_B$ receptor antagonists, but are not suppressed by NCS-382, an antagonist at [^{3}H] GHB binding sites (Maitre *et al.* 1990). Moreover, some pharmacological properties of exogenous GHB are shared by GABA$_B$ receptor agonists such as *R*-(−)-baclofen (Xie and Smart 1992; Engberg and Nissbrandt 1993; Williams *et al.* 1995; Madden and Johnson 1998; Tables 3.2a and b). These findings, suggesting that exogenous GHB can act as an agonist at GABA$_B$ receptors, were confirmed by binding studies demonstrating that GHB is a weak, but selective, agonist of GABA$_B$ receptors (Bernasconi *et al.* 1992; Mathivet *et al.* 1997). More recently, Lingenhoehl *et al.* (1999) have demonstrated that GHB activates recombinant heteromeric GABA$_B$R1/R2 receptors expressed in *Xenopus* oocytes. However, these data do not imply that GABA$_B$ receptors mediate all the actions of endogenous or exogenous GHB (Feigenbaum and Howard 1996; Maitre 1997; Mathivet *et al.* 1997). It is unlikely that endogenous GHB activates

Table 3.1 Displacement of various radioligands by GHB

Radioligand	Binding site	GHB affinity	Reference
[³H]GHB	GHB	$K_{d1} = 30\text{--}580\ nM^a$ $K_{d2} = 2.3\text{--}16\ \mu M^a$	Maitre (1997)
[³H]baclofen	GABA$_B$	$K_i = 125\ \mu M^b$	Bernasconi *et al.* (1992)
[³H]baclofen	GABA$_B$	$K_i = 398\ \mu M^{b,c}$	Ito *et al.* (1995)
[³H]baclofen	GABA$_B$	$K_i = 630\ \mu M^{b,c}$	Ishige *et al.* (1993, 1996)
[³H]CGP 27492	GABA$_B$	$K_i = 80\text{--}120\ \mu M^{b,d}$	Mathivet *et al.* (1997)
[³H]GABA^e	GABA$_B$	$K_i = 80\text{--}120\ \mu M^{b,d}$	Mathivet *et al.* (1997)
[³H]CGP 54626	GABA$_B$	$K_i = 3,300\ \mu M^f$	Mathivet *et al.* (1997)
[³H]muscimol	GABA$_A$	—	Bernasconi *et al.* (1992)
[³⁵S]TBPS^g	Cl⁻ channel	—	Serra *et al.* (1991)
[³H]flunitrazepam^h	Benzodiazepine	—	Serra *et al.* (1991)
[³H]naloxone	μ-Opioid	—	Bernasconi *et al.* (1992)
[³H]DAGO	μ-Opioid	—	Feigenbaum and Simantov (1996)
[³H]diprenorphine	δ-Opioid	—	Feigenbaum and Simantov (1996)
[³H]etophirne	κ-Opioid	—	Feigenbaum and Simantov (1996)
[³H]prazosin	α_1-Adrenoceptor	—	Bernasconi *et al.* (1992)
[³H]clonidine	α_2-Adrenoceptor	—	Bernasconi *et al.* (1992)
[³H]DHA	β-Adrenoceptor	—	Bernasconi *et al.* (1992)
[³H]5-HT	5-HT$_1$	—	Bernasconi *et al.* (1992)
[³H]doxepine	Histamine H$_1$	—	Bernasconi *et al.* (1992)
[³H]QNB	Muscarinic	—	Bernasconi *et al.* (1992)
[³H]CHA	Adenosine A$_1$	—	Bernasconi *et al.* (1992)
[³H]substance P	Substance P	—	Bernasconi *et al.* (1992)
[³H]MK801	NMDA	—	Gessa *et al.* (1993)

The methods used for the [³H]GHB binding and for the GABA$_B$ receptor assay using agonist and antagonist radioligand are different. Crude membrane fractions prepared from total rat brain bind [³H]GHB in a saturable and reversible manner (Benavides *et al.* 1982a). The binding reaches equilibrium within 15–20 min at 0 °C in a buffered medium. Maximum binding occurs at pH 5.5. Absence of Ca^{2+} in the incubation buffer does not change the binding characteristics. The GABA$_B$ receptor-binding assays are performed using frozen and thawed crude membranes prepared from rat cerebral cortex (Bittiger *et al.* 1996). The radioreceptor assay is performed at pH 7.4 and at 20 °C for 40 min. Ca^{2+} is necessary for the high affinity of GABA on native and recombinant GABA$_B$ receptors (Galvez *et al.* 2000).

Notes
a K_{d1} = high-affinity and K_{d2} = low-affinity GHB binding sites.
b K_i values were calculated from the IC$_{50}$ values by means of the Cheng–Prusoff equation (Leff and Douglas 1993).
c Ito *et al.* (1995) and Ishige *et al.* (1996) use cultured cerebellar granule cells and mouse cerebellar membrane, respectively.
d Similar K_i of GABA$_B$ receptors were measured in membranes prepared from rat hippocampus or cerebellum.
e This binding assay is performed in the presence of the GABA$_A$ agonist isoguvacine hydrobromide.
f The IC$_{50}$ values of GABA$_B$ receptor agonists are much higher in the antagonist [³H]CGP 54626 assay than in the [³H]baclofen and [³H]CGP 27492 agonist radioligand assays (Bittiger *et al.* 1992).
g GHB does not modify *in vitro* muscimol-stimulated^{36} Cl⁻ uptake. *In vivo* administration of sedative and hypnotic doses of GHB (300–700 mg/kg) fail to induce any significant changes in the [³⁵S]TBPS binding. Moreover, GHB also does not antagonize the increase in [³⁵S]TBPS binding induced by isoniazid.
h GHB fails to modify the modulatory action of GABA on [³H]flunitrazepam binding.

TBPS: *t*-butylbicyclophosphorothionate; DAGO: (D-Ala², N-Me-Phe⁴, glycinol⁵)-enkephalin; DHA: dihydro-alprenolol; QNB: quinuclidinyl benzylate; CHA: cyclohexyl-adenosine; —: no displacement of specific binding by GHB up to 10^{-4}M.

Table 3.2a Pharmacological *in vivo* actions of GHB attributed either to stimulation of GHB binding sites or GABA$_B$ receptors

Antagonized by	GHB (GBL) doses	Pharmacological responses	Reference
NCS-382	5–10 mg/kg	Spontaneous firing rate	Godbout *et al.* (1995)
NCS-382	5–30 mg/kg	Hyperthermia	Kaufman *et al.* (1990, see 4.1)
NCS-382	500 mg/kg	Increase of cGMP[a]	Vayer *et al.* (1987, 1989)
GABA$_B$ antagonists	100–400 mg/kg	Decrease of cGMP[b]	Bernasconi *et al.* (1992, see 3.3)
GABA$_B$ antagonists	100 mg/kg	Increase in CRE and AP-1 DNA-binding activities[c]	Ishige *et al.* (1996)
GABA$_B$ antagonists	12.5–1,600 mg/kg	Decrease in firing rate of DA neurons[d]	Erhardt *et al.* (1998, see 3.8)
NCS-382 GABA$_B$ antagonists	200–400 mg/kg	Absence-like seizures	Snead (1996, see 4.2)
NCS-382 GABA$_B$ antagonists	250–1,500 µM	GABA and glutamate release[e]	Banerjee *et al.* (1995) Hu *et al.* (2000, see 3.5)
GABA$_B$ antagonists	400–600 mg/kg	Hypothermia, loss of righting reflex	Kaufman *et al.* (1990, see 4.1)

Notes

a GHB induces an increase of cGMP content in the hippocampus of rats, which was prevented by valproate and naloxone. Neither NCS-382 nor GABA$_B$ receptor antagonists has been tested.

b GHB decreases cGMP content in the cerebellum and thalamus of rats and mice. Administration of NCS-382 (200–400 mg/kg i.p.), suggested to act as a specific antagonist of GHB sites, does not affect the decrease of cGMP induced by GHB.

c GBL increases dose-dependently nuclear cyclic AMP-responsive element (CRE) and activator protein 1 (AP-1) DNA-binding activities in mouse brain.

d GHB induces a decrease in the firing rate and a regularization of the firing rhythm with a reduction of burst activity of nigral dopamine (DA) neurons (Engberg and Nissbrandt 1993; Nissbrandt *et al.* 1994). There are several reasons to believe that GHB produced its effects on the firing of nigral DA neurons specifically via activation of GABA$_B$ receptors and not via stimulation of GHB binding sites. First, baclofen which is devoid of affinity to GHB binding sites (Snead 1996), produces almost identical effects on the firing of nigral DA neurons as GHB (Engberg *et al.* 1993). Second, pretreatment with specific GABA$_B$ receptor antagonists, which are deprived of affinity for GHB binding sites (Snead 1996), antagonizes the effects of GHB and baclofen on the firing of nigral DA neurons. Third, NCS-382, which is reported to selectively block GHB binding sites with no antagonistic action at GABA$_B$ receptors (Maitre *et al.* 1990; Snead 1996), fails to antagonize the effects of GHB and baclofen on nigral DA activity. Fourth, baclofen markedly increases DA synthesis early after administration and later reduces it (Waldmeier 1991). During this later phase, it causes a marked accumulation of the levels of DA and its deaminated metabolites. GHB closely reproduces this sequence of effects, indicating that both drugs share a similar mode of action, namely stimulation of GABA$_B$ receptors. GABA$_B$ receptor antagonists attenuate the increase in rat striatal dopamine synthesis elicited by graded doses of GHB (Waldmeier 1991).

e Basal extracellular release of GABA, but not that of glutamate (GLU), is inhibited by GHB (250–1500 µM of GHB added to the perfusion solution) in rat thalamic ventrobasal nucleus (Banerjee and Snead 1995). On the other hand, the K^+-evoked release of both GABA and GLU in the thalamic ventrobasal nucleus is significantly decreased by all concentrations of GHB tested. Both the basal and the K^+-evoked release of GABA are significantly decreased in the frontal cortex at the onset of GHB-induced absence seizures. In contrast, neither the basal nor the K^+-evoked release of GLU is altered in superficial laminae of cerebral cortex at any time during absence seizures elicited by GHB (Hu *et al.* 2000). The effects of GHB in thalamic ventrobasal nucleus and in cerebral cortex are partially reversed by NCS-382 and GABA$_B$ receptor antagonists. R-(−)-baclofen also produces a concentration-dependent decrease in basal and K^+-evoked release of GABA and GLU in the thalamic ventrobasal nucleus. Unfortunately, Hu *et al.* (2000) have not investigated the effects of GABA$_B$ receptor agonists on the basal and K^+-evoked release of GABA and GLU in the cerebral cortex. According to Gobaille *et al.* (1999), 500 mg/kg GHB increases GABA content in dialysates of the frontal cortex whereas 250 mg/kg decreases GABA release. Peripheral administration of NCS-382 blocks both the increase and the decrease of GABA content in dialysates. GABA$_B$ receptor antagonists have not been used in this study.

Table 3.2b Pharmacological *in vitro* actions of GHB attributed either to stimulation of GHB binding
sites or GABA$_B$ receptors

Antagonized by	GHB concentrations	Biological responses	Reference
NCS-382	25–50 μM	Inhibition of Ca^{2+} conductance[a]	Kemmel *et al.* (1998)
NCS-382	2.5–10 μM	Decrease of NO synthase[b]	Cash *et al.* (1999)
GABA$_B$ antagonists	0.1–3 mM	Increase of [Ca^{2+}]$_i$[c]	Ito *et al.* (1995)
GABA$_B$ antagonists	EC$_{50}$ = 0.88 mM	Increase of K$^+$ conductance[d]	Madden and Johnson (1998)
GABA$_B$ antagonists	IC$_{50}$ = 10.85 mM	Decrease of EFPs in the CA1 region[e]	King *et al.* (1997)

Notes

a A GABA$_B$ receptor antagonist does not block the GHB-induced inhibition of Ca^{2+} conductances in neurohybridoma cells (NCB-20).

b GHB produces a reduction of NO synthase activity at doses in the range 2.5–10 μM.

c GHB induces a concentration-dependent increases in [Ca^{2+}]$_i$ in primary cultures of mouse cerebellar granule cells.

d GHB hyperpolarizes DA neurons in the ventral tegmental area of the rat midbrain slice by a GABA$_B$ receptor-mediated increase in K$^+$ conductance and reduces depolarization-activated Ca^{2+} spikes. None of these effects are prevented by NCS-382.

e GHB produces a concentration-dependent depression of evoked synaptic field potentials (EFPs) recorded extracellularly in the CA1 region of the rat hippocampal slice. Baclofen also depresses the EFP, whereas the weak GABA$_B$ receptor antagonist 2-hydroxysaclofen blocks most of this effect. No antagonism of GHB-induced depression of EFP by NCS-382 could be detected.

GABA$_B$ receptors. Concentrations of GHB that are necessary to bind to or stimulate GABA$_B$ receptors are one order of magnitude higher than the brain levels of GHB (Benavides *et al.* 1982a; Snead and Liu 1984; Bernasconi *et al.* 1992). [^{3}H]GHB binding sites have no affinity for GABA or *R*-(−)-baclofen (Snead and Liu 1984) and differ from GABA$_B$ receptor binding sites, both in their distribution (Hechler *et al.* 1992) and developmental appearance (Snead 1994). Furthermore, some specific responses produced by low doses of exogenous GHB (5–50 mg/kg) are reversed by NCS-382 but not by GABA$_B$ receptor antagonists, whereas some behavioral and EEG effects of GHB cannot be reproduced by GABA$_B$ receptor agonists (Diana *et al.* 1991; Godbout *et al.* 1995; Maitre 1997; Kemmel *et al.* 1998; Tables 3.2a and b).

An attractive hypothesis is that GHB can bind to both GHB and GABA$_B$ receptor sites, and thus can induce different biological responses depending on whether GHB activates one or both of these receptors. The action of low doses of GHB (5–50 mg/kg) that produce no change in GHB brain levels (Kaufman *et al.* 1990) is probably due to stimulation of the high-affinity GHB binding sites only. The effects of higher doses (100–400 mg/kg), which increase GHB brain content to concentrations sufficient to interact with GABA$_B$ receptors (Snead 1991), may be related to stimulation of GABA$_B$ receptors alone, or to simultaneous activation of both GHB and GABA$_B$ receptors. At very large doses (over 500–750 mg/kg) the action of GHB may not be selective.

Despite the significant amount of knowledge that has been accumulated about GHB, much remains unknown. This review will attempt to describe the current knowledge on the GABA$_B$ receptors and GHB binding sites, the putative interactions between both sites, and the physiological significance of the stimulation of GABA$_B$ receptors by GHB. The results presented in this chapter suggest that $GABA_B$ receptors play an important role in the effects induced by systemic administration of exogenous GHB at doses between 100 and 400 mg/kg.

$GABA_B$ receptors and GHB binding sites

$GABA_B$ receptors: a seven-transmembrane heterodimer

The neurotransmitter GABA activates both the ionotropic $GABA_A$ and $GABA_C$ receptors and the G protein-coupled metabotropic $GABA_B$ receptors. $GABA_A$ and $GABA_C$ receptors are permeable to Cl^- ions and mediate, in most cases, fast inhibition of postsynaptic neurons. $GABA_B$ receptors couple to pertussis toxin-sensitive G proteins (G_i or G_o) and accordingly inhibit the activity of adenylate cyclase and voltage-gated Ca^{2+} channels or activate G protein-regulated inwardly rectifying K^+ channels (Kir3) (Bowery 2000).

In rodent brain, $GABA_B$ binding sites appear to be present at an early stage of life, and peak at regionally specific times during the first three postnatal weeks, and then decrease to levels detected in adult animals (Knott *et al.* 1993; Turgeon and Albin 1993). The characteristics of the receptors and their signaling in young animals seem to be similar to those in adults, although some indications of pharmacological differences have been found (Turgeon and Albin 1993).

In adult mammalian brain, the ubiquitous distribution of $GABA_B$ receptor sites, as determined by receptor autoradiography (Bowery *et al.* 1987; Chu *et al.* 1990; Bischoff *et al.* 1999), is consistent with the presence of functional $GABA_B$ receptors on many neurons throughout the CNS (Mott and Lewis 1994; Misgeld *et al.* 1995). Expression in glial cells, if any, is marginal (Bischoff *et al.* 1999; but see Kang *et al.* 1998 and Fritschy *et al.* 1999a). Recently, Otten *et al.* (2000) have demonstrated that $GABA_B$ receptors are present in primary cultures of rat astrocytes and that activation of these receptors induces specific neurotrophin expression. $GABA_B$ receptors play important roles in physiological and pathological phenomena such as pain transmission, absence epilepsy, spasticity and drug addiction (Mott and Lewis 1994; Kerr and Ong 1995; Misgeld *et al.* 1995).

Data from biochemical and electrophysiological studies indicate that $GABA_B$ receptors are present on presynaptic terminals and at postsynaptic sites (Alford and Grillner 1991; Thompson and Gähwiler 1992). Activation of postsynaptic $GABA_B$ receptors generates the late inhibitory postsynaptic potential (IPSP) that is important for the fine-tuning of inhibitory neurotransmission (Crunelli and Leresche 1991; Deisz 1997). Characteristically, the late IPSP caused by an increase in membrane K^+ conductance is slower in onset and has a prolonged duration compared with the fast IPSP, which derives from $GABA_A$ receptors. Presynaptic $GABA_B$ receptor activation decreases neurotransmitter and neuropeptide release, presumably by diminution of a Ca^{2+} conductance. Presynaptic $GABA_B$ receptors are found not only on GABAergic terminals (autoreceptors), but also on non-GABAergic (e.g. glutamatergic) terminals (heteroreceptors) (Davies and Collingridge 1996; Marshall *et al.* 1999).

The structure of $GABA_B$ receptors was first identified in 1997 when two isoforms $GABA_{B(1a)}$ and $GABA_{B(1b)}$ (molecular weight 130 and 92 kDa, respectively) were isolated using an expression cloning approach (Kaupmann *et al.* 1997) (Fig. 3.2). This was dependent on the development of a high-affinity radiolabeled iodinated antagonist ligand [^{125}I]CGP 64213. The topology of the receptor is indicative of seven transmembrane domain receptors with an amino-terminus facing the extracellular space and a C-terminus protruding into the cell (Kaupmann *et al.* 1997). The sequence homology classifies $GABA_B$ receptors as

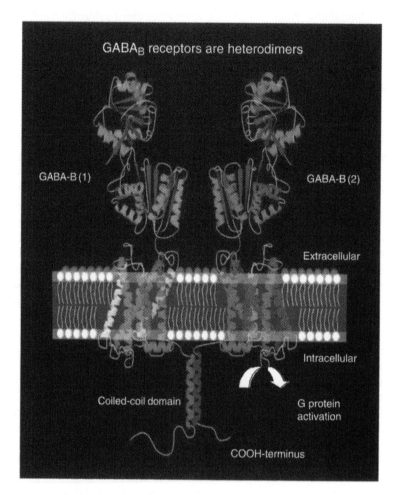

Figure 3.2 Model of the GABA$_B$ receptor heterodimer formed by association of the subunits GABA$_{B(1)}$ and GABA$_{B(2)}$. Features supported by experimental evidence include binding sites for all known competitive GABA$_B$ antagonists, GABA and GHB on GABA$_{B(1)}$ and coiled-coil domains at the intracellular C-termini that are important for surface trafficking of the receptor. The G proteins are coupled to the GABA$_{B(2)}$ receptor (Galvez *et al.* 2001; Margeta-Mitrovic *et al.* 2001; Robbins *et al.* 2001; Duthey *et al.* 2002). A new class of compounds, the positive modulators, binds to the GABA$_{B(2)}$ subunit (Urwyler *et al.* 2001). These compounds do not activate the receptor on their own; rather they potentiate the efficacy and affinity of agonists on the GABA$_B$ receptor, as observed with the benzodiazepines on the GABA$_A$ receptor. GHB is not a positive modulator of the GABA$_B$ receptor. (See Color Plate I.)

a member of family 3 G protein-coupled receptors (GPCRs). The two most prominent splice variants GABA$_{B(1a)}$ and GABA$_{B(1b)}$ differ in their extreme amino-terminal sequence, but share ligand-binding and effector domains. The first 163 amino acids of GABA$_{B(1a)}$ differ from the N-terminus of GABA$_{B(1b)}$ and contain so-called 'Sushi domains', known to be involved in protein–protein interactions (Hawrot *et al.* 1998). This amino-terminal extracellular domain of GABA$_B$ receptors contains all the structural information that is necessary and sufficient for ligand binding (Malitschek *et al.* 1999). $GABA_{B(1a)}$ and $GABA_{B(1b)}$, when expressed alone, display an affinity for antagonists and an overall distribution corresponding to native $GABA_B$ receptors. However, recombinant $GABA_{B(1a)}$ and $GABA_{B(1b)}$ exhibit a 100–150-fold reduced binding potency for agonists as compared to native receptors (Kaupmann *et al.* 1997). Furthermore, they couple only weakly to adenylate cyclase and their coupling to other effector systems, such as K^+ and Ca^{2+} channels, remained elusive. In addition, when expressed in mammalian cells, $GABA_{B(1)}$ is not transported to the cell membrane. These features were explained with the identification of $GABA_{B(2)}$ receptor protein, a second component of $GABA_B$ receptors (Fig. 3.2).

The $GABA_{B(2)}$ protein is a seven-transmembrane domain protein that displays a sequence homology of about 35% with the $GABA_{B(1)}$ receptor (Jones *et al.* 1998; Kaupmann *et al.* 1998a; White *et al.* 1998; Kuner *et al.* 1999; Ng *et al.* 1999). Its distribution corresponds largely to that of $GABA_{B(1)}$ as demonstrated by *in situ* hybridization (Fig. 3.3). $GABA_{B(2)}$ fulfills several functions. First, $GABA_{B(2)}$ traffics $GABA_{B(1)}$ to the cell surface where both proteins form a heterodimer as shown by co-immunoprecipitation (Jones *et al.* 1998; Kaupmann *et al.* 1998a; White *et al.* 1998). Second, these two receptors were found to be tightly associated via an interaction through coiled-coil domains at the C-terminal (White *et al.* 1998; Kuner *et al.* 1999). This interaction is essential for surface trafficking, but not for assembly of the heterodimer *per se* (Calver *et al.* 2001; Pagano *et al.* 2001). Further experiments showed that the interactions between $GABA_{B(1)}$ and $GABA_{B(2)}$ were specific and that neither subunit forms homodimers (Kammerer *et al.* 1999). Third, the heterodimerization of the two subunits is a prerequisite for a robust functional activity and is necessary for coupling to all three known effectors: Ca^{2+} channels, adenylate cyclase and inward rectifier K^+ channels (Kir3.1) (Filippov *et al.* 2000). Fourth, the $GABA_{B(2)}$ protein influences ligand affinity: the heterodimer displays an agonist affinity that is higher than that of the individually expressed $GABA_{B(1)}$ subunit and close to that of native receptors (Kaupmann *et al.* 1998a; White *et al.* 1998).

Thus, it is now apparent that functional $GABA_B$ receptors are a heterodimeric complex made up of two closely related transmembrane proteins that interact at the C-terminus in a stoichiometry of $1:1$ (Fig. 3.2). Although receptor homodimers had previously been described for GPCRs (Hébert and Bouvier 1998; Angers *et al.* 2000), $GABA_B$ receptors are the first clear evidence for heterodimerization among GPCRs. The $GABA_B$ receptor dimer appears to be the basic building block for most, if not all, $GABA_B$ receptors in the brain. *In situ* hybridization and light microscopic immunohistochemistry revealed their co-localization in agreement with the distribution of $GABA_B$ receptor binding sites previously identified by autoradiography (Bischoff *et al.* 1999; Fritschy *et al.* 1999b; Margeta-Mitrovic *et al.* 1999).

Interestingly, there are several areas of the CNS (e.g. the striatum, corpus callosum and spinal cord) where there is little $GABA_{B(2)}$ mRNA relative to $GABA_{B(1)}$ (Calver *et al.* 2000; Fig. 3.3). It is possible that $GABA_{B(1)}$ can function as a monomer in some discrete regions or in association with proteins other than $GABA_{B(2)}$. Unlike $GABA_{B(1)}$, $GABA_{B(2)}$ is expressed predominantly at the cell surface in recombinant cell lines, even when expressed alone (Marshall *et al.* 1999).

GABA_B receptor subtypes

Pharmacological evidence supports the existence of pharmacological subtypes of $GABA_B$ receptors (Bonanno and Raiteri 1993). However, differences in the potency of a particular drug in functional assays could relate to distinct neuronal effector systems and do not necessarily

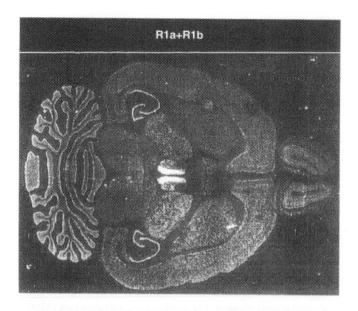

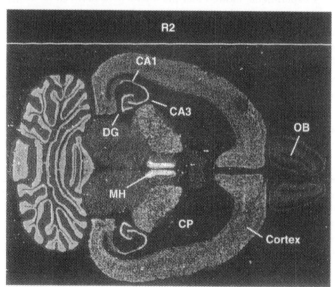

Figure 3.3 Cellular and subcellular distribution of GABA$_B$ receptor mRNA in rat brain by *in situ* hybridization analysis. Transcripts of both GABA$_{B(1)}$ and GABA$_{B(2)}$ are abundant in the medial habenula (MH), the pyramidal cells of the CA1–CA3 subfields of the hippocampus, the granular layer of the dentate gyrus (DG) and the cortex. GABA$_{B(2)}$ transcripts are less abundant than GABA$_{B(1)}$ transcripts in the caudate putamen (CP) and the olfactory bulb (OB). GABA$_{B(1)}$ and GABA$_{B(2)}$ transcript expression in cerebellar cortex is higher in Purkinje cells than in the granule cell layer. (See Color Plate II.)

reflect the existence of multiple forms of binding sites. Currently we have no molecular explanation for the proposed $GABA_B$ receptor subtypes. Although multiple splice variants of $GABA_{B(1)}$ and $GABA_{B(2)}$ appear to exist (Calver *et al.* 2000), no differences in their pharmacological properties have been observed (Kaupmann *et al.* 1998a). The rank order of agonist and antagonist binding affinities at recombinant $GABA_B$ and native receptors is identical (Kaupmann *et al.* 1998a). Moreover, the distribution of $GABA_{B(1)}$ and $GABA_{B(2)}$ transcripts in the brain, as studied by *in situ* hybridization, qualitatively parallels that of $GABA_B$ radioligand binding sites (Bischoff *et al.* 1999; Fritschy *et al.* 1999b; Margeta-Mitrovic *et al.* 1999). On the other hand, the expression of $GABA_{B(1)}$ and $GABA_{B(2)}$ receptor subunits in the peripheral tissues differs from that in the CNS. $GABA_{B(1)}$ subunit splice variants are expressed throughout the CNS and peripheral tissues, whereas $GABA_{B(2)}$ subunit splice variants are neural specific (Calver *et al.* 2000). Analysis by immunoblotting using antisera specific to individual $GABA_B$ receptor subunits has confirmed these findings at the protein level and demonstrates the presence of the $GABA_{B(1)}$ subunit, but not the $GABA_{B(2)}$ subunit, in uterus and spleen. Therefore, areas of non-overlap between $GABA_{B(1)}$ and $GABA_{B(2)}$ subunit expression in tissues known to contain functional $GABA_B$ receptors have been identified. Such areas are of interest as they may well contain novel $GABA_B$ subunit isoforms, expression of which would enable the $GABA_{B(1)}$ subunit to reach the cell surface and form functional $GABA_B$ receptors. Finally, there is always the possibility that there are still other $GABA_B$ receptors yet to be identified.

GHB binding sites

Binding sites for GHB were first described by Benavides *et al.* (1982a) soon after Hill and Bowery (1981) had observed that [^{3}H]baclofen and [^{3}H]GABA bind to a bicuculline-insensitive GABA site in rat brain. The distribution of GHB binding sites in rat does not match that of $GABA_A$ and $GABA_B$ receptors and appears to be specific (Maitre 1997). The ontogeny of [^{3}H]GHB binding sites has been studied by Snead (1994) and differs from that of $GABA_B$ receptors (Knott *et al.* 1993; Turgeon and Albin 1993). GABA and baclofen, two $GABA_B$ receptor agonists, do not displace [^{3}H]GHB from its binding sites (Maitre 1997). These results indicate that GHB binding sites differ from $GABA_B$ receptor (Mathivet *et al.* 1997). It has been shown that guanine nucleotide binding proteins (G proteins) are coupled to the GHB binding protein, modifying the high-affinity GHB binding (Ratomponirina *et al.* 1995; Snead 2000). This suggests that GHB binding sites belong to the family of G protein-linked receptors and that a pertussis-sensitive (G_i or G_o family) G protein is involved in the coupling to the cellular response.

The solubilization of the GHB binding sites was undertaken by Cash *et al.* (1996) as a first step to their molecular characterization (see also Chapter 2 of this volume). After solubilization, the GHB binding sites were purified by affinity chromatography and preparative electrophoresis. A dominant protein band of 54 kDa was purified and sequenced to define specific primers for polymerase chain reaction (PCR). A cDNA was obtained coding for a protein of 50 kDa (Andriamampandry *et al.* 1999). The cDNA and amino acid sequences of the GHB receptor protein were described in a patent application by Andriamampandry and Maitre (2000). Surprisingly, the deduced GHB protein sequence shares similarity with the tetraspanin family of proteins but not with $GABA_{B(1)}$ and $GABA_{B(2)}$ or with any other known GPCRs. After transfection of the tetraspanin-like cDNA into CHO cells, significant high-affinity binding of [^{3}H]GHB (K_d = 425 nM, B_{max} = 4.8 pmol/mg protein) was observed. GHB and other GHB receptor agonists, but not GABA, baclofen and glutamate, displace [^{3}H]GHB from this recombinant-binding site. Binding data with the GHB receptor antagonist

NCS-382 are not yet available. Although the native receptor has been claimed to be coupled to adenylate cyclase (Snead 2000), no coupling of the protein receptor stably expressed in CHO cells to adenylate cyclase and to other effector systems such as K^+ and Ca^{2+} channels has been reported. However, more studies are required to confirm that GHB is a natural ligand of tetraspanin-like proteins.

The tetraspanin superfamily (TM4SF), first recognized in 1990, has grown to over thirty different genes (Bronstein 2000). TM4SF is a family of cell-surface transmembrane glyco-proteins that share a common protein structure (Maecker *et al.* 1997). All tetraspanins possess four highly conserved hydrophobic, putative transmembrane domains, forming a small and a large extracellular domain with short intracellular amino and carboxyl tails (Maecker *et al.* 1997).

The high-affinity GHB binding sites as defined by Maitre (1997) and Snead (2000) are absent from peripheral tissues, including heart, kidney, liver and skeletal muscle, although GHB is present in significant amounts in these organs (Snead and Liu 1984). Thus, pharma-cological effects produced by GHB in peripheral organs (e.g. the anti-ischemic cardioprotec-tive activity (Kolin *et al.* 1993)) are unlikely to be mediated by these high-affinity GHB binding sites. Some tetraspanins are found in virtually all tissues (CD81, CD82, CD9, CD63), whereas others are highly tissue restricted, such as Tspan-2 or Tspan-5, which are highly expressed in the CNS (Garcia-Frigola *et al.* 2000). Thus, binding of GHB to peripheral tetraspanin-like proteins may explain the pharmacological effects of GHB in peripheral organs.

GHB receptor subtypes

Like many other GPCRs, GHB receptors are likely to be heterogeneous, but at present it is unclear whether receptor subtypes are present in rodent brains. The data obtained from the elucidation of the structure of the receptor have not provided any clear basis for receptor heterogeneity (Andriamampandry *et al.* 1999; Andriamampandry and Maitre 2000). However, some biochemical and physiological studies in rodent brains suggest that there may be differences between GHB receptors. Thus, the binding characteristics of *trans*-hydroxycrotonic acid (THCA), an endogenous agonist (Table 3.1, Maitre 1997), suggest that subclasses of GHB binding sites may be present in rat brain. In the dorsal hippocampus, GHB is a better displacer of [^{3}H]GHB than THCA, whereas in the striatum, THCA is more affine for the GHB high-affinity sites than GHB itself (Hechler *et al.* 1990). These results may be in line with the existence of several high-affinity subclasses of GHB binding sites in rat brain and that the natural ligand of some receptor isoforms could be THCA. However, these data do not explain that, in contrast to GHB, THCA does not induce absence seizures in non-epileptic rats and does not increase the total number of spike–wave discharges (SWD) in rats with spontaneous absence epilepsy. On the other hand, evidence on the roles of GHB and $GABA_B$ receptors in the induction of long-term potentiation (LTP) of the mouse hippocampal CA1 region and in the genesis of absence seizures suggest that baclofen and GHB have different effects on LTP, although they have a common mode of action on the thalamocortical function related to the genesis of absence seizures in lethargic mice (Aizawa *et al.* 1997). Thus, hippocampal GHB receptors may be different from those in thalamus. These results are supported by a recent study of Snead (2000) showing an effect of GHB on adenylate cyclase in cortex and hippocampus but not in thalamus. Alternatively, GHB might couple to a second messenger in the thalamus that differs from the GHB-coupled second-messenger system in the cortex and hippocampus.

It remains to be seen how these apparent functional distinctions can be equated with the lack of diversity in receptor structure. A major problem in defining and establishing any differences in pharmacological characteristics is the lack of ligands with specificity for the proposed GHB receptor subtypes. Therefore, although subtypes have been postulated, the effects of pharmacological agents do not seem robust enough to make unequivocal decisions about the status of multiple GHB receptors in the brain.

Exogenous GHB binds to GABA$_B$ receptors

The data from Bernasconi *et al.* (1992), Ishige *et al.* (1993, 1996), Ito *et al.* (1995), Mathivet *et al.* (1997) and Lingenhoehl *et al.* (1999) have characterized GHB as a weak agonist of GABA$_B$ receptors (Table 3.1). In rats, the K_i values of GHB are similar in three GABA$_B$ agonist radioligand assays (80–100 μM) and are much lower than the K_i (~3,000 μM) measured with GABA$_B$ antagonists (Table 3.1). The K_i values of GHB observed with three agonist radioligands and the higher K_i values obtained in a GABA$_B$ antagonist assay are similar in the cortex, hippocampus and cerebellum (Mathivet *et al.* 1997). The observation that the K_i values for the binding of GHB at GABA$_B$ receptors are identical in brain regions with a high density (cortex, hippocampus) and in a brain area where GHB binding sites are absent (cerebellum), suggests strongly that the GABA$_B$ and GHB receptors are two different proteins which do not appear to interact with each other.

This large discrepancy in the K_i values between agonist and antagonist radioligand assays is a characteristic of all GABA$_B$ receptor agonists (Bittiger *et al.* 1992). On the contrary, GABA$_B$ receptor antagonists have similar K_i values in the antagonist assay as in the agonist assay. In addition, the slope of the agonist inhibition curves is much flatter than in the agonist assay (Hill factor about 0.5). Taking into account that the antagonist ligand labels two to three times more receptor sites in a cortical membrane preparation, the agonist inhibition curves are explained by the fact that the antagonist radioligand labels high- and low-affinity states of the receptor and that high-affinity sites are measured with radioligand agonists (Bittiger *et al.* 1992).

The K_i values assessed with GABA$_B$ agonist and antagonist radioligands (Mathivet *et al.* 1997) (Table 3.1) indicate that GHB is an agonist at GABA$_B$ receptors. The fact that Gpp(NH)p produces a significant decrease of GHB affinity for GABA$_B$ receptors in antagonist binding assays confirms that GHB shares the same binding characteristics as GABA, *R*-(−)-baclofen and other GABA$_B$ receptor agonists. These results also indicate that GHB-sensitive GABA$_B$ receptors are receptors that interact with heterotrimeric G proteins (Mathivet *et al.* 1997).

In mice, Ito *et al.* (1995) and Ishige *et al.* (1996) have confirmed that GHB inhibits the binding of [^{3}H]baclofen in primary cultures of cerebellar granule cells and in cerebellar membranes with K_i values of 398 and 630 μM, respectively. The discrepancy between the K_i values reported by Ishige *et al.* (1996) and Ito *et al.* (1995) and those observed by Mathivet *et al.* (1997) may result from differences in animal species (mouse versus rat) or in membrane preparations. The K_i values of the two other GABA$_B$ receptor agonists GABA and baclofen measured by Ito *et al.* (1995) and Ishige *et al.* (1996) are also higher than those observed by Mathivet *et al.* (1997).

Despite its relatively high K_i value, the binding of GHB to GABA$_B$ receptors is selective. On the one hand, close analogs of GHB do not display any affinity for GABA$_B$ receptors. Thus, γ-butyrolactone (GBL) the cyclized derivative and prodrug of GHB (Fig. 3.1), does not interact with GHB or with GABA$_B$ binding sites. THCA, a close structural analog of GHB, which can be considered as a semi-rigid, non-lactonizable analog of GHB

(Fig. 3.1), is as potent as GHB in displacing the binding of [^{3}H]GHB *in vitro* (Benavides *et al.* 1982a), but does not displace specifically bound [^{3}H]GABA and [^{3}H]CGP 27492 from GABA$_B$ receptors (Mathivet *et al.* unpublished results).

On the other hand, another compound, the structure of which is not related to GHB, (*S*)-(−)-HA 966 (Fig. 3.1), displaces both the GABA$_B$ agonist [^{3}H]CGP 27492 ($K_i = 250\,\mu M$) and [^{3}H]GHB ($K_i = 250$ nM) from their binding sites, whereas its enantiomer, (*R*)-(+)-HA 966, has no affinity for GABA$_B$ receptors (Mathivet *et al.* 1996). Lingenhoehl *et al.* (1999) have measured the affinity of GHB for recombinant GABA$_B$ receptors co-expressed with Kir3 channels in *Xenopus* oocytes. A concentration of 30 mM GHB significantly displaces the GABA$_B$ antagonist [^{125}I]CGP 64213 binding at homomeric GABA$_{B(1a)}$ and GABA$_{B(1b)}$ receptors expressed in COS cells. This high concentration of GHB is in line with two observations: (1) GHB is thirty times less potent as a displacer of GABA$_B$ antagonists from native receptors than as a displacer of GABA$_B$ agonists (Mathivet *et al.* 1997), and (2) agonists have a 100–150-fold lower affinity at GABA$_{B(1)}$ receptor when compared to native receptors (Kaupmann *et al.* 1997). In this regard, GHB behaves as other GABA$_B$ receptor agonists.

Endogenous brain levels of GHB (2–5 nmol/g) are too low to stimulate GABA$_B$ receptors (K_i between 80 and 120 μM), and physiological effects of endogenous GHB are likely to be mediated by receptors or signaling pathways other than GABA$_B$. However, little is known about the micro- and ultrastructural distribution of endogenous GHB or its receptor *in situ*, and the possibility that synaptically released GHB reaches millimolar concentration in the synaptic cleft cannot be ruled out. Alternatively, other GABA$_B$ receptor subtypes may exist that differ in their sensitivity to GHB. On the other hand, after administration of a dose of 100–400 mg/kg of exogenous GHB, the cortical GHB concentrations reached ~250 nmol/g and are sufficient to stimulate GABA$_B$ receptors (Snead 1991). Thus, the pharmacological effects observed after injection of non-anesthetic doses of GHB may be mediated through stimulation of GABA$_B$ receptors.

Exogenous GHB is not reconverted into GABA during the receptor assay

About 1–2% of the metabolic flux through the cerebral GABA shunt pathway leads to the synthesis of GHB (Maitre 1997). The reversed pathway, that is, the reconversion of GHB into GABA, is still a matter of controversy. Whereas Möhler *et al.* (1976) and Doherty and Roth (1978) have failed to detect the formation of GABA from GHB, Vayer *et al.* (1985) found that [^{3}H]GHB can be converted into [^{3}H]GABA, both *in vivo* and *in vitro*. Subsequently, the converted GABA could bind to GABA$_A$ and GABA$_B$ receptors, thus simulating the displacement of the radioligand by GHB. A conversion of only 0.1% of GHB into GABA would be sufficient to explain the results of the binding experiments. It was of central importance, therefore, to test whether GHB could be reconverted into GABA during radioligand incubation. According to Vayer *et al.* (1985), the conversion of GHB into GABA is catalyzed by a cytosolic GHB dehydrogenase, an enzyme that is completely blocked by some antiepileptic drugs including valproate and ethosuximide (Kaufman and Nelson 1991). The K_i values of GHB dehydrogenase for valproate and ethosuximide range between 60 and 80 μM (Vayer *et al.* 1985; Kaufman and Nelson 1991). A concentration of 3 mM of the two antiepileptic drugs corresponds approximately to thirty- to fiftyfold the K_i. Valproate and ethosuximide have no affinity for GABA$_B$ receptors. The IC$_{50}$ values and the number of binding sites occupied by GHB were similar in the presence or absence of valproate and ethosuximide (Mathivet *et al.* 1997). The results of these experiments demonstrate that the affinity of GHB for GABA$_B$ receptors is not due to GHB-derived GABA acting selectively at GABA$_B$ receptors.

At the same time, Hechler *et al.* (1997) reported the opposite results, that is, that the conversion of GHB into GABA induces the displacement of GABA$_B$ binding and that inhibition of the reconversion of GHB into GABA by valproate and ethosuximide blocks the ability of GHB to interfere with GABA$_B$ and also with GABA$_A$ binding. It is difficult to reconcile the results of these two studies; however, a number of lines of evidence do not support the hypothesis that GHB is a prodrug of GABA that induces GABA-like effects at GABA$_A$ and GABA$_B$ receptors during binding assays:

1   Newly synthesized GABA should have the same affinity for GABA$_A$ as for GABA$_B$ receptors (Bowery *et al.* 1987; Chu *et al.* 1990). However, several reports have demonstrated that GHB does not displace agonist radioligand at GABA$_A$ receptors in the rat and in human brain (Enna and Maggi 1979; Lloyd and Dreksler 1979; Serra *et al.* 1991; Bernasconi *et al.* 1992; Snead and Liu 1993). Finally, GHB elicits a membrane inhibition of nigral and neocortical cells that is resistant to GABA$_A$ antagonists (Olpe and Koella 1979), but is inhibited by GABA$_B$ receptor antagonists (Engberg and Nissbrandt 1993; Ito *et al.* 1995; Nissbrandt and Engberg 1996; Erhardt *et al.* 1998). In addition, GHB induces a membrane hyperpolarization that is selectively blocked by GABA$_B$ receptor antagonists (Xie and Smart 1992; Williams *et al.* 1995).

2   The amount of GABA formed from GHB depends on the activity of the coupled action of GHB dehydrogenase and NADP, which oxidizes GHB to succinic semialdehyde (SSA), and then of GABA-transaminase which transaminates SSA into GABA. It is unlikely that the activity of these enzymes is kept constant in the different batches of membrane preparations that we used (Bernasconi *et al.* 1992; Mathivet *et al.* 1997), leading to a large heterogeneity of K_i. However, the K_i values of GHB for the displacement of different GABA$_B$ receptor agonist radioligands are highly reproducible (79–126 μM) and are not dependent on the batch of membrane preparation.

3   Gobaille *et al.* (1999) have observed that the conversion of [^{3}H]GHB to [^{3}H]GABA does not take place in two-thirds of the brain regions investigated and is highly variable in the other remaining structures. This regionally selective difference in the mechanisms of GHB reconversion into GABA should be reflected in the binding parameters. However, the K_i values of GHB observed with the different GABA$_B$ receptor agonist radioligands are similar in all brain structures examined (Mathivet *et al.* 1997).

4   According to Gobaille *et al.* (1999), 1–2% of GHB is converted *in vivo* into GABA after a rather long delay (160 min), and this long time lapse may explain why Möhler *et al.* (1976) as well as Doherty and Roth (1978) have failed to detect the formation of GABA. Under the conditions used for binding at GABA$_B$ receptors, the incubation lasts 40 min (20 °C) and is too short to allow a significant transformation of GHB into GABA.

5   It is unlikely that GHB is converted into GABA under conditions of continuous perfusion of the intact oocyte (Lingenhoehl *et al.* 1999). It is also highly improbable that GABA is synthesized from GHB in membranes prepared from CHO cells stably expressing rat GABA$_{B(1a)}$ and GABA$_{B(1b)}$ receptors. Most important is the observation of Lingenhoehl *et al.* (1999) that GHB is a partial agonist at GABA$_B$ receptors, whereas GHB synthesized from GABA should act as a full agonist at GABA$_B$ receptors.

Pharmacological properties of exogenous GHB

The evaluation of the functional relevance of *in vitro* binding results requires *in vivo* paradigms (Table 3.2a) in which the stimulation of GABA$_B$ receptors by GHB can be demonstrated. To enhance the reproducibility and predictability of the biological responses induced by GHB, the drug has to be injected intravenously or more conveniently by using the GHB precursor GBL (Snead 1992a). GBL is used because of consistency and rapidity of onset of its effects. It has been shown to produce exactly the same EEG and behavioral changes as that of GHB. GBL is a cyclized derivative of GHB (Fig. 3.1) that does not displace [^{3}H]GHB from its binding sites and is inactive *per se*. However, GBL is rapidly and irreversibly hydrolyzed into GHB by peripheral γ-lactonase and the fast onset of bilaterally synchronous SWD produced by GBL in rats correlated with a rapid appearance of GHB in brain.

GHB modulates potassium currents mediated through GABA$_B$ receptors

In *Xenopus* oocytes expressing heteromeric GABA$_{B(1a)}$/GABA$_{B(2)}$ and heteromultimeric G proteins activated inwardly rectifying potassium channels (Kir3), the application of high-potassium Ringer induces a large inward current (I_{kir}). Activation of GABA$_B$ receptor by *R*-(−)-baclofen (10 μM) increases this current by a factor of 1.5–3. Similarly, GHB increases I_{kir} in a dose-dependent and reversible way (EC$_{50}$ = 4.4 mM) (Lingenhoehl *et al.* 1999). This action of GHB is mediated by GABA$_B$ receptors because oocytes expressing Kir3 channels alone do not show an increase of I_{kir} upon application of *R*-(−)-baclofen or GHB. Currents activated by GHB and *R*-(−)-baclofen application are inwardly rectifying and reverse at approximately −20 mV in high-potassium Ringer. GABA$_{B(1b)}$/GABA$_{B(2)}$, a heteromeric combination containing a different splice variant of GABA$_{B(1)}$, yields the same increase in I_{kir} upon GHB application (EC$_{50}$ = 4.7 mM), indicating that GHB (as GABA and *R*-(−)-baclofen) does not discriminate between the GABA$_{B(1)}$ splice variants. At the highest concentration tested (30 mM), GHB does not induce a saturating current response and produces 80% of the response induced by 10 μM *R*-(−)-baclofen. Furthermore, the asymptotic maximal activation of I_{kir} by GHB as calculated from the curve fit is significantly lower (70%) than that by *R*-(−)-baclofen, suggesting that GHB is a partial agonist at the GABA$_B$ receptor. Lingenhoehl *et al.* (1999) have also analyzed the effect of co-application of *R*-(−)-baclofen and GHB. When 10 mM GHB is co-applied to oocytes with increasing concentrations of *R*-(−)-baclofen, the resulting concentration–response curve crosses the one obtained with *R*-(−)-baclofen alone. This suggests again that GHB may be acting as a partial agonist that functionally competes with *R*-(−)-baclofen at the GABA$_B$ receptor.

Competitive GABA$_B$ receptor antagonists inhibit the GHB-induced response completely, and the IC$_{50}$ values are similar to those found in other studies (Kaupmann *et al.* 1998a). On the other hand, the GHB antagonist NCS-382 does not change GHB-induced inward currents (Lingenhoehl *et al.* 1999).

GHB reproduced GABA$_B$ inhibitory postsynaptic potentials

Electrophysiological studies in isolated lateral geniculate (Williams *et al.* 1995), CA1 pyramidal (Xie and Smart 1992) and nigrostriatal neurons (Harris *et al.* 1989) have demonstrated that GHB induces hyperpolarization and rebound Ca^{2+} spikes in a manner similar to *R*-(−)-baclofen. The selective GABA$_B$ receptor antagonist CGP 35348 blocks the action of GHB, but not of tetrodotoxin, demonstrating that the effects of GHB are mediated by a direct action on postsynaptic GABA$_B$ receptors. Moreover, the electrophysiological effects of GHB on the postsynaptic GABA$_B$ receptor are not blocked by NCS-382, indicating strongly that GHB binding sites are not implicated in the hyperpolarization-mediated oscillation (see Chapter 5 of this volume).

Effects of GHB and the GABA$_B$ receptor agonist R-(−)-baclofen on brain cGMP content

The nucleotide guanosine 3′,5′-cyclic monophosphate (cGMP) has been shown to be an intracellular second-messenger system within the brain, especially in the cerebellum where high concentrations of cGMP, the synthetic enzyme guanylate cyclase, the degradative enzyme cGMP phosphodiesterase and cGMP-dependent protein kinases are present (Wood 1991). The cerebellum is a unique CNS region for biochemical studies in that a large number of neuronal projections are biochemically well characterized (Fig. 3.4). Both neurochemical and anatomical data suggest that a balance between excitatory and inhibitory pathways control cGMP levels within the cerebellum.

GABA$_B$ receptors are abundant in the cerebellum as reflected by the high level of mRNA transcripts for GABA$_{B(1a)}$/GABA$_{B(1b)}$ and GABA$_{B(2)}$ (Fig. 3.3) as well as the high density of [³H]CGP 54626 binding sites. However, the GABA$_{B(1a)}$ and GABA$_{B(1b)}$ splice variants exhibit distinct localization within the cerebellum (Fig. 3.4). The GABA$_{B(1b)}$ is almost

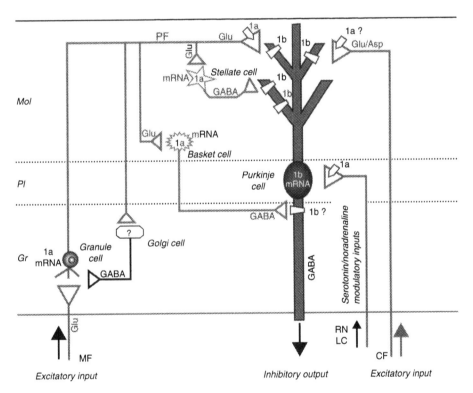

Figure 3.4 Basic circuitry diagram of the cerebellum with the neurons of the molecular (Mol), Purkinje cell (Pl) and granular cell layers (Gr). The possible localization of the GABA$_{B(1a)}$ and GABA$_{B(1b)}$ receptor variants on the cerebellar circuitry is also indicated. This model is based on the data of Bischoff *et al.* (1999) on the localization of the mRNA transcripts and the autoradiographic distribution of [³H]CGP 54626-labeled binding sites. This model is in line with the first immunohistochemical data with ultrastructural localization of antibodies in the Mol, Pl and Gr of the cerebellum (Kaupmann *et al.* 1998). (CF: climbing fibers; LC: locus coeruleus; MF: mossy fibers; PF: parallel fibers; RN: raphe nucleus; 1a: GABA$_{B(1a)}$; 1b: GABA$_{B(1b)}$ (modified from Bischoff *et al.* 1999).) (See Color Plate III.)

exclusively expressed in Purkinje cells and the receptor protein is mainly localized on the dendritic arborization within the molecular layer. Thus the $GABA_{B(1b)}$ subunit may play a key role in the inhibitory output of the cerebellum. In contrast, the $GABA_{B(1a)}$ mRNA is mainly (though modestly) expressed in the granule cells giving rise to receptors located on presynaptic terminals of the glutamate-containing parallel fibers. Weak expression of $GABA_{B(1a)}$ is also seen in discrete neurons of the molecular layer that might correspond to stellate and basket cells (Bischoff *et al.* 1999; Fig. 3.4). $GABA_{B(2)}$ is strongly expressed in the Purkinje cells and is weaker in the granule cells, stellate and basket cells (Kaupmann *et al.* 1998a). As would be predicted by the localization of the $GABA_{B(1a)}/GABA_{B(1b)}$ receptor variants on the cerebellar circuitry, systemic administration of the $GABA_B$ receptor agonist R-(−)-baclofen dose- and time-dependently decreases cerebellar cGMP levels in rodents (Bernasconi *et al.* 1992). Specific $GABA_B$ receptor antagonists at the doses used do not alter cerebellar cGMP *per se*, but antagonize the decrease of cGMP content induced by R-(−)-baclofen (Bernasconi *et al.* 1992). The effects of GHB (or GBL) on cerebellar cGMP show a remarkable similarity to those produced by R-(−)-baclofen. Exogenous administration of GBL (200 mg/kg) in rodents induces a dose- and time-dependent decrease of cerebellar cGMP levels that is partially, but significantly, antagonized by pretreatment with $GABA_B$ receptor antagonists (Fig. 3.5)

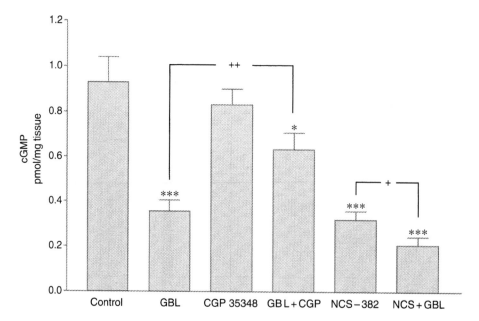

Figure 3.5 Antagonistic effect of CGP 35348 on the decrease in levels of cGMP in the cerebellum induced by GBL compared to the lack of antagonistic effect of NCS-382. CGP 35348 (200 mg/kg) was injected 25 min before GBL (200 mg/kg) and rats were sacrificed 60 min after treatment with the $GABA_B$ receptor antagonist. NCS-382 (400 mg/kg) was given 30 min before GHB and rats were killed by fast focused microwave irradiation of the head 30 min later. Each group represents the mean ± SEM of eight animals. * $p < 0.05$, *** $p < 0.001$ relative to controls, + $p < 0.05$, ++ $p < 0.01$ compared to animals receiving GHB or NCS-382 alone (Dunnett's test). cGMP assays were performed using a radioimmunoassay kit with [^{3}H]cGMP obtained from Amersham or Anawa Trading S.A. (Zurich). Additional experiments were carried out with an acetylated [^{125}I]cGMP RIA kit of Advanced Magnetic (Cambridge, MA).

(Bernasconi *et al.* 1992). It is important to recall that any biochemical effect that GHB may have in the cerebellum is not related to stimulation of the GHB binding sites, which are not present in the cerebellum (Maitre 1997).

Surprisingly, NCS-382 *per se* induces changes of cerebellar cGMP content similar to those induced by GHB: NCS-382 alone (400 mg/kg) does not antagonize the action of GHB but it dose-dependently and significantly decreases cGMP levels in the cerebellum of rats (34% of control value) (Fig. 3.5). This effect is additive to the decrease produced by 200 mg/kg GBL, suggesting that GHB and NCS-382 decrease cerebellar cGMP through different mechanisms (Fig. 3.5). NCS-382 displays no affinity for GABA$_B$ receptors (Maitre *et al.* 1990; Snead 1996; and our own data $K_i > 10$ mM) and there are no GHB binding sites in the cerebellum (Maitre 1997), so that the similar effects of the agonist GHB and the antagonist NCS-382 are not mediated by GHB binding sites. NCS-382 was administered at a dose as high as that required to fully antagonize GHB-induced absence-like seizures or to inhibit SWD in a genetic model of generalized non-convulsive epilepsy (Maitre *et al.* 1990). These results strongly suggest that the action of NCS-382 in the cerebellum is unspecific and cannot be explained in terms of mechanisms implicating GHB- or GABA$_B$-receptors and that this compound is not a selective antagonist for GHB binding sites.

We have measured the content of cGMP in three different brain regions of adult male rats after administration of 300 mg/kg of GHB (Fig. 3.6A). GHB induces a decline of cGMP in the cerebellum and the thalamus, whereas a slight but significant increase (116%) is observed in the frontal cortex. GABA$_B$ receptor antagonists attenuate significantly the decrease of cGMP in the cerebellum and in the thalamus (results not shown). NCS-382 has not been tested. For comparison we have measured cGMP concentrations in the same regions of the brain of male rats after injection of R-(−)-baclofen. As shown in Fig. 3.6B, the similarities of response with GHB are striking: the content of cGMP is markedly reduced in the cerebellum and in the thalamus, whereas cGMP levels are not changed in the frontal cortex. Taken

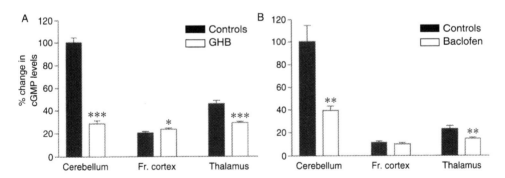

Figure 3.6 Effect of GHB (A) and R-(−)-baclofen (B) on cGMP content in the cerebellum, frontal cortex and thalamus of male rats. Rats were killed by microwave irradiation and cGMP levels were determined by radioimmunoassay and expressed as mean ± SEM of eight rats. Results are expressed in percent of control values. cGMP content in cerebellum of controls was 0.381 pmol/mg tissue in experiment A and 0.639 pmol/mg tissue in experiment B. Statistical significance was calculated by Dunnett's test. * $p < 0.05$, ** $p < 0.01$, *** $p < 0.001$. (A) Levels of cGMP 30 min after injection of 300 mg/kg GHB. (B) Concentrations of cGMP 60 min after administration of 6 mg/kg R-(−)-baclofen. A marked and significant decrease of hippocampal cGMP was also observed after treatment of R-(−)-baclofen (data not shown).

together, these results indicate that the effects of GHB on cerebellar cGMP are partially mediated through activation of $GABA_B$ receptors. GABAergic-induced hyperpolarization is known to depress cGMP levels (Wood 1991). Thus, the decrease in cGMP content observed after administration of GHB supports the notion that GHB, like baclofen, stimulates $GABA_B$ receptors. However, the observation that $GABA_B$ receptor antagonists only partially attenuate the decrease of cGMP induced by GHB suggests that GHB may interact with other neurotransmitter systems.

Effects of GHB and R-(−)-baclofen on adenylate cyclase

$GABA_B$ receptors are coupled via G proteins to membrane K^+ and Ca^{2+} channels, as well as to adenylate cyclase (Bowery 2000). $GABA_B$ receptor agonists can either inhibit or stimulate adenylate cyclase in brain tissue, modifying adenosine $3',5'$-cyclic monophosphate (cAMP) production. $GABA_B$-mediated inhibition of forskolin-activated neuronal adenylate cyclase is produced via G protein (G_i/G_o) coupling (Bowery 2000), whereas $GABA_B$ receptor agonists enhance cAMP formation induced by G_s-coupled receptor agonists (e.g. isoprenaline) in brain slice preparations. This dual action of $GABA_B$ receptor agonists is also manifest *in vivo* (Bowery 2000). Snead (2000) has used the well-documented response of $GABA_B$ receptor agonists on adenylate cyclase to test the hypothesis that GHB binds to a specific GPCR that is different from the $GABA_B$ receptor. The effects of GHB on regional and subcellular brain adenylate cyclase were determined and compared to those of R-(−)-baclofen. Neither GHB nor R-(−)-baclofen alter basal cAMP concentrations. GHB significantly decreases forskolin-stimulated cAMP levels by 40–50% in cortex and hippocampus $(ED_{50} = 4\,\mu M)$ but not in thalamus or cerebellum, whereas R-(−)-baclofen has an effect throughout the brain $(ED_{50} = 2\,\mu M)$ (Snead 2000). The effect of GHB on adenylate cyclase is observed in presynaptic synaptosomal and not in postsynaptic synaptoneurosomal preparations. On the other hand, the effect of R-(−)-baclofen is observed in both subcellular preparations. The GHB-induced alteration in forskolin-induced cAMP formation is blocked by NCS-382 but not by a specific $GABA_B$ receptor antagonist. The decrease in forskolin-induced cAMP formation produced by R-(−)-baclofen is blocked by a specific $GABA_B$ receptor antagonist but not by NCS-382. GHB and R-(−)-baclofen both stimulate guanosine $5'$-O-(thiotriphosphate) (GTPγS) binding and low-K_m guanosine $5'$-triphosphate (GTP)-ase activity by 40–50%. The GHB-induced effect is blocked by NCS-382 but not by a $GABA_B$ receptor antagonist, and occurs only in cortex and hippocampus. The effects elicited by R-(−)-baclofen are blocked by a $GABA_B$ receptor antagonist but not by NCS-382, and are present throughout the brain. These data support the hypothesis that GHB induces a G protein-mediated decrease in adenylate cyclase via a GHB-specific G protein-coupled presynaptic receptor that is different from the $GABA_B$ receptor. The data of Snead (2000) suggest that GHB acts primarily at a presynaptic site. Alternatively, there may be postsynaptic GHB receptors that are not negatively coupled to adenylate cyclase. If this hypothesis is valid, any postsynaptic GHB-mediated effects would have been missed in the study of Snead (2000).

However, the results of Snead (2000) are in contradiction with those described by Andriamampandry and Maitre (2000). The high similarity between the GHB receptor protein and the tetraspanin family of receptors, which are not coupled to G proteins, speaks against an inhibition or stimulation of adenylate cyclase by GHB. Another possibility is that tetraspanin-like proteins, similar to the receptor-activity-modifying proteins (RAMPs), associate to the GPCRs (McLatchie *et al.* 1998). The effects of GHB on adenylate cyclase should be re-evaluated in more-representative biological systems than synaptosomes or synaptosoneurosomes.

Effects of GHB and R-(−)-baclofen on neurotransmitter release

One of the most widely used *in vitro* methods to study presynaptic modulation of transmitter release from nerve terminals is superfusion and electrical field stimulation of slices preloaded with radiolabeled neurotransmitter or by measuring the concentration of endogenous released neurotransmitter by high-pressure liquid chromatography. Waldmeier *et al.* (1994) have used this technique to investigate the role of presynaptic GABA$_B$ receptors on the release of GABA via autoreceptors and on the release of glutamate and biogenic amines via heteroreceptors. Their results showed that R-(−)-baclofen inhibits dose-dependently the release of GABA and glutamate, but does not change the release of dopamine, acetylcholine, noradrenaline and serotonin. Using the same experimental protocol, we have investigated the effects of GHB on the release of neurotransmitters. The results summarized in Table 3.3 show that, like R-(−)-baclofen, GHB has no effect on basal GABA and glutamate release, but it dose-dependently decreases electrically evoked release of [^{3}H]GABA and of endogenous glutamate from cortical slices. On the other hand, GHB failed to alter the release of other neurotransmitters (Table 3.3, Bernasconi *et al.* 1995). The threshold dose for the release of [^{3}H]GABA is 3 mM, and at 30 mM the inhibition reaches approximately 50%. Thus, at the highest dose used (30 mM), GHB does not completely inhibit the electrically induced evoked release of GABA and produces only 50% of the response elicited by 10 μM R-(−)-baclofen, suggesting that GHB is a partial agonist at the GABA$_B$ autoreceptors.

The addition of valproate to the superfusion medium has no effect on the decrease of [^{3}H]GABA and endogenous glutamate, demonstrating that these effects of GHB are

Table 3.3 Inhibition on the electrically evoked release of some neurotransmitters from rat cortical slices by GHB (10 mM) and R-(−)-baclofen (10 μM)

Neurotransmitter released	*Release in % of controls*	
	GHB	*R-(−)-baclofen*
[^{3}H]GABA	74.2 ± 2.3**	10 ± 7.5**
Glutamate	68.9 ± 5.5**	70 ± 3**
[^{3}H]dopamine	79.9 ± 10.1	68.6 ± 10.5
[^{3}H]noradrenaline	>90	>90
[^{3}H]acetylcholine	>90	>90
[^{3}H]serotonin	>90	>90

For each concentration of GHB or R-(−)-baclofen, a separate experiment with its own control was carried out ($n = 4$ for both drug and control) in which the slices were stimulated twice at 0.125 Hz for 8 min (60 pulses) for monitoring the release of [^{3}H]GABA, and at 1 Hz for 4 min for the release of endogenous glutamate, which was measured by high-pressure liquid chromatography. The release experiments were performed as described by Waldmeier *et al.* (1994). GHB or R-(−)-baclofen were added to the superfusion medium two fractions after the first stimulation (S_1). The effects of GHB or R-(−)-baclofen were evaluated by expressing the corresponding S_2/S_1 ratios as a percentage of those of the controls. The IC$_{50}$ values of GHB and R-(−)-baclofen for the release of GABA were 30 mM and 0.37 μM, respectively. GBL (10 mM) did not modify GABA release evoked by electrical stimulation from rat cortical slices.

The values given for [^{3}H]GABA and glutamate are the means of eight and three different experiments, respectively. Statistical calculations were done by one-factor ANOVA (repeated measures) followed by Dunnett's test. ** $p < 0.01$ compared with controls.

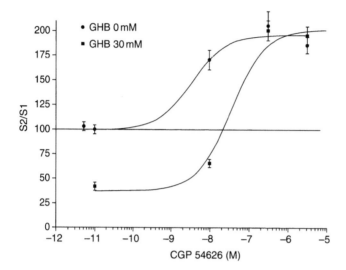

Figure 3.7 Effect of CGP 54626 on the release of [³H]GABA in the absence and in the presence of GHB. At a concentration of 10 nM CGP 54626, the increase in [³H]GABA release compared to the controls is 170% in the absence of GHB and 35% in the presence of 30 mM GHB. This is a significant difference, $p < 0.001$, Dunnett's test.

not related to a conversion of GHB into GABA but are due to a stimulation of GABA$_B$ autoreceptors by GHB. GABA$_B$ receptor antagonists inhibit the effects of GHB on the release of the amino acid neurotransmitters (Fig. 3.7). Thus, the stimulation of presynaptic GABA$_B$ auto- or heteroreceptors by either GHB or R-(−)-baclofen shows similar effects on neurotransmitter release.

The doses of GHB used in the release experiments refer to the concentrations in the superfusion medium and not to concentrations in brain slices, which have not been measured. It is possible that the active uptake of GHB by a transporter reduces its potency in the slice (Benavides *et al.* 1982b; Hechler *et al.* 1985). Just as with GHB, higher concentrations of baclofen are needed to induce an inhibitory effect on the release of [³H]GABA from prelabeled slices ($IC_{50} = 0.37$ μM versus $K_i = 12$ nM) (Waldmeier *et al.* 1994).

Banerjee and Snead (1995) as well as Hu *et al.* (2000) have measured the effects of GHB and baclofen on the *in vivo* extracellular release of GABA and glutamate by using brain microdialysis. Their results are shown in Table 3.2a (see also Timmerman and Westerink 1997).

Effects of exogenous GHB and R-(−)-baclofen on GABA synthesis

One approach to the problem of determining whether GHB alters GABA synthesis has been to administer a compound such as aminooxyacetic acid, which preferentially inhibits GABA catabolism *in vivo*, and to measure whether GHB (or GBL) affects the subsequent accumulation of GABA (Pericic *et al.* 1978). GBL (150–600 mg/kg) produces a dose- and time-dependent decrease in aminooxyacetic acid-induced accumulation of GABA in the striatum, globus pallidus, cerebellum and cortex of rats. The decrease of *in vivo* GABA synthesis elicited by GBL is not blocked by picrotoxin, an antagonist at GABA$_A$ receptors (Pericic *et al.* 1978).

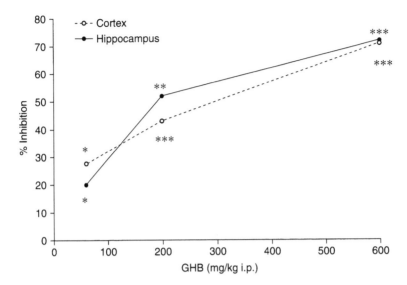

Figure 3.8 Effect of GHB on the gabaculine-induced accumulation of GABA in the cortex and hippocampus of mice. Similar results were observed in the striatum and cerebellum (data not shown). GHB was administered i.p. 65 min before animals were killed by microwave irradiation and gabaculine (150 mg/kg i.p.) was injected 5 min after GHB. GHB has no effect on basal GABA content. The results are the means for six mice. SEM are less than 5% and have been omitted to improve the clarity of the figure. Statistical significance of differences was calculated by Dunnett's test. * $p < 0.05$, ** $p < 0.01$, *** $p < 0.001$ compared with the group treated with gabaculine alone.

We have confirmed the above results by using gabaculine, a more selective inhibitor of GABA transaminase (Bernasconi *et al.* 1981). GHB decreases dose-dependently the accumulation of GABA levels produced by gabaculine in the striatum, hippocampus, cerebellum and cortex of mice (Fig. 3.8). NCS-382 does not antagonize the decrease in the rate of GABA synthesis induced by GHB, whereas GABA$_B$ receptor antagonists have not been tested. *R*-(−)-baclofen produces a marked dose-dependent and stereoselective decrease of gabaculine-induced GABA accumulation. GABA$_B$ receptor antagonists inhibit this effect, suggesting that the GABA synthesis is controlled to a significant extent by presynaptic GABA$_B$ autoreceptors. Thus, GHB, like *R*-(−)-baclofen, decreases GABA release and GABA synthesis.

Effects of GHB and GABA$_B$ receptor agonists on addictive conducts

Drug discrimination studies have shown that baclofen elicits discriminative stimulus effects similar to those produced by GHB (Colombo *et al.* 1998). A large part of the pharmacological effects of GHB, including the discriminative stimulus effects, are blocked by GABA$_B$ receptor antagonists (Colombo *et al.* 1998; Bernasconi *et al.* 1999).

A beneficial role of GHB in the treatment of alcoholism has been documented both in rodents and humans (Poldrugo and Addolorato 1999). The similarities between the pharmacological effects of baclofen and GHB would predict that ethanol self-administration and withdrawal syndrome might also be reduced by baclofen. This hypothesis has recently been confirmed by Colombo *et al.* (2000), who have shown that baclofen dose-dependently decreases the intensity of ethanol withdrawal syndrome and protects ethanol-dependent rats from audiogenic seizures. Moreover, baclofen selectively and dose-dependently reduces ethanol intake in ethanol-preferring rats. A preliminary clinical study by Addolorato *et al.* (2000) has validated these experimental results. They investigated the effect of short-term baclofen administration on craving for alcohol, ethanol intake and abstinence from ethanol in alcoholic patients. This preliminary clinical study supports the evidence of the effect of baclofen in reducing alcohol intake. The anticraving properties of baclofen suggest a possible use of GABA$_B$ receptor agonists in the treatment of individuals with alcohol problems. GABA$_B$ receptor agonists have been proved to possess antimotivational effects and to inhibit cocaine, heroin and nicotine reinforcing effects in addition to alcohol in rats (Brebner *et al.* 1999; Fattore *et al.* 2000).

Effects of GHB on striatal dopamine synthesis

Like many drugs with abuse potential (Di Chiara 1995; Chapter 6 of this volume, and Tables 3.2a and b), GHB significantly alters dopaminergic neurotransmission in the brain. Systemic injections of GHB to rats increase dopamine synthesis early after administration and later reduce it. During this later phase, it causes a marked accumulation of the levels of dopamine in the striatum (Walters and Roth 1972) and other brain regions (Da Prada and Keller 1976; Nissbrandt and Engberg 1996). This sequence of effects closely resembles that caused by baclofen (Walters and Roth 1972). It was interpreted as a consequence of the interruption of impulse flow in nigrostriatal dopaminergic neurons caused by GHB and baclofen (Nissbrandt *et al.* 1994; Feigenbaum and Howard 1997). GABA$_B$ receptor antagonists do not alter dopamine synthesis on their own up to the highest doses tested (e.g. 500 mg/kg i.p. CGP 36742). However, they antagonize the increase elicited by graded doses of GHB and baclofen (Waldmeier 1991). The *in vivo* effects of GABA$_B$ receptor agonists on rat striatal dopamine synthesis have been corroborated by *in vitro* studies. Perfusion of midbrain slices with GHB has been shown to cause a concentration-dependent membrane hyperpolarization and a reduction in input resistance in dopamine neurons recorded intracellularly (Harris *et al.* 1989; Nissbrandt and Engberg 1996; Madden and Johnson 1998). Voltage-clamp recordings demonstrate that GHB hyperpolarizes dopaminergic neurons by a GABA$_B$ receptor-mediated increase in K$^+$ conductance because the GABA$_B$ receptor antagonist CGP 35348 (300 μM) prevents GHB from causing hyperpolarization. In addition, NCS-382 (10 μM) does not significantly alter the increase in conductance produced by 10 mM GHB. These results agree with those of Da Prada and Keller (1976) and Engberg and Nissbrandt (1993), who have performed extracellular recordings of midbrain dopaminergic neurons *in vivo* and intracellular studies of hippocampal neurons *in vitro* (Xie and Smart 1992). These data support the notion that GHB is an agonist at GABA$_B$ receptors and inhibits dopamine release probably by causing K$^+$-dependent membrane hyperpolarization. Such a mode of action of GHB is supported by biochemical and behavioral data of Nissbrandt and Engberg (1996), who have shown that the GABA$_B$ receptor antagonist CGP 35348 is able to antagonize similar effects of GHB and baclofen both on striatal dopamine release and locomotor activity.

GHB and GABA$_B$ receptors present in mammalian heart

GHB has also been found in non-neuronal tissue such as the heart, in which its concentration may be five- to six-fold that of brain (Nelson *et al.* 1981). The surprisingly high tissue content, together with anti-ischemic and protective effects of GHB in the heart (Kolin *et al.* 1993), raises the question of a possible influence of GHB on excitable cardiac cells. Lorente *et al.* (2000) have provided electrophysiological evidence in isolated rat ventricular myocytes that high doses of GHB (2 mM), corresponding to blood levels reached during anesthesia, can activate Kir3 current. This effect is mimicked by baclofen, reversibly inhibited by GABA$_B$ antagonists and prevented by pertussis toxin pretreatment. Because GHB binding sites are absent from the heart (Snead and Liu 1984), it can be assumed that this effect is produced through GABA$_B$ receptors present in rat ventricular cell membranes. Indeed, Castelli *et al.* (1999) have shown, by using reverse transcription with polymerase chain reaction (RT-PCR), that GABA$_{B(1a)}$ and GABA$_{B(1b)}$ transcripts are expressed not only in the brain but also in the heart and in a wide range of peripheral organs. However, the yield of PCR products indicated that the levels of mRNA encoding GABA$_{B(1a)}$ and GABA$_{B(1b)}$ receptors are considerably lower in the heart than in the brain. Furthermore, both GABA$_{B(1)}$ and GABA$_{B(2)}$ are detected in cardiomyocytes by Western blotting and are shown to co-immunoprecipitate. Laser scanning confocal microscopy discloses an even distribution of the two receptors in the sarcolemma and along the transverse tubular system of the cardiomyocyte.

These results indicate that heteromeric GABA$_{B(1)}$/GABA$_{B(2)}$ receptors are present in mammalian heart and appear to be functionally coupled to G proteins of the G$_{\alpha i}$ class. Endogenous GHB might assume its cardioprotective function by stimulating GABA$_{B(1)}$/GABA$_{B(2)}$ receptors in cardiomyocytes. Thus, GHB elicits specific GABA$_B$-type responses in cardiomyocytes in the same way as baclofen, but with lower affinity at GABA$_B$Rs.

Some pharmacological properties of GHB may require activation of both GHB and GABA$_B$ receptors

Opposite pharmacological responses induced by low and high doses of GHB

Two studies have shown that low and high doses of GHB induce different pharmacological responses. Godbout *et al.* (1995) have observed that low doses of GHB (5–10 mg/kg i.p.) increase the spontaneous firing rate in the prefrontal cortex neurons tested in urethane-anesthetized rats (Chapter 7 of this volume and Table 3.2a). This effect is blocked by NCS-382 (10 or 320 mg/kg i.p.), which at these doses does not have any effect on cell firing activity. On the other hand, high doses of GHB (160–320 mg/kg i.p.) inhibit the firing rate in the prefrontal cortex neurons. This response is not modified by NCS-382. The results of Godbout *et al.* (1995) indicate that the excitatory effects of low doses of GHB may be mediated by activation of GHB binding sites, whereas the inhibitory effects of high doses involve a mechanism in which GHB binding sites are not implicated. GABA$_B$ receptor antagonists have not been tested in this study.

In another study, Kaufman *et al.* (1990) have found that low doses of GHB (5–10 mg/kg) increase, whereas high doses (300–500 mg/kg) decrease, body temperature (Table 3.2a). Brain levels of GHB are not altered by the low doses, whereas large doses (>200 mg/kg i.p.) produce an increase in the brain concentration (see also Snead (1991) for a correlation between doses of GHB and brain levels). Thus, hyperthermia occurs when there have been no or very small changes in GHB content in the brain, whereas hypothermia correlates with a large increase in GHB concentration in the brain.

The results of Kaufman *et al.* (1990) parallel those of Godbout *et al.* (1995) and indicate that the different responses to the low and high doses of GHB are probably related to two different mechanisms of action. Neither NCS-382 nor GABA$_B$ receptor antagonists have been used in the study of Kaufman *et al.* (1990). We have therefore re-investigated the effects of graded doses of GHB on the body temperature of adult male rats and tested the action of NCS-382 and a GABA$_B$ receptor antagonist (CGP 36742) on the hyper- and hypothermia induced by low and high doses of GHB, respectively. Our results confirm that graded doses of GHB produce biphasic effects on body temperature. The hyperthermia ($\Delta t = 0.29$ °C, $p < 0.05$) caused by low doses of 10 mg/kg GHB is partly antagonized by 70 mg/kg of NCS-382, but not by 200 mg/kg of the GABA$_B$ antagonist CGP 36742. High doses of GHB (100–600 mg/kg) produce a sharp drop in body temperature, which is not attenuated by NCS-382 (70 mg/kg), but is antagonized by 200 mg/kg CGP 36742 ($p < 0.01$), which has by itself no effect on body temperature in rats (Fig. 3.9). Thus, GABA$_B$ receptors are implicated in the control of thermoregulation. This concept was confirmed by the observation that baclofen produces hypothermia in mice (Gray *et al.* 1987) and that selective GABA$_B$ receptor antagonists completely abolish the hypothermic effects of baclofen (Jackson and Nutt 1991).

These findings suggest that GABA$_B$ receptors are probably not tonically involved in the control of body temperature. In addition, the reversal of baclofen- and GHB-induced hypothermia by GABA$_B$ antagonists indicates that this response is mediated by GABA$_B$ receptors. In contrast to GHB, baclofen does not induce hyperthermia at any doses. Taken

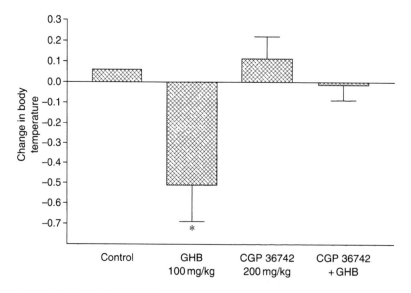

Figure 3.9 Effect of CGP 36742 on the hypothermia induced by 100 mg/kg i.p. GHB. Adult male Wistar rats (weight range 350–400 g) were housed in groups of four at an ambient temperature of 23 ± 2 °C for at least 7 days before use. Body temperature was recorded using a rat rectal probe (inserted 2 cm). Rectal temperatures were taken 60 min before injection, immediately before injection and 60 min after drug treatment. The first reading was taken to measure baseline body temperature and to familiarize animals with the experimental procedure, which can produce an initial increase in body temperature. In the interaction studies, CGP 36742 was given to animals 15 min before GHB. Results are expressed as means $\pm$ SEM for groups of eight to twelve animals. A significant change in core temperature between experimental and saline-treated control rats is indicated as * $p < 0.05$ (one-way ANOVA followed by Dunnett's test for each experimental group compared to control).

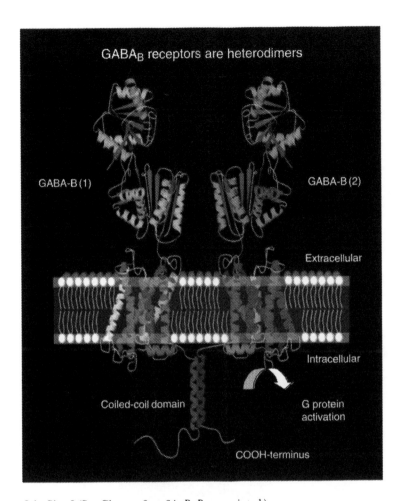

Color Plate I (See Chapter 3, p. 34. *R. Bernasconi* et al.)

Figure 3.2 Model of the GABA$_B$ receptor heterodimer formed by association of the subunits GABA$_{B(1)}$ and GABA$_{B(2)}$. Features supported by experimental evidence include binding sites for all known competitive GABA$_B$ antagonists, GABA and GHB on GABA$_{B(1)}$ and coiled-coil domains at the intracellular C-termini that are important for surface trafficking of the receptor. The G proteins are coupled to the GABA$_{B(2)}$ receptor (Galvez *et al.* 2001; Margeta-Mitrovic *et al.* 2001; Robbins *et al.* 2001; Duthey *et al.* 2002). A new class of compounds, the positive modulators, binds to the GABA$_{B(2)}$ subunit (Urwyler *et al.* 2001). These compounds do not activate the receptor on their own; rather they potentiate the efficacy and affinity of agonists on the GABA$_B$ receptor, as observed with the benzodiazepines on the GABA$_A$ receptor. GHB is not a positive modulator of the GABA$_B$ receptor.

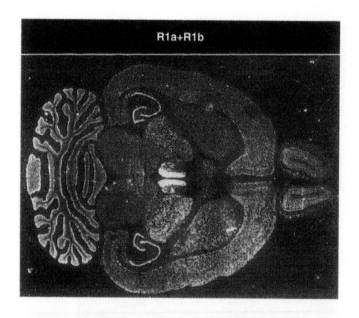

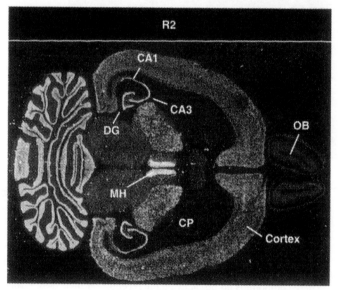

Color Plate II (See Chapter 3, p. 36. *R. Bernasconi* et al.)

Figure 3.3 Cellular and subcellular distribution of GABA$_B$ receptor mRNA in rat brain by *in situ* hybridization analysis. Transcripts of both GABA$_{B(1)}$ and GABA$_{B(2)}$ are abundant in the medial habenula (MH), the pyramidal cells of the CA1–CA3 subfields of the hippocampus, the granular layer of the dentate gyrus (DG) and the cortex. GABA$_{B(2)}$ transcripts are less abundant than GABA$_{B(1)}$ transcripts in the caudate putamen (CP) and the olfactory bulb (OB). GABA$_{B(1)}$ and GABA$_{B(2)}$ transcript expression in cerebellar cortex is higher in Purkinje cells than in the granule cell layer.

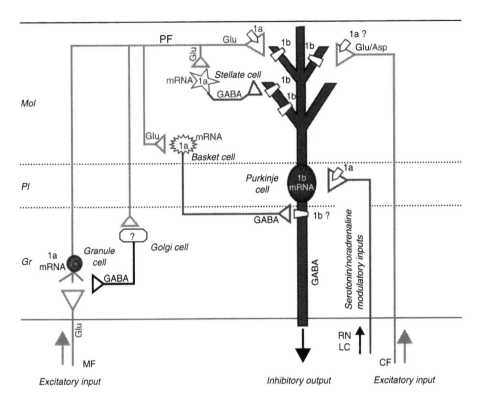

Color Plate III (See Chapter 3, p. 43. *R. Bernasconi* et al.)

Figure 3.4 Basic circuitry diagram of the cerebellum with the neurons of the molecular (Mol), Purkinje cell (Pl) and granular cell layers (Gr). The possible localization of the $GABA_{B(1a)}$ and $GABA_{B(1b)}$ receptor variants on the cerebellar circuitry is also indicated. This model is based on the data of Bischoff *et al.* (1999) on the localization of the mRNA transcripts and the autoradiographic distribution of [^{3}H]CGP 54626-labeled binding sites. This model is in line with the first immunohistochemical data with ultrastructural localization of antibodies in the Mol, Pl and Gr of the cerebellum (Kaupmann *et al.* 1998). (CF: climbing fibers; LC: locus coeruleus; MF: mossy fibers; PF: parallel fibers; RN: raphe nucleus; 1a: $GABA_{B(1a)}$; 1b: $GABA_{B(1b)}$ (modified from Bischoff *et al.* 1999).)

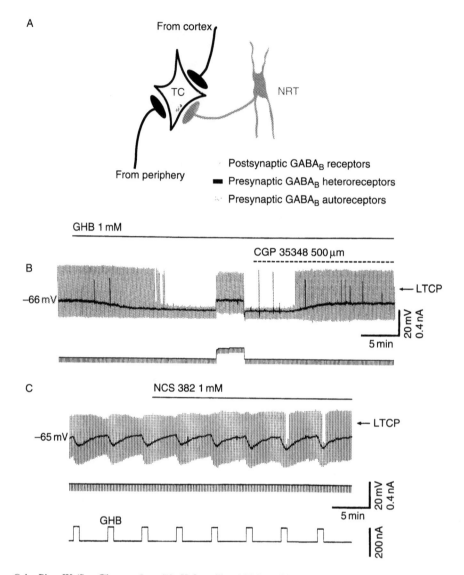

Color Plate IV (See Chapter 5, p. 76. *V. Crunelli and N. Leresche*)

Figure 5.1 Postsynaptic effect of GHB on TC neurons in brain slices. (A) Schematic diagram of sensory, cortical and nucleus reticularis (NRT) synapses on a TC neuron showing the location of presynaptic and postsynaptic GABA$_B$ receptors on which GHB has been shown to act. Note that NRT neurons are GABAergic. (B) Intracellular voltage record (top trace) shows the hyperpolarization elicited by GHB to be abolished by the subsequent application of CGP 35348. Upward deflections in the trace represent low threshold Ca^{2+} potentials (LTCPs). Downward deflections represent the voltage response to hyperpolarizing current pulses (bottom trace) that were used to measure the input resistance of the neuron. The reduction in input resistance associated with the response to GHB persists during repolarization of the membrane potential (by steady dc injection) to the pre-GHB value. (C) Repeated iontophoretic applications of GHB elicit a rapid-onset hyperpolarization (uppermost trace) that is not blocked by bath application of NCS 382. Note how during continuous perfusion of NCS 382 there is a clear, though small, increase in the response to GHB, which in turn leads to a failure of evoking LTCPs, as indicated by the absence of upward voltage deflections accompanying the peak of the last two responses to GHB. The middle trace is the current injected via the intracellular electrode, whereas the bottom trace represents the iontophoretic current used to eject GHB. (B) and (C) Reproduced with permission from Williams *et al.* (1995).

together, these results suggest that GHB binding sites may be implicated in the hyperthermia response induced by low doses of GHB, whereas GABA$_B$ receptors are involved in the hypothermic action of high doses of GHB.

GHB binding sites, GABA$_B$ receptors and absence seizures

Absence seizures are characterized by bilateral synchronous and rhythmic spike-and-wave discharges (SWD), which are associated with brief unresponsiveness to environmental stimuli and cessation of behavioral activity. EEG and cellular records, as well as lesion experiments, performed in various animal models, have demonstrated that absences are generated in specific neuronal networks involving sensory-motor cortex as well as reticular and relay thalamic nuclei from both hemispheres (Crunelli and Leresche 1991). The reticular nucleus is exclusively composed of GABAergic interneurons that project massively to relay thalamo-cortical neurons. Absences are, at least in part, related to an enhanced GABAergic influence on the thalamic relay nuclei. The reinforcement of inhibitory inputs leads to excessive synchronized rhythmic oscillations within the thalamocortical circuitry. The respective roles of GHB and GABA$_B$ receptors in absence epilepsy have been extensively investigated in various pharmacological or genetic models (Table 3.4).

In non-epileptic rats, systemic administration of GHB or GBL at doses high enough to interact with both GHB and GABA$_B$ receptors provokes absence-like seizures consisting in irregular SWD accompanied by immobility and myoclonic jerks. GHB-induced seizures are responsive to classical antiabsence drugs (Snead 1992a). *trans*-Hydroxycrotonic acid (THC) is a potent endogenous GHB receptor agonist that does not bind to GABA$_B$ receptors (Fig. 3.1; Benavides *et al.* 1982a). Administration of THCA (250–1,000 mg/kg) does not modify the EEG activity in non-epileptic control rats (Depaulis *et al.* 1988). Sedation is observed at the largest dose, suggesting that significant amounts of THCA cross the blood–brain barrier. The effects of higher doses of THCA on EEG activity have not been investigated. *R*-(−)-baclofen, which has no affinity for GHB binding sites (table 3.4; Snead 1994), does not produce absence seizures in non-epileptic control rats, but sometimes irregular small-amplitude oscillations are observed after administration of high doses of this potent GABA$_B$ receptor agonist (Aizawa *et al.* 1997).

Genetic absence epilepsy rats from Strasbourg (GAERS) closely reproduce human absence epilepsy. In GAERS, systemic administration of *R*-(−)-baclofen, or of GHB or GBL, induces

Table 3.4 Drugs involved in the control of absence seizure in GAERS versus non-epileptic control rats

Drugs	GHB binding sites k_i	GABA$_B$ receptors k_i	Non-epileptic rats	GAERS with spontaneous SWD
GBL (GHB)	30–580 nM	80–120 μM	Induction of SWD	Increase in SWD duration
(S)-(−)-HA966	250 nM	250 μM	—	Increase in SWD duration
trans-Hydroxy-crotonic acid	7 nM	No affinity	—	—
R-(−)-baclofen	No affinity	12–32 nM	—	Increase in SWD duration

GAERS: Genetic absence epilepsy rats from Strasbourg = rats with spontaneous bilaterally synchronous spike and wave discharges (SWD).

a dose-dependent increase in the duration of SWD evolving gradually into 'absence status' (Marescaux *et al.* 1992a,b). Similar results are obtained when the same drugs are injected bilaterally into thalamic relay nuclei (Liu *et al.* 1992; Marescaux *et al.* 1992a). Reciprocally, systemic or intrathalamic injections of $GABA_B$ receptor antagonists dose-dependently and completely suppress spontaneous SWD in GAERS (Marescaux *et al.* 1992a,b). On the contrary, the GHB antagonist NCS-382 produces only a partial and dose-independent decrease of SWD (Liu *et al.* 1991). In addition, THCA does not change in GAERS the cumulative duration of SWD at any stage after injection (Depaulis *et al.* 1988; Table 3.4). Thus, stimulation of only one of the GHB or $GABA_B$ receptors using potent but selective agonists is not sufficient to induce absence seizures in control rats, and simultaneous activation of both receptors may be required for the generation of absence seizures (Table 3.4).

Since $GABA_B$ receptors are coupled to G proteins, the effects of pertussis toxin (PTX) on SWD duration has been used to determine the role of the $GABA_B$–G protein receptor complex in the generation of absence seizures (Bowery *et al.* 1999). The injection of PTX into relay nuclei of the thalamus induces a time-dependent reduction in the duration of SWD in GAERS. After 6 days the SWD duration of the rats treated with PTX decreases by 96% compared with vehicle saline or denatured-PTX-injected rats, and $GABA_B$ but not $GABA_A$ receptor binding is significantly reduced by 70–80% in the ventrolateral and ventral posteriolateral thalamic nuclei. The decrease in $GABA_B$ binding sites by PTX, presumably by reducing high-affinity binding in the ventrolateral and ventral posteriolateral thalamic nuclei, could account for the virtual abolition of SWD. These data implicate G protein mechanisms in the generation of SWD in GAERS and support the role of $GABA_B$ receptors in their induction within the thalamus. The role of PTX-sensitive G proteins in the generation of SWD has also been investigated in the GHB and low-dose pentylenetetrazol models of absence epilepsy in the rat. Pretreatment with PTX administered i.c.v. induces a significant decrease in the duration of seizures in both models (Snead 1992b,c). Unfortunately, no binding experiments have been performed in these studies, but the doses of GBL correspond to activation of both GHB and $GABA_B$ receptors.

To investigate a possible involvement of the gene encoding the $GABA_{B(1)}$ receptor in idiopathic generalized epilepsies (IGE) comprising juvenile myoclonic epilepsy (JME) and idiopathic absence epilepsies (IAE), Kaupmann *et al.* (1998b) have determined the chromosomal localization of the $GABA_{B(1)}$ receptor gene. This gene maps close to the human major histocompatibility complex (HLA) locus on chromosome 6p21.3 in the same region to which a major susceptibility locus for IGE, EJM1, has been localized (Sander 1996; Kaupmann *et al.* 1998b). Mutation screening of the entire coding sequence of the $GABA_{B(1)}$ receptor gene by single-strand conformation polymorphism analysis has been performed in eighteen patients with JME. These patients are all members of families with positive evidence for linkage to the 6p21.3 region. Analyses have revealed several DNA polymorphisms, two of which result in amino acid changes occurring in all IGE-affected patients. However, clinical unaffected relatives do carry the same variations, excluding these amino acid substitutions as the cause for IGE in these families (Peters *et al.* 1998). This study has been continued and extended to 248 unrelated probands by Sander *et al.* (1999). The results reveal no evidence for an allelic association to any of the $GABA_{B(1)}$ receptor sequence variants with either JME or IAE. Therefore, there is little evidence to date to suggest that the $GABA_{B(1)}$ receptor gene is a candidate gene for IGE.

The potential role of $GABA_{B(2)}$ receptors in generalized epilepsies has not been examined so far. An evaluation of the involvement of the gene encoding the human GHB receptor in IGE probands has not been described to date.

Conclusions

The mechanism by which GHB produces its central effects remains a matter of controversy. According to radioligand binding studies, some actions of GHB have been suggested to be mediated via activation of specific membrane-binding sites without affinity for GABA or GABAergic agonists and antagonists (Maitre 1997). These GHB high-affinity binding sites have a different regional distribution and ontogeny in rat brain as compared to GABA$_A$ and GABA$_B$ receptors (Hechler *et al.* 1992; Snead 1994). On the other hand, many effects elicited by high doses of exogenous GHB are suppressed by administration of GABA$_B$ receptor antagonists, whereas the GHB antagonist NCS-382 is either weakly active or ineffective. Similarly, many effects of GHB on the CNS show a marked similarity to those elicited by the GABA$_B$ receptor agonist *R*-(−)-baclofen. To reconcile these apparently divergent results, we have introduced in this chapter the notion that exogenous GHB may stimulate both GHB and GABA$_B$ binding sites independently. We have provided evidence that, in addition to being an agonist at its specific binding sites, GHB is a weak, partial agonist both at pre- and postsynaptic GABA$_B$ receptors. The reduction of GABA and glutamate release is in line with a presynaptic stimulation of GABA$_B$ auto- and heteroreceptors. The activation of Kir3 channels indicates that GHB is a partial agonist at postsynaptic GABA$_B$ receptors (Misgeld *et al.* 1995; Kaupmann *et al.* 1998b; Lingenhoehl *et al.* 1999).

Brain levels of GHB after systemic administration of doses between 5 and 30 mg/kg are probably too low to activate GABA$_B$ receptors, and the pharmacological responses may correspond to a stimulation of GHB binding sites. In addition, these effects are completely suppressed by NCS-382, confirming that they are induced by a stimulation of GHB binding sites only. On the contrary, brain concentrations of GHB (240 μM) measured after injection of single high doses of GHB (100–400 mg/kg) are in the same order of magnitude as the K_i (80–120 μM) of GHB for GABA$_B$ receptors. The effects seen after administration of these doses are the result of the stimulation of both GHB and GABA$_B$ binding sites.

In agreement with the different regional distribution and ontogeny of the two receptors, we do not postulate a co-localization of the two binding sites. The mechanisms underlying a possible cross talk between GHB binding sites and GABA$_B$ receptors are unknown, and the answer to this question has to wait the cloning of the GHB receptor. The generation of GABA$_{B(1)}$ and GABA$_{B(2)}$ knockout mice will be an invaluable tool to investigate the effects of GHB in the absence of GABA$_B$ receptors. Many questions concerning the role of endogenous GHB and of [^{3}H] GHB binding sites have not been resolved. The results presented do not explain the biological role of endogenous GHB and its receptor in the CNS: no clear functional responses such as modulation of adenylate cyclase and K$^+$ or Ca^{2+} channels, of phospholipase C, phospholipase A$_2$ or stimulation of GTPγS binding have been reported after exposure to low doses of GHB corresponding to endogenous levels. Molecular cloning of the GHB receptor should characterize the physiological significance of GHB and further clarify the functional role of the endogenous agonist *trans*-hydroxycrotonic acid. On the other hand, the need for high concentrations of GHB to produce changes in membrane conductance or to inhibit GABA and glutamate release is consistent with a pharmacological action rather than with a physiological role of GHB as a GABA$_B$ receptor agonist.

In summary, the data presented in this chapter support the hypothesis that GHB is a weak partial agonist at GABA$_B$ receptors and that the effects induced in animals by administration of exogenous GHB at doses of approximately 100–400 mg/kg are due in part to a stimulation of GABA$_B$ receptors.

Acknowledgments

We thank Therese Leonhardt, Verena Rubio, Nicole Reymann, Jakob Heid and Pierre Martin for their skilful experimental work. We are grateful to Wolfgang Froestl, Nathalie Leresche, Vincenzo Crunelli and Klemenz Kaupmann for their comments on the manuscripts. The EU community, contract European BMH 4CT972093, supported this work.

References

Addolorato G., Caputo F., Capristo E., *et al.* (2000) Ability of baclofen in reducing alcohol craving and intake: II – preliminary clinical evidence. *Alcoholism: Clin. Exp. Res.* 24, 67–71.

Aizawa M., Ito Y. and Fukuda H. (1997) Roles of γ-aminobutyric acid$_B$ (GABA$_B$) and γ-hydroxybutyric acid receptors in hippocampal long-term potentiation and pathogenesis of absence seizures. *Biol. Pharm. Bull.* 20, 1066–70.

Alford S. and Grillner S. (1991) The involvement of GABA$_B$ receptors and coupled G-proteins in spinal GABAergic presynaptic inhibition. *J. Neurosci.* 11, 3718–26.

Andriamampandry C. and Maitre M. (2000) Cloning, expression and characterisation of a cDNA coding for a rat brain gamma-hydroxybutyrate (GHB) receptor. WO 00/78948 A2.

Andriamampandry C., Kemmel V., Hechler V., *et al.* (1999) Characterization of the receptor for the neurotransmitter and addictive drug γ-hydroxybutyrate (GHB). *Biochimie* 81 (Suppl. 6), 107.

Angers S., Salahpour A., Joly E., *et al.* (2000) Detection of β_2-adrenergic receptor dimerization in living cells using bioluminescence resonance energy transfer (BRET). *Proc. Natl Acad. Sci. USA* 97, 3684–89.

Banerjee P. K. and Snead O. C. (1995) Presynaptic gamma-hydroxybutyric acid (GHB) and gamma-aminobutyric acidB (GABA$_B$) receptor-mediated release of GABA and glutamate (GLU) in rat thalamic ventrobasal nucleus (VB): a possible mechanism for the generation of absence-like seizures induced by GHB. *J. Pharmacol. Exp. Ther.* 273, 1534–43.

Benavides J., Rumigny J. F., Bourguignon J. J., *et al.* (1982a) High affinity binding sites for γ-hydroxybutyric acid in rat brain. *Life Sci.* 30, 953–61.

Benavides J., Rumigny J. F., Bourguignon J. J., *et al.* (1982b) A high-affinity, Na$^+$-dependent uptake system for γ-hydroxybutyrate in membrane vesicles prepared from rat brain. *J. Neurochem.* 38, 1570–75.

Bernasconi R., Maitre L., Martin P., *et al.* (1981) The use of inhibitors of GABA-transaminase for the determination of GABA turnover in mouse brain regions: an evaluation of aminooxyacetic acid and gabaculine. *J. Neurochem.* 38, 57–66.

Bernasconi R., Lauber J., Marescaux C., *et al.* (1992) Experimental absence seizures: potential role of γ-hydroxybutyric acid and GABA$_B$ receptors. *J. Neural Transm. (Suppl.)* 35, 155–77.

Bernasconi R., Mathivet P., Marescaux P., *et al.* (1995) Similar effects of (−)-baclofen and γ-hydroxybutyric acid (GHB) on presynaptic GABA$_B$ receptors. *Soc. Neurosci. Abstr.* 21, 778.

Bernasconi R., Mathivet P., Bischoff S., *et al.* (1999) Gamma-hydroxybutyric acid: an endogenous neuromodulator with abuse potential. *Trends Pharmacol. Sci.* 20, 135–41.

Bischoff S., Leonhard S., Reymann N., *et al.* (1999) Spatial distribution of GABA$_B$R1b receptor mRNA and binding sites in the rat brain. *J. Comp. Neurol.* 412, 1–16.

Bittiger H., Reymann N., Froestl W., *et al.* (1992) ^{3}H-CGP 54626: a potent antagonist radioligand for GABA$_B$ receptors. *Pharmacol. Commun.* 2, 23.

Bittiger H., Bellouin C., Froestl W., *et al.* (1996) [^{3}H] CGP 62349: a new potent GABA$_B$ receptor antagonist radioligand. *Pharmacol. Rev. Commun.* 8, 97–98.

Bonanno G. and Raiteri M. (1993) Multiple GABA$_B$ receptors. *Trends Pharmacol. Sci.* 14, 259–61.

Bowery N. G. (2000) GABA$_B$ receptors. Structure and function. In *GABA in the Nervous System: The View at Fifty Years* (Martin D. L. and Olsen R. W., eds), pp. 233–44. Lippincott Williams and Wilkins, Philadelphia, PA.

Bowery N. G., Hudson A. L. and Price G. W. (1987) GABA$_A$ and GABA$_B$ receptor site distribution in the rat central nervous system. *Neuroscience* 20, 365–83.

Bowery N. G., Parry K., Boehrer A., *et al.* (1999) Pertussis toxin decreases absence seizures and GABA$_B$ receptor binding in thalamus of a genetically prone rat (GAERS). *Neuropharmacology* 38, 1691–97.

Brebner K., Froestl W., Andrews M., *et al.* (1999) The GABA$_B$ agonist CGP 44532 decreases cocaine self-administration in rats: demonstration using a progressive ratio and a discrete trials procedure. *Neuropharmacology* 38, 1797–804.

Bronstein J. M. (2000) Function of tetraspan proteins in the myelin sheath. *Curr. Opin. Neurobiol.* 10, 552–55.

Calver A. R., Medhurst A. D., Robbins M. J., *et al.* (2000) The expression of GABA$_{B1}$ and GABA$_{B2}$ receptor subunits in the CNS differs from that in peripheral tissues. *Neuroscience* 100, 155–70.

Calver A. R., Robbins M. J., Cosio C., *et al.* (2001) The C-terminal domains of the GABA$_B$ receptor subunits mediate intracellular trafficking but are not required for receptor signalling. *J. Neurosci.* 21, 1203–10.

Cash C. D. (1994) Gamma-hydroxybutyrate: an overview of the pros and cons for it being a neurotransmitter and/or a useful therapeutic agent. *Neurosci. Biobehav. Rev.* 18, 291–304.

Cash C. D., Hechler V., Mersel M., *et al.* (1996) Kinetic characterisation and solubilisation of γ-hydroxybutyrate receptors from rat brain. *Neurosci. Lett.* 209, 25–28.

Cash C. D., Gobaille S., Kemmel V., *et al.* (1999) γ-Hydroxybutyrate receptor function studied by the modulation of nitric oxide synthase activity in rat frontal cortex punches. *Biochem. Pharmacol.* 58, 1815–19.

Castelli M. P., Ingianni A., Stefanini E., *et al.* (1999) Distribution of GABA$_B$ receptor mRNA in the rat brain and peripheral organs. *Life Sci.* 64, 1321–28.

Chu D. C. M., Albin R. L., Young A. B., *et al.* (1990) Distribution and kinetics of GABA$_B$ binding sites in rat central nervous system: a quantitative autoradiographic study. *Neuroscience* 34, 341–57.

Colombo G., Agabio R., Lobina C., *et al.* (1998) Involvement of GABA$_A$ and GABA$_B$ receptors in the mediation of discriminative stimulus effects of gamma-hydroxybutyric acid. *Physiol. Behav.* 64, 293–302.

Colombo G., Agabio R., Carai M. A. M., *et al.* (2000) Ability of baclofen in reducing alcohol intake and withdrawal severity: I – preclinical evidence. *Alcoholism: Clin. Exp. Res.* 24, 58–66.

Crunelli V. and Leresche N. (1991) A role for GABA$_B$ receptors in excitation and inhibition of thalamo-cortical cells. *Trends Neurosci.* 14, 16–21.

Da Prada M. and Keller H. H. (1976) Baclofen and gamma-hydroxybutyrate: similar effects on cerebral dopamine neurones. *Life Sci.* 19, 1253–64.

Davies C. H. and Collingridge G. L. (1996) Regulation of EPSPs by the synaptic activation of GABA$_B$ autoreceptors in rat hippocampus. *J. Physiol.* 496, 451–70.

Deisz R. A. (1997) Electrophysiology of GABA$_B$ receptors. In *The GABA Receptors* (Enna S.J. and Bowery N. G., eds), pp. 157–207. Humana Press, Totowa, NJ.

Depaulis A., Bourguignon J. J., Marescaux C., *et al.* (1988) Effects of gamma-hydroxybutyrate and gamma-butyrolactone derivatives on spontaneous generalized non-convulsive seizures in the rat. *Neuropharmacology* 27, 683–89.

Diana M., Mereu G., Mura A., *et al.* (1991) Low doses of gamma-hydroxybutyric acid stimulate the firing rate of dopaminergic neurons in unanesthetized rats. *Brain Res.* 566, 208–11.

Di Chiara G. (1995) The role of dopamine in drug abuse viewed from the perspective of its role in motivation. *Drug Alcohol Depend.* 38, 95–137.

Doherty J. D. and Roth R. H. (1978) Metabolism of γ-hydroxy-[1-^{14}C] butyrate by rat brain: relationship to the Krebs cycle and metabolic compartmentation of amino acids. *J. Neurochem.* 30, 1305–09.

Duthey B., Caudron S., Perroy J., *et al.* (2002) A single subunit (GB2) is required for G-protein activation by the heterodimeric GABA$_B$ receptor. *J. Biol. Chem.* 277, 3236–41.

Emri Z., Antal K. and Crunelli V. (1996) Gamma-hydroxybutyric acid decreases thalamic sensory excitatory postsynaptic potentials by an action on presynaptic GABA$_B$ receptors. *Neurosci. Lett.* 216, 121–24.

Engberg G. and Nissbrandt H. (1993) Gamma-hydroxybutyric acid (GHBA) induces pacemaker activity and inhibition of substantia nigra dopamine neurons by activating GABA$_B$ receptors. *Naunyn-Schmied. Arch. Pharmacol.* 348, 491–97.

Engberg G., Kling-Petersen T. and Nissbrandt H. (1993) GABA$_B$ receptor activation alters the firing pattern of dopamine neurons in the rat substantia nigra. *Synapse* 15, 229–38.

Enna S. J. and Maggi A. (1979) Biochemical pharmacology of GABAergic agonists. *Life Sci.* 24, 1717–35.

Erhardt S., Andersson B., Nissbrandt H., *et al.* (1998) Inhibition of firing rate and changes in the firing pattern of nigral dopamine neurons by γ-hydroxybutyric acid (GHBA) are specifically induced by activation of GABA$_B$ receptors. *Naunyn-Schmied Arch. Pharmacol.* 357, 611–19.

Fattore L., Cossu G., Martellota M. C., *et al.* (2000) Effect of baclofen on intravenous self-administration in mice and rats. *Eur. J. Neurosci.* 12 (Suppl. 11), 204.

Feigenbaum J. J. and Howard S. G. (1996) Gamma-hydroxybutyrate is not a GABA agonist. *Prog. Neurobiol.* 50, 1–7.

Feigenbaum J. J. and Howard S. G. (1997) Naloxone reverses the inhibitory effect of γ-hydroxybutyrate on central DA release *in vivo* in awake animals: a microdialysis study. *Neurosci. Lett.* 224, 71–74.

Feigenbaum J. J. and Simantov R. (1996) Lack of effect of γ-hydroxybutyrate on μ, δ and κ opioid receptor binding. *Neurosci. Lett.* 212, 5–8.

Filippov A. K., Couve A., Pangalos M. N., *et al.* (2000) Heterodimeric assembly of GABA$_B$R1 and GABA$_B$R2 receptor subunits inhibits Ca^{2+} current in sympathetic neurons. *J. Neurosci.* 20, 2867–74.

Fritschy J. M., Giorgetta O. and Benke D. (1999a) Differential maturation and cellular distribution of the GABA$_B$-receptor splice variants in developing brain. In *GABA$_B$ Receptors: The 8th Neuropharmacology Conference* (Bettler B., Bowery N. G. and Collingridge G. L., eds), p. 51. Pergamon Elsevier Science, New York.

Fritschy J. M., Meskenaite V., Weinmann O., *et al.* (1999b) GABA$_B$-receptor splice variants GB1a and GB1b in rat brain: developmental regulation, cellular distribution and extrasynaptic localization. *Eur. J. Neurosci.* 11, 761–68.

Galvez T., Urwyler S., Prézeau L., *et al.* (2000) Ca^{2+} requirement for high-affinity γ-aminobutyric acid (GABA) binding at GABA$_B$ receptors: involvement of serine 269 of the GABA$_B$R1 subunit. *Mol. Pharmacol.* 57, 419–26.

Galvez T., Duthey B., Kniazeff J., *et al.* (2001) Allosteric interactions between GB1 and GB2 subunits are required for optimal GABA$_B$ receptor function. *EMBO J.* 20, 2152–59.

Garcia-Frigola C., Burgaya F., Calbet M., *et al.* (2000) Mouse Tspan-5, a member of the tetraspanin superfamily, is highly expressed in brain cortical structures. *NeuroReport* 11, 3181–85.

Gessa G. L., Diana M., Fadda F., *et al.* (1993) Gamma-hydroxybutyric acid (GHB) for treatment of ethanol dependence. *Eur. Neuropsychopharmacol.* 3, 224–25.

Gobaille S., Hechler V., Andriamampandry C., *et al.* (1999) γ-Hydroxybutyrate modulates synthesis and extracellular concentration of γ-aminobutyric acid in discrete rat brain regions *in vivo*. *J. Pharmacol. Exp. Ther.* 290, 303–09.

Godbout R., Jelenic P., Labrie C., *et al.* (1995) Effects of gamma-hydroxybutyrate and its antagonist NCS-382 on spontaneous cell firing in the prefrontal cortex of the rat. *Brain Res.* 673, 157–60.

Gray J. A., Goodwin G. M., Heal D. J., *et al.* (1987) Hypothermia induced by baclofen, a possible index of GABA$_B$ receptor function in mice, is enhanced by antidepressant drugs and ECS. *Br. J. Pharmacol.* 92, 863–70.

Harris N. C., Webb C. and Greenfield S. A. (1989) The effects of gamma-hydroxybutyrate on the membrane properties of guinea pig pars compacta neurons in the substantia nigra *in vitro*. *Neuroscience* 31, 363–70.

Hawrot E., Xiao Y., Shi Q. L., *et al.* (1998) Demonstration of a tandem pair of complement protein modules in GABA$_B$ receptor 1a. *FEBS Lett.* 432, 103–08.

Hébert T. E. and Bouvier M. (1998) Structural and functional aspects of G protein-coupled receptor oligomerization. *Biochem. Cell Biol.* 76, 1–11.

Hechler V., Bourguignon J. J., Wermuth C. G., *et al.* (1985) γ-Hydroxybutyrate uptake by rat brain slices. *Neurochem. Res.* 10, 387–96.

Hechler V., Schmitt M, Bourguignon J. J., *et al.* (1990) *trans*-Hydroxycrotonic acid binding sites in brain: evidence for a subpopulation of γ-hydroxybutyrate sites. *Neurosci. Lett.* 110, 204–09.

Hechler V., Gobaille S. and Maitre M. (1992) Selective distribution pattern of gamma-hydroxybutyrate receptors in the rat forebrain and midbrain as revealed by quantitative autoradiography. *Brain Res.* 572, 345–48.

Hechler V., Ratomponirina C. and Maitre M. (1997) γ-Hydroxybutyrate conversion into GABA induces displacement of GABA$_B$ binding that is blocked by valproate and ethosuximide. *J. Pharmacol. Exp. Ther.* 281, 753–60.

Hill D. R. and Bowery N. G. (1981) ^{3}H-baclofen and ^{3}H-GABA bind to bicuculline-insensitive GABA$_B$ sites in rat brain. *Nature* 290, 149–52.

Hu R. Q., Banerjee P. K. and Snead O. C. (2000) Regulation of γ-aminobutyric acid (GABA) release in cerebral cortex in the γ-hydroxybutyric acid (GHB) model of absence seizures in rat. *Neuropharmacology* 39, 427–39.

Ishige K., Ito Y. and Fukuda H. (1993) Differential effects of Triton Y-100 on ligand binding to GABA$_B$ receptors in mouse cerebral cortex, cerebellum and whole brain. *Gen. Pharmacol.* 6, 1533–40.

Ishige K., Aizawa M., Ito Y., *et al.* (1996) γ-Butyrolactone-induced absence-like seizures increase nuclear CRE- and AP-1 DNA-binding activities in mouse brain. *Neuropharmacology* 35, 45–55.

Ito Y., Ishige K., Zaitsu E., *et al.* (1995) γ-Hydroxybutyric acid increases intracellular Ca^{2+} concentration and nuclear cyclic AMP-responsive element- and activator protein 1 DNA-binding activities through GABA$_B$ receptor in cultured cerebellar granule cells. *J. Neurochem.* 65, 75–83.

Jackson H. C. and Nutt D. J. (1991) Inhibition of baclofen-induced hypothermia in mice by the novel GABA$_B$ antagonist CGP 35348. *Neuropharmacology* 30, 535–38.

Jones K. A., Borowsky B., Tamm J. A., *et al.* (1998) GABA-B receptor function as a heteromeric assembly of the subunits GABA B R1 and GABA B R2. *Nature* 396, 674–79.

Kammerer R. A., Frank S., Schulthess T., *et al.* (1999) Heterodimerization of a functional GABA$_B$ receptor is mediated by parallel coiled-coil α-helices. *Biochemistry* 38, 13263–69.

Kang J., Jiang L., Goldman S. A., *et al.* (1998) Astrocyte-mediated potentiation of inhibitory synaptic transmission. *Nature Neurosci.* 1, 683–92.

Kaufman E. E. and Nelson T. (1991) An overview of γ-hydroxybutyrate catabolism: the role of the cytosolic NAPD$^+$-dependent oxidoreductase EC 1.1.1.19 and of a mitochondrial hydroxy-acid–oxoacid transhydrogenase in the initial, rate-limiting step in this pathway. *Neurochem. Res.* 16, 965–74.

Kaufman E. E., Porrino L. J. and Nelson T. (1990) Pyretic action of low doses of γ-hydroxybutyrate in rats. *Biochem. Pharmacol.* 40, 2637–40.

Kaupmann K., Huggel K., Heid J., *et al.* (1997) Expression cloning of GABA$_B$ receptors uncovers similarity to metabotropic glutamate receptors. *Nature* 386, 239–46.

Kaupmann K., Malitschek B., Schuler V., *et al.* (1998a) GABA$_B$-receptor subtypes assemble into functional heteromeric complexes. *Nature* 396, 683–87.

Kaupmann K., Schuler V., Mosbacher J., *et al.* (1998b) Human γ-aminobutyric acid type B receptors are differentially expressed and regulate inwardly rectifying K$^+$ channels. *Proc. Natl Acad. Sci. USA* 95, 14991–96.

Kemmel V., Taleb O., Perard A., *et al.* (1998) Neurochemical and electrophysiological evidence for the existence of a functional γ-hydroxybutyrate system in NCB-20 neurons. *Neuroscience* 86, 989–1000.

Kerr D. I. B. and Ong J. (1995) GABA$_B$ receptors. *Pharmacol. Ther.* 67, 187–246.

King M. A., Thinschmidt J. S. and Walker D. W. (1997) Gammahydroxybutyrate (GHB) receptor ligand effects on evoked synaptic field potentials in CA1 of the rat hippocampal slice. *J. Neural Transm.* 104, 1177–93.

Knott C., Maguire J. J. and Bowery N. G. (1993) Age-related regional sensitivity to pertussis toxin-mediated reduction in $GABA_B$ receptor binding in rat brain. *Mol. Brain Res.* 18, 353–57.

Kolin A., Brezina A., Mamelak M., *et al.* (1993) Cardioprotective action of sodium gamma-hydroxybutyrate against isoproterenol induced myocardial damage. *Int. J. Exp. Pathol.* 74, 275–81.

Kuner R., Kohr G., Grunewald S., *et al.* (1999) Role of heteromer formation in $GABA_B$ receptor function. *Science* 283, 74–77.

Leff P. and Douglas G. (1993) Further concerns over Cheng–Prusoff analysis. *Trends Pharmacol. Sci.* 14, 110–12.

Lingenhoehl K., Brom R., Heid J., *et al.* (1999) γ-Hydroxybutyrate is a weak agonist at recombinant $GABA_B$ receptors. *Neuropharmacology* 38, 1667–73.

Liu Z., Snead O. C., Vergnes M., *et al.* (1991) Intrathalamic injections of γ-hydroxybutyric acid increase genetic absence seizures in rats. *Neurosci. Lett.* 125, 19–21.

Liu Z., Vergnes M., Depaulis A., *et al.* (1992) Involvement of intrathalamic $GABA_B$ neurotransmission in the control of absence seizures in the rat. *Neuroscience* 48, 87–93.

Lloyd K. G. and Dreksler S. (1979) An analysis of [^{3}H]-gamma-aminobutyric acid (GABA) binding in human brain. *Brain Res.* 163, 77–87.

Lorente P., Lacampagne A., Pouzeratte Y., *et al.* (2000) γ-Aminobutyric acid type B receptors are expressed and functional in mammalian cardiomyocytes. *Proc. Natl Acad. Sci. USA* 97, 8664–69.

McLatchie L. M., Fraser N. J., Main M. J., *et al.* (1998) RAMPs regulate the transport and ligand specificity of the calcitonin-receptor-like receptor. *Nature* 393, 333–39.

Madden T. E. and Johnson S. W. (1998) Gamma-hydroxybutyrate is a $GABA_B$ receptor agonist that increases a potassium conductance in rat ventral tegmental dopamine neurons. *J. Pharmacol. Exp. Ther.* 287, 261–65.

Maecker H. T., Todd S. C. and Levy S. (1997) The tetraspanin superfamily: molecular facilitators. *FASEB J.* 11, 428–42.

Maitre M. (1997) The gamma-hydroxybutyrate signalling system in the brain: organization and functional implications. *Prog. Neurobiol.* 51, 337–61.

Maitre M., Hechler V., Vayer P., *et al.* (1990) A specific gamma-hydroxybutyrate receptor ligand possesses both antagonistic and anticonvulsant properties. *J. Pharmacol. Exp. Ther.* 255, 657–63.

Malitschek B., Schweizer C., Keir M., *et al.* (1999) The N-terminal domain of $GABA_B$ receptors is sufficient to specify agonist and antagonist binding. *Mol. Pharmacol.* 56, 448–54.

Marescaux C., Liu Z., Bernasconi R., *et al.* (1992a) $GABA_B$ receptors are involved in the occurrence of absence seizures in rats. *Pharmacol. Commun.* 2, 57–62.

Marescaux C., Vergnes M. and Bernasconi R. (1992b) $GABA_B$ receptor antagonists: potential new-antiabsence drugs. *J. Neural Transm. (Suppl.)* 35, 179–88.

Margeta-Mitrovic M., Mitrovic I., Riley R. C., *et al.* (1999) Immunohistochemical localization of $GABA_B$ receptors in the rat central nervous system. *J. Comp. Neurol.* 405, 299–321.

Margeta-Mitrovic M., Jan Y.N. and Jan L.Y. (2001) Function of GB1 and GB2 subunits in G protein coupling of $GABA_B$ receptors. *Proc. Natl. Acad. Sci. USA* 98, 14649–54.

Marshall F. H., Jones K. A., Kaupmann K., *et al.* (1999) $GABA_B$ receptors – the first 7TM heterodimers. *Trends Pharmacol. Sci.* 20, 396–99.

Mathivet P., Boehrer A., Reymann N., *et al.* (1996) Effects of (−)-HA966 and (+)-HA966 on $GABA_A$, $GABA_B$ and γ-hydroxybutyric acid (GHB) receptors. *Pharmacol. Rev. Commun.* 8, 209.

Mathivet P., Bernasconi R., de Barry J., *et al.* (1997) Binding characteristics of γ-hydroxybutyric acid as weak but selective $GABA_B$ receptor agonist. *Eur. J. Pharmacol.* 321, 67–75.

Misgeld U., Bijak M. and Jarolimek W. (1995) A physiological role for $GABA_B$ receptors and the effects of baclofen in the mammalian central nervous system. *Prog. Neurobiol.* 46, 423–62.

Möhler H., Patel A. J. and Balazs R. (1976) Gamma-hydroxybutyrate degradation in the brain *in vivo*: negligible direct conversion to GABA. *J. Neurochem.* 27, 253–58.

Mott D. D. and Lewis D. L. (1994) The pharmacology and function of $GABA_B$ receptors. *Int. Rev. Neurobiol.* 36, 97–223.

Nelson T., Kaufman E., Kline J., *et al.* (1981) The extraneural distribution of γ-hydroxybutyrate. *J. Neurochem.* 37, 1345–48.

Ng G. K., Clark J., Coulombe N., *et al.* (1999) Identification of a $GABA_B$ receptor subunit, GB2, required for functional $GABA_B$ receptor activity. *J. Biol. Chem.* 274, 7607–10.

Nissbrandt H. and Engberg G. (1996) The $GABA_B$-receptor antagonist CGP 35348 antagonises γ-hydroxybutyrate and baclofen-induced alterations in locomotor activity and forebrain dopamine levels in mice. *J. Neural Transm.* 103, 1255–63.

Nissbrandt H., Elverfors A. and Engberg G. (1994) Pharmacologically induced cessation of burst activity in nigral dopamine neurons: significance for the terminal dopamine efflux. *Synapse* 17, 217–24.

Olpe H. R. and Koella W. (1979) Inhibition of nigral and neocortical cells by gamma-hydroxybutyrate: a microiontophoretic investigation. *Eur. J. Pharmacol.* 53, 359–64.

Otten U., Dimitriades-Schmutz, Bettler B., *et al.* (2000) Modulation by $GABA_B$ receptor agonists and antagonists of neurotrophin levels in primary cultures of rat astrocytes. *Soc. Neurosci. Abstr.* 26, 585.

Pagano A., Rovelli G., Mosbacher J., *et al.* (2001) C-terminal interaction is essential for surface trafficking but not for heteromeric assembly of $GABA_B$ receptors. *J. Neurosci.* 21, 1189–202.

Pericic D., Eng N. and Walters J. R. (1978) Post-mortem and aminooxyacetic acid-induced accumulation of GABA: effect of gamma-butyrolactone and picrotoxin. *J. Neurochem.* 30, 767–73.

Peters H. W., Kämmer G., Volz A., *et al.* (1998) Mapping, genomic structure, and polymorphisms of the human $GABA_B$R1 receptor gene: evaluation of its involvement in idiopathic generalized epilepsy. *Neurogenetics* 2, 47–54.

Poldrugo F. and Addolorato G. (1999) The role of gamma-hydroxybutyric acid in the treatment of alcoholism: from animal to clinical studies. *Alcohol Alcohol.* 34, 15–24.

Ratomponirina C., Hodé Y., Hechler V., *et al.* (1995) γ-Hydroxybutyrate receptor binding in rat brain is inhibited by guanyl nucleotides and pertussis toxin. *Neurosci. Lett.* 189, 51–53.

Robbins M.J., Calver A.R., Filipov A.K., *et al.* (2001) $GABA_{B2}$ is essential for G-protein coupling of the $GABA_B$ receptor heterodimer. *J. Neurosci.* 21, 8043–52.

Sander T. (1996) The genetics of idiopathic generalised epilepsy: implications for the understanding of its aetiology. *Mol. Med. Today* 2, 173–80.

Sander T., Peters C., Kämmer G., *et al.* (1999) Association analysis of exonic variants of the gene encoding the $GABA_B$ receptor and idiopathic generalized epilepsy. *Am. J. Med. Genet.* 88, 305–10.

Serra M., Sanna E., Foddi C., *et al.* (1991) Failure of γ-hydroxybutyrate to alter the function of the $GABA_A$ receptor complex in the rat cerebral cortex. *Psychopharmacology* 104, 351–55.

Snead O. C. (1991) The γ-hydroxybutyrate model of absence seizures: correlation of regional brain levels of γ-hydroxybutyric acid and γ-butyrolactone with spike wave discharges. *Neuropharmacology* 30, 161–67.

Snead O. C. (1992a) Pharmacological models of generalized absence seizures in rodents. *J. Neural Transm. (Suppl.)* 35, 7–19.

Snead O. C. (1992b) $GABA_B$ receptor mediated mechanisms in experimental absence seizures in rat. *Pharmacol. Commun.* 2, 63–69.

Snead O. C. (1992c) Evidence for G protein modulation of experimental-generalized absence seizures in rat. *Neurosci. Lett.* 148, 15–18.

Snead O. C. (1994) The ontogeny of [³H]γ-hydroxybutyrate and [³H] $GABA_B$ binding sites: relation to the development of experimental absence seizures. *Brain Res.* 659, 147–56.

Snead O. C. (1996) Relation of the [³H]γ-hydroxybutyric acid (GHB) binding site to the γ-aminobutyric acid$_B$ ($GABA_B$) receptor in the rat brain. *Biochem. Pharmacol.* 52, 1235–43.

Snead O. C. (2000) Evidence for a G protein-coupled γ-hydroxybutyric acid receptor. *J. Neurochem.* 75, 1986–96.

Snead O. C. and Liu C. C. (1984) Gamma-hydroxybutyric acid binding sites in rat and human brain synaptosomal membranes. *Biochem. Pharmacol.* 33, 2587–90.

Snead O. C. and Liu C. C. (1993) GABA$_A$ receptor function in the γ-hydroxybutyrate model of generalized absence seizures. *Neuropharmacology* 32, 401–09.

Thompson S. M. and Gähwiler B. H. (1992) Comparison of the actions of baclofen at pre- and postsynaptic receptors in the rat hippocampus *in vitro. J. Physiol.* 451, 329–45.

Timmerman W. and Westerink B. H. C. (1997) Brain microdialysis of GABA and glutamate: what does it signify? *Synapse* 27, 242–61.

Turgeon S. M. and Albin R. L. (1993) Pharmacology, distribution, cellular localization, and development of GABA$_B$ binding in rodent cerebellum. *Neuroscience* 55, 311–23.

Urwyler S., Mosbacher J., Lingenhoehl K., *et al.* (2001) Positive allosteric modulation of native and recombinant γ-aminobutyric acid$_B$ receptors by 2,6-di-*tert*-butyl-4-(3-hydroxy-2,2-dimethyl-propyl)-phenol (CGP7930) and its aldehyde analog CGP 13501. *Mol. Pharmacol.* 60, 963–71.

Vayer P. and Maitre M. (1989) γ-Hydroxybutyrate stimulation of the formation of cyclic GMP and inositol phosphates in rat hippocampal slices. *J. Neurochem.* 52, 1382–87.

Vayer P., Mandel P. and Maitre M. (1985) Conversion of γ-hydroxybutyrate to γ-aminobutyrate *in vitro. J. Neurochem.* 45, 810–14.

Vayer P., Gobaille S., Mandel P., *et al.* (1987) 3′–5′ cyclic-guanosine monophosphate increase in rat brain hippocampus after gamma-hydroxybutyrate administration. Prevention by valproate and naloxone. *Life Sci.* 41, 605–10.

Waldmeier P. (1991) The GABA$_B$ antagonist, CGP 35348, antagonizes the effects of baclofen, γ-butyrolactone and HA 966 on rat striatal dopamine synthesis. *Naunyn-Schmied Arch. Pharmacol.* 343, 173–78.

Waldmeier P., Wicki P., Feldtrauer J. J., *et al.* (1994) GABA and glutamate release affected by GABA$_B$ receptor antagonists with similar potency: no evidence for pharmacologically different presynaptic receptors. *Br. J. Pharmacol.* 113, 1515–21.

Walters J. R. and Roth R. H. (1972) Effect of γ-hydroxybutyrate on dopamine and dopamine metabolites in the rat striatum. *Biochem. Pharmacol.* 21, 2111–21.

White J. H., Wise A., Main M. J., *et al.* (1998) Heterodimerisation is required for the formation of a functional GABA$_B$ receptor. *Nature (London)* 396, 679–82.

Williams S. R., Turner J. P. and Crunelli V. (1995) Gamma-hydroxybutyrate promotes oscillatory activity of rat and cat thalamocortical neurons by a tonic GABA$_B$ receptor-mediated hyperpolarization. *Neuroscience* 66, 133–41.

Wood P. (1991) Pharmacology of the second messenger, cyclic guanosine 3′, 5′-monophosphate, in the cerebellum. *Pharmacol. Rev.* 43, 1–25.

Xie X. and Smart T. G. (1992) Gamma-hydroxybutyrate hyperpolarizes hippocampal neurones by activating GABA$_B$ receptors. *Eur. J. Pharmacol.* 212, 291–94.

Addendum

GABA$_{B(1)}$ deficient mice

Mice lacking the gene for the GABA$_{B(1)}$ subunit have been developed by two different groups (Prosser *et al.*, 2001; Schuler *et al.*, 2001). Null-mutant mice generated on the 129Sv background survive for 3–4 weeks (Prosser *et al.*, 2001), whereas those generated on the Balb/c background have a normal viability (Schuler *et al.*, 2001). The latter null-mutant mice have been used to characterize GABA$_B$ receptor paradigms. Administration of GABA$_B$ agonists to Balb/c null-mutants failed to produce the typical muscle relaxation, hypothermia or delta EEG waves observed in wild-type animals. These behavioral findings were paralleled by a loss of all biochemical and electrophysiological GABA$_B$ responses. Importantly, the

GABA$_{B(2)}$ subunit is heavily down regulated in GABA$_{B(1)}$ knockout mice. This demonstrates that GABA$_{B(1)}$ subunits are an essential component of pre- and postsynaptic GABA$_B$ receptors. Accordingly, Balb/c null-mutant mice offer the possibility to tackle the question, which specific GHB actions depend on GABA$_B$ receptors.

Effects of GHB on GABA$_B$ receptors

Several recent studies add further support to the hypothesis, that GABA$_B$ receptors constitute a central site of action of GHB. Thus, the sedative/hypnotic effects of GHB (1000 mg/kg) in DBA mice are completed prevented by GABA$_B$ receptor antagonists, whereas NCS-382 reinforces these effects of GHB (Carai *et al.* 2001).

The action of GABA$_B$ receptor agonists on acetylcholine release has been examined by microdialysis in freely moving rats (Nava *et al.* 2001). GHB (200 and 500 mg/kg) and baclofen (20 mg/kg) reduce extracellular acetylcholine levels in the hippocampus by 25% to 50%. The GABA$_B$ receptor antagonist SCH 50911 suppresses these effects, while the putative GHB receptor antagonist NCS 382 is ineffective. These data indicate that GHB- and baclofen-induced reduction of hippocampal acetylcholine release is mediated by GABA$_B$ receptors. Another study by Ferraro *et al.* (2001) demonstrates that pharmacological effects of low and high doses of GHB are mediated by GHB and GABA$_B$ receptors, respectively. These authors examined the modulation of hippocampal glutamate release by GHB both *in vitro* and *in vivo*. The results of this study indicate that GHB exerts a concentration-dependent regulation of hippocampal glutamate transmission via two opposing mechanisms: a direct GHB binding site mediated facilitation is observed at nanomolar GHB concentrations, whereas a GABA$_B$ receptor mediated inhibition predominates at millimolar concentrations.

Jensen and Mody (2001) studied the electrophysiological effects of GHB on cortical neurons in brain slices from 12- to 33-day-old mice. Using whole-cell patch-clamp recordings they found that GHB activates both pre- and postsynaptic GABA$_B$ receptors in neocortical neurons participating in fast synaptic transmission, leading to a powerful depression of cortical network activity.

References

Carai M.A.M., Colombo G., Brunetti G., *et al.* (2001) The role of GABA$_B$ receptors in the sedative/hypnotic effect of γ-hydroxybutyric acid. *Eur. J. Pharmacol.* 428, 315–21.

Ferraro L., Tanganelli S., O'Connor W.T., *et al.* (2001) γ-Hydroxybutyrate modulation of glutamate levels in the hippocampus: an *in vivo* and *in vitro* study. *J. Neurochem.* 78, 929–39.

Jensen K. and Mody I. (2001) GHB depresses fast excitatory and inhibitory synaptic transmission via GABA$_B$ receptors in mouse neocortical neurons. *Cerebral Cortex* 11, 424–29.

Nava F., Carta G., Bortolato M., *et al.* (2001) γ-Hydroxybutyric acid and baclofen decrease extracellular acetylcholine levels in the hippocampus via GABA$_B$ receptors. *Eur. J. Pharmacol.* 430, 261–63.

Prosser H.M., Gill C.H., Hirst W.D., *et al.* (2001) Epileptogenesis and enhanced prepulse inhibition in GABA$_{B1}$-deficient mice. *Molecular and Cellular Neuroscience* 10, 1–10.

Schuler V., Lüscher C., Blanchet C., *et al.* (2001) Epilepsy, hyperalgesia, impaired memory and loss of pre- and postsynaptic GABA$_B$ responses in mice lacking GABA$_{B(1)}$. *Neuron* 31, 47–58.

4   Membrane transport of γ-hydroxybutyrate

G. Tunnicliff

Introduction

Soon after the discovery of γ-aminobutyric acid (GABA) in brain tissue and its possible role in central nervous system (CNS) function (Roberts and Frankel 1950), researchers became interested in the systemic administration of GABA to study its pharmacological effects in the intact animal. Unfortunately, the blood–brain barrier prevented most of the amino acid getting into the brain. Because of its structural resemblance to GABA, the fatty acid derivative γ-hydroxybutyrate (GHB) was synthesized as a potential GABA mimetic (Laborit 1964). At that time, it was known that the intraperitoneal injection of sodium *n*-butyrate and other short-chain fatty acids had a hypnotic action in rabbits (White and Samson 1956). Since butyric acid was known to be rapidly metabolized by β-oxidation, Laborit felt that GHB might undergo a reduced metabolic degradation and consequently exhibit a prolonged pharmacological effect.

Early pharmacological experiments revealed that GHB induced a sleep-like state in experimental animals when treated intraperitoneally (Drakontides *et al.* 1962; Laborit 1964). This apparent CNS depressant action was later exploited in human subjects where it was used as an induction agent for general anesthesia (Aldrete and Barnes 1968; Blumenfeld *et al.* 1962). Like sodium thiopental, GHB offered a poor analgesic profile, although this could be remedied by providing supplementary analgesia with opioids or nitrous oxide (Solway and Sadove 1965; Vickers, 1969). Several animal studies have provided clear evidence that GHB can induce EEG patterns of activity reminiscent of petit mal seizures (Winters and Spooner 1965; Godschalk *et al.* 1977). Despite these observations in experimental animals, no such evidence exists for this type of seizure activity in humans following GHB treatment (e.g. Entholzner *et al.* 1995). Nowadays GHB is used infrequently in general anesthesia. Other potential therapeutic uses of the drug include treatment of alcoholism (Addolorato *et al.* 1999) and other drug addictions (Gallimberti *et al.* 2000), as well as treatment of fibromyalgia, narcolepsy and additional sleep disorders (Scharf *et al.* 1985, 1998; Scrima *et al.* 1990; Lapierre *et al.* 1990).

In addition to its pharmacological actions, GHB could well have a physiological role in brain function. Relatively low levels have been discovered in the CNS (Bessman and Fishbein 1963; Anderson *et al.* 1977), and it has an uneven distribution (Snead and Morley 1981; Vayer *et al.* 1988). Evidently, succinic semialdehyde is a precursor after its production from GABA (Roth and Giarman 1969). Enzymes involved in its biosynthesis and degradation have been described. The enzyme responsible for its formation is cytosolic succinic semialdehyde reductase (Cash *et al.* 1979). Evidence suggests that two oxidative enzymes are involved in the metabolic degradation of GHB. One is GHB dehydrogenase and the other is

GHB keto-acid transhydrogenase (Kaufman and Nelson 1991). Two possible plasma membrane-located targets for the mediation of GHB action have been identified – the GHB receptor and the GABA$_B$ receptor. Both GHB recognition sites are linked to G-protein activation (Snead 2000; Thalmann 1988). GHB has a higher affinity for the GHB receptor than for the GABA$_B$ receptor (Bernasconi *et al.* 1999). It is plausible that the pharmacological actions of GHB are mediated via both receptors whereas the physiological effects are mediated via the GHB receptor.

Transport of GHB by nervous tissue

Transport of molecules and ions across biomembranes is a well-studied phenomenon. These membranes are generally impermeable to solutes, but there are occasions when they will allow transport. The structure of the biomembrane contributes to this selective permeability. Such membranes are rich in lipids, although the protein content can be substantial. Some lipophilic substances can readily pass through membranes by simple diffusion, for example, ethanol. However, all other examples of transport involve membrane proteins acting as channels or carriers. Water also can penetrate biomembranes despite its inability to dissolve in lipids. The mechanism involved seems to rely on aquaporins within the membranes (Engel *et al.* 2000). Several different ions, for instance Na^+, K^+ or Cl^-, can pass through membranes via specific channels. Molecules can be transported by two types of specific membrane proteins: those involved in facilitated diffusion and those involved in active transport. In both cases there is a recognition site on the carrier protein for the transported molecule. In facilitated diffusion transport is along a concentration gradient, but with active transport the solute is carried against its concentration gradient, usually with the co-transport of inorganic anions or cations (Stein 1990).

In the brain, neurotransmitters are transported by four different families of carrier proteins. One of the most important types of membrane transport is the uptake of neurotransmitters from the synaptic gap by plasma membrane carriers. This process biologically inactivates the neurotransmitter by rapidly removing it from the vicinity of its receptor and depends on the co-transport of Na^+ and Cl^- (Nelson 1998). Proteins belonging to this family exhibit a topography of twelve transmembrane helices. Four subfamilies belong to this group: (a) GABA transporters, (b) monoamine transporters, (c) amino acid transporters and (d) NTT4 (orphan) transporters. Ironically the first neurotransmitter to be identified, acetylcholine, was found to be inactivated not by an uptake process but by the hydrolytic action of acetylcholinesterase. Subsequently, almost all neurotransmitters since the discovery of acetylcholine have been shown to be removed from the synaptic gap by plasma membrane transport proteins rather than undergo enzymatic degradation. These transporters are located either in nerve cells or glial cells and remove their ligands from the synaptic cleft by a secondary active process. A second type of transporter concentrates neurotransmitters in synaptic vesicles (Schuldiner 1994). In fact, new data from Muller *et al.* (2002) demonstrate the existence of an ATP-dependent uptake process for GHB by such vesicles. A third type of transporter is dependent on Na^+ and K^+ and is responsible for glutamate uptake (Kanner 1993). Finally, general amino acid transport processes regulate amino acid neurotransmitters extracellularly (McGivan and Pastor-Anglada 1994).

Brain tissue contains a secondary active uptake system for GHB which is both Na^+- and Cl^--dependent (Benavides *et al.* 1982; McCormick and Tunnicliff 1998). Thus, there is good reason to believe that the transport protein responsible belongs to the Na/Cl family. According to hydropathic analysis, these proteins have twelve membrane-spanning regions. Using a vesicular preparation from rat brain, Benavides *et al.* (1982) observed that [^{3}H]GHB was taken up by an Na^+-dependent process, though in the absence of Na^+ there was still a measurable amount of uptake (about 30 pmol/min/mg protein). However, at a constant Cl^- concentration of 120 mM, when Na^+ was gradually increased to 120 mM, uptake of [^{3}H]GHB systematically increased till a threefold increase was observed. When uptake was measured at 140 mM Na^+ but in the absence of Cl^-, uptake was substantially reduced, though not eliminated. Thus, in the absence of both Na^+ and Cl^-, significant uptake occurred but a marked stimulation was noted with an increase in either ion. When the concentration of GHB was progressively increased, the rate of uptake also increased in a manner exhibiting saturation kinetics. Results plotted by the Eadie–Hofstee method showed a biphasic curve. One component of uptake yielded a K_m (transport constant) of 46.4 µM and a V_{max} of 155 pmol/min/mg protein, whereas the other component had a K_m of 325 µM and a V_{max} of 746 pmol/min/mg protein. This means two transport systems were present in the vesicular preparation, one having a high affinity for the ligand and the other having a considerably lower affinity. When the data for the high-affinity system were treated by the method of Feldman (1972), a K_m of 34.3 µM was calculated. This method assumes the presence of two binding sites for the ligand. The authors found that there were regional differences in the rate of GHB uptake in brain. For instance, of the five areas studied, striatum had the greatest uptake capacity while the pons medulla exhibited almost half as much. The authors measured the potential inhibitory effects of several GHB structural analogues. The most effective inhibitors were 3-hydroxypropane sulfonate, GABA, trans-4-hydroxycrotonic acid and 3-methylGHB, each of which produced about a 63% inhibition of 20 µM [^{3}H]GHB when present at 250 µM during the assay.

In addition to transport by a vesicular preparation, brain slices have also been shown to transport GHB. Hechler et al. (1985) reported that striatal slices from rat brain when incubated in a buffer containing [^{3}H]GHB took up the radiolabelled substrate in a manner consistent with secondary active transport. The uptake system displayed saturation kinetics, with a K_m of 0.7 mM and a V_{max} of 48 pmol/min/mg. Uptake was stimulated by Na^+ in a concentration-dependent manner. Uptake was also dependent upon K^+ since its omission resulted in a 40% inhibition. A series of GHB analogues were tested as potential inhibitors. The four most potent were γ-phenylGHB, cyclohexanol 2-acetic acid, trans-4-hydroxycrotonic acid and β-phenyl-trans-4-hydroxycrotonic acid which, against 50 µM GHB, yielded IC_{50} values ranging from 0.86 to 0.95 mM.

Mouse brain also has been shown to possess an uptake mechanism for GHB (Tunnicliff 1994). A synaptosomal fraction prepared from male ICR mice transported [^{3}H]GHB by a system that showed saturation kinetics when the concentration of the ligand was increased. The K_m was calculated as 48 µM and the V_{max} as 0.32 nmol/min/mg protein. Intriguingly, the uptake was found to be markedly stimulated by pyridoxal 5'-phosphate. In an attempt to characterize the binding site of GHB on its transport protein, pyridoxal 5'-phosphate was added to the assay system. The rationale was that pyridoxal 5'-phosphate can inhibit certain biologically active proteins by binding to essential lysine residues (e.g. Anderson et al. 1966; Tunnicliff 1979). Indeed, pyridoxal 5'-phosphate has previously been found to inhibit GABA transport by both cat and mouse brain (Tunnicliff 1980a,b). The presence of pyridoxal 5'-phosphate in the GHB uptake assay led to an increasing stimulation of transport up to about 3.5 mM pyridoxal 5'-phosphate. At this point, uptake was enhanced by about 330%. Higher concentrations of pyridoxal 5'-phosphate produced decreasing amounts of stimulation until finally there was an inhibition of GHB uptake at 7 mM pyridoxal 5'-phosphate. The significance of these observations is not known, and whether there are any physiological implications is also poorly understood.

The stimulation of GHB transport by Na^+ and Cl^- in rat brain, first reported by Benavides *et al.* (1982), has been shown to occur in mouse brain (McCormick and Tunnicliff 1998). Very little uptake by synaptosomes was observed when no Na^+ was added to the assay (less than 10 pmol/min/mg protein). A steady increase in transport occurred as Na^+ was added to the assay system in increasing amounts. At 100 mM Na^+, there was more than a ten fold stimulation of uptake. Mouse brain differs from rat brain in that, in the latter, there is substantial uptake even in the absence of Na^+ (about 30 pmol/min/mg protein). In mouse brain, even in the absence of Cl^- there was significant transport (about 60 pmol/min/mg protein). Addition of 100 mM Cl^- increased this by about 50%. The influence of Na^+ on GHB transport is supported by the observation that harmaline was found to inhibit uptake (McCormick and Tunnicliff 1998). It is established that this inhibitor binds to Na^+ sites on sodium-dependent transporters (Young *et al.* 1988). Thus, GHB uptake appears to rely on a transporter belonging to the family of Na/Cl carriers in which Na^+ is transported downhill while the substrate is transported uphill via a secondary active process.

A functional GHB system has been shown to exist in NCB-20 neuronal cells (Kemmel *et al.* 1998). These are a cell line derived from the fusion of mouse neuroblastoma N18TG2 and embryonic Chinese brain cells that possess properties of nerve cells (MacDermot *et al.* 1979). These NCB-20 cells possess a specific GHB receptor binding site and exhibit Ca^{2+} conductances which were inhibited by GHB. Further, this cell line possesses an Na^+-dependent uptake process exhibiting a K_m of 35 μM and a V_{max} of 80 pmol/min/mg protein. A Hill plot derived from increasing Na^+ concentrations had a slope approaching 1.0, which indicated that one Na^+ ion is transported with each molecule of GHB.

No potent inhibitors of GHB transport have yet been discovered, although a few moderately effective competitive inhibitors have been reported. When a series of structural analogues was tested in mouse brain synaptosomes, 2-hydroxycinnamic acid and 3-(2-furyl)acrylic acid inhibited [^{3}H]GHB uptake, with K_i values of 144 and 152 μM , respectively (McCormick and Tunnicliff 1998). 3-Hydroxypropane sulfonic acid also inhibited, but much more weakly ($K_i = 4$ mM). This compound was also tested against rat brain GHB transport (Benavides *et al.* 1982). It produced a 38% inhibition at 250 μM against 20 μM substrate, which was the most effective inhibitor out of a total of twenty-two monitored in that investigation. In mouse, GABA and certain GABA transport inhibitors had no appreciable effect on brain [^{3}H]GHB uptake (McCormick and Tunnicliff 1998). This strongly suggests that GHB is not transported by the GABA transporter.

The regional distribution of GHB uptake sites in brain is not well studied. However, Benavides *et al.* (1982) measured initial rates of transport in a P_2 fraction of five different areas of rat brain and found that striatum and hypothalamus possessed the highest values, that is 146 and 140 pmol/min/mg protein, respectively. The pons medulla yielded a value of 87 pmol/min/mg protein.

GHB transporter binding

Recently, a study has examined the binding of [^{3}H]GHB to its transporter (Tunnicliff, unpublished observations). In mouse brain homogenates, under conditions highly unfavorable to transport, binding showed saturability with an IC_{50} of 40 μM (Fig. 4.1). Binding was proportional to Na^+ concentration up to 100 mM, and in its absence no appreciable binding was observed. The effect of increasing Cl^- ions was similar, except that in their absence the degree of [^{3}H]GHB binding was still measurable (Fig. 4.2). A number of structural analogues of GHB were able to inhibit binding (data not shown), the most effective

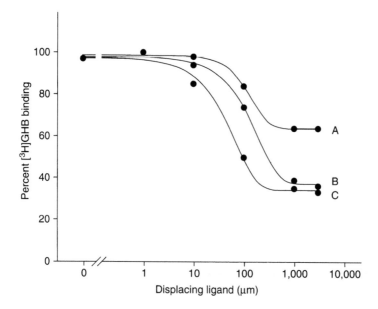

Figure 4.1 Displacement of [³H]GHB, by ligands, from transporter binding sites in membranes from cerebral cortex of adult ICR male mice. Membranes were incubated with 12 nM [³H]GHB and inhibiting ligand in 50 mM Tris-citrate buffer containing 100 mM NaCl, pH 7.4, for 30 min at 4 °C. After centrifugation and pellet washing, the amount of radioactivity was measured in the membranes. Each point is the mean of four determinations. Error bars are smaller than the symbols. Binding of 100% represents 13 pmol/mg protein. A: harmaline; B: 2-hydroxycinnamic acid; C: unlabeled GHB.

being 2-hydroxycinnamic acid. When this inhibition was studied in more detail, the effect resembled competitive inhibition and an IC_{50} of 110 μM was calculated (Fig. 4.1) Harmaline, known to interfere with Na^+ binding, also inhibited [³H]GHB binding in a manner reminiscent of noncompetitive inhibition (Fig. 4.1). This was further evidence that binding and subsequent transport depends on the presence of Na^+ ions.

The binding of [³H]GHB was also found to be irreversibly inhibited by phenylglyoxal (Fig. 4.3). Inactivation of the GHB recognition site on the transporter was both time- and concentration-dependent. Since this inhibitor has been used extensively to suggest the presence of essential arginyl residues in enzymes and other biologically active proteins (Takahashi 1968; Tunnicliff and Ngo 1978; Franks *et al.* 1980), these data indicate that such residues might play a role in the binding of GHB to its carrier protein.

Release of GHB

Nerve cells communicate with each other by extruding neurotransmitter molecules, stored at presynaptic sites, into the synaptic gap where they are then in close proximity to their receptors located on plasma membranes of neighboring cells. The primary process for the release of neurotransmitter is by regulated exocytosis. The underlying mechanisms responsible for this release involve a complicated series of neuronal protein–protein interactions that are far from understood (Fernández-Chacón and Südhof 1999). The neurotransmitter molecules

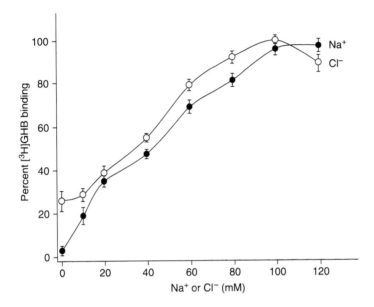

Figure 4.2 Effects of Na$^+$ and Cl$^-$ on [^{3}H]GHB binding to transporter sites in membranes from cerebral cortex of adult ICR male mice. Membranes were incubated with 12 nM [^{3}H]GHB in 50 mM Tris-citrate buffer, pH 7.4, for 30 min at 4 °C. When Na$^+$ was varied, tubes contained a mixture of NaCl and KCl in proportions that yielded the desired Na$^+$ concentration. When Cl$^-$ was varied, tubes contained a mixture of NaCl and sodium acetate in proportions that yielded the desired Cl$^-$ concentration. After centrifugation and pellet washing, the amount of radioactivity was measured in the membranes. Each point is the mean ($\pm$SD) of four determinations. Binding of 100% represents 14.6 pmol/mg protein.

are contained in vesicles that fuse with the plasma membrane. Apparently, SNARE proteins in the vesicular membrane dock with SNARE proteins in the plasma membrane at the end bouton. Further, recently described cysteine-string proteins (Chamberlain and Burgoyne 2000) are emerging as an additional key element in the initiation of the subsequent fusion of the vesicular and plasma membranes, leading to the release of neurotransmitter (Hanson *et al.* 1997). Intracellular Ca^{2+} ions also are necessary for this presynaptic exocytosis (Knight and Baker 1982), but their exact role has yet to be elucidated (Avery *et al.* 1999). With this in mind, it is interesting that new data have emerged suggesting that a target for initial Ca^{2+} action is the synaptic protein synaptotagmin I which seems to act as a Ca^{2+} sensor (Fernández-Chacón *et al.* 2001).

In addition to exocytosis, certain neurotransmitters might undergo release at the nerve ending by a reversal of the transporter responsible for uptake from the synaptic cleft (see the review by Levi and Raiteri 1993). Typically this release is independent of Ca^{2+} ions but does require Na$^+$ ions. It is attenuated by transport inhibitors. Unlike exocytosis, this type of release does not involve vesicles as a source of the neurotransmitter. Instead the transmitter substances are free in the cytoplasm.

GHB exhibits certain properties of a neurotransmitter. For example, it shows an uneven distribution throughout the brain; it activates specific receptor sites linked to G-protein

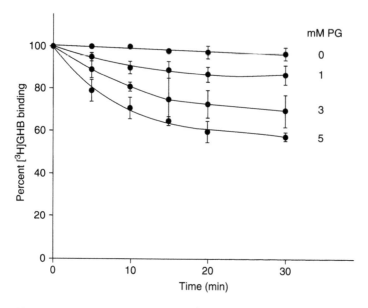

Figure 4.3 Effects of phenylglyoxal on [³H]GHB binding to transporter sites in membranes from cerebral cortex of adult ICR male mice. Membranes suspended in 50 mM Tris-citrate buffer, pH 7.4, were preincubated with various concentrations of phenylglyoxal for different periods of time at 4 °C. Aliquots of the membrane suspensions were then incubated in buffer containing 12 nM [³H]GHB and 100 mM NaCl for 30 min at 4 °C. This also had the effect of producing a fiftyfold dilution of the phenylglyoxal. After centrifugation and pellet washing, the amount of radioactivity was measured in the membranes. Each point is the mean (±SD) of four determinations. Binding of 100% represents 14.1 pmol/mg protein.

function; the presence of anabolic and catabolic enzymes has been demonstrated; and a saturable, Na⁺-dependent uptake system has been established. Demonstrating that a putative neurotransmitter can be released by nerve endings upon depolarization is further evidence. Maitre *et al.* (1983) reported that preloaded striatal slices from rat brain released [³H]GHB in the presence of either K⁺ or veratridine. Both are traditional depolarizing agents, and in either case release was concentration-dependent. Tetrodotoxin is known to block voltage-sensitive Na⁺ channels, and at a concentration of 10 μM, this drug abolished [³H]GHB release, indicating that the process is triggered by action potentials secondary to depolarization. In a Ca²⁺-free medium, release was decreased by up to 60%. Release could also be reduced by the Ca²⁺ channel blocker verapamil (100 μM), further supporting the notion that [³H]GHB release was Ca²⁺-dependent. Vayer and Maitre (1988) measured regional differences in [³H]GHB release using rat brain slices. Veratridine-induced release was greatest in cerebral cortex, hippocampus and striatum, but remarkably low in medulla and cerebellum.

In another study, the anticonvulsant drugs valproate and ethosuximide were both observed to inhibit the depolarization-evoked [³H]GHB release from hippocampal and striatal slices (Vayer *et al.* 1987). Both drugs have been reported to suppress seizures induced by GHB in experimental animals (Godschalk *et al.* 1976).

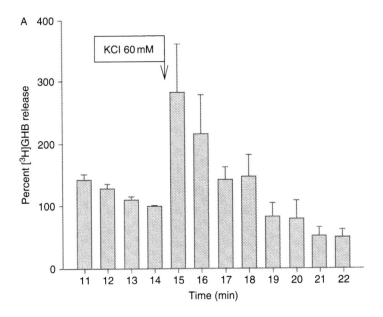

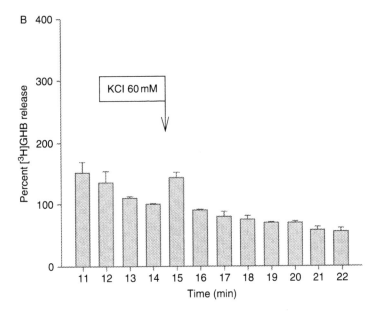

Figure 4.4 K$^+$-evoked [^{3}H]GHB release in differentiated NCB-20 neurons. (A) Release in the presence of 1.3 mM Ca^{2+}. Spontaneous [^{3}H]GHB release $=3.8 \pm 0.04$ pmol/min/mg protein. Peak [^{3}H]GHB release $=11.2 \pm 1.1$ pmol/min/mg protein. (B) [^{3}H]GHB release in the presence of 1 mM EGTA and the absence of Ca^{2+}. These conditions led to a 91% decrease in release. Results expressed as mean $\pm$ SEM. Redrawn from Kemmel *et al.* (1998).

In NCB-20 neurons grown in culture (Kemmel *et al.* 1998), undifferentiated cells were unable to release preloaded [^{3}H]GHB upon depolarization by KCl. Cells differentiated by dibutyryl cyclic AMP, however, released [^{3}H]GHB in a time-dependent manner. This release was drastically attenuated in the absence of Ca^{2+} in the assay medium (Fig. 4.4).

Details of the GHB release mechanism are unknown. Yet, because of the dependence on Ca^{2+} ions, it seems clear that the process relies on exocytosis rather than on a reversal of the Na^+-dependent uptake system.

Summary

Two important characteristics of a neurotransmitter system are (1) a means of releasing transmitter molecules from nerve endings and (2) a means of inactivating these molecules immediately after receptor activation. Typically, this latter mechanism relies on the rapid removal of neurotransmitter from the synaptic gap by an initial binding to a specific transporter, followed by uptake into adjacent cells. This active transport process is dependent on Na^+ and Cl^- ions. There is substantial evidence that systems for both release and active uptake exist for GHB in mammalian brain, providing support for the idea that GHB functions as a neurotransmitter in the CNS. The transporter probably belongs to the Na/Cl family of carrier proteins. Initial evidence suggests that arginyl residues at the GHB recognition site are important for ligand binding.

References

Addolorato G., Balducci G., Capristo E., Attilia M. L., Taggi G., Gasbarrini G. and Ceccanti M. (1999) Gamma-hydroxybutyric acid (GHB) in the treatment of alcohol withdrawal syndrome: a randomized comparative study versus benzodiazepines. *Alcoholism: Clin. Exp. Res.* 23, 1596–1604.

Aldrete J. A. and Barnes D. P. (1968) 4-Hydroxybutyrate anesthesia for cardiovascular surgery. *Anesthesia* 23, 558–65.

Anderson B. M., Anderson C. D. and Churchich J. E. (1966) Inhibition of glutamic dehydrogenase by pyridoxal 5'-phosphate. *Biochemistry* 5, 2893–900.

Anderson R. A., Ritzmann R. F. and Tabakoff B. (1977) Formation of gamma-hydroxybutyrate in brain. *J. Neurochem.* 28, 633–39.

Avery J., Jahn R. and Edwardson J. M. (1999) Reconstitution of regulated exocytosis in cell-free systems: critical approach. *Annu. Rev. Physiol.* 61, 777–807.

Benavides J., Rumigny J. F., Bourguignon J. J., Wermuth C. G., Mandel P. and Maitre M. (1982) A high-affinity, Na^+-dependent uptake system for γ-hydroxybutyrate in membrane vesicles prepared from rat brain. *J. Neurochem.* 38, 1570–75.

Bernasconi R., Mathivet P., Bischoff S. and Marescaux C. (1999) Gamma-hydroxybutyric acid: an endogenous neuromodulator with abuse potential? *Trends Pharmacol. Sci.* 20, 135–41.

Bessman S. P. and Fishbein W. M. (1963) Gamma-hydroxybutyrate, a normal brain metabolite. *Nature* 200, 1207–08.

Blumenfeld M., Suntay R. G. and Harmel M. H. (1962) Sodium gamma-hydroxybutyric acid: a new anesthetic adjuvant. *Anesth. Analg. Curr. Res.* 41, 721–26.

Cash C. D., Maitre M. and Mandel P. (1979) Purification from human brain and some properties of two NADPH-linked aldehyde reductases which reduce succinic semialdehyde to 4-hydroxybutyrate. *J. Neurochem.* 33, 1169–75.

Chamberlain L. H. and Burgoyne R. D. (2000) Cysteine-ring protein: the chaperone at the synapse. *J. Neurochem.* 74, 1781–89.

Drakontides A. B., Schneider J. A. and Funderburk W. H. (1962) Some effects of sodium gamma-hydroxybutyrate on the central nervous system. *J. Pharmacol. Exp. Ther.* 135, 275–84.

Engel A., Fujiyoshi Y. and Agre P. (2000) The importance of aquaporin water channel protein structures. *EMBO J.* 19, 800–06.

Entholzner E., Mielke L., Pichlmeier R., Weber F. and Schneck H. (1995) EEG changes during sedation with gamma-hydroxybutyric acid. *Anaesthetist* 44, 345–50.

Feldman H. A. (1972) Mathematical theory of complex-ligand binding systems at equilibrium: some methods for parameter fitting. *Anal. Biochem.* 48, 317–38.

Fernández-Chacón R. and Südhof T. C. (1999) Genetics of synaptic vesicle function: toward the complete functional anatomy of an organelle. *Annu. Rev. Physiol.* 61, 753–76.

Fernández-Chacón R., Königstorfer A., Gerber S. H., Garcia J., Matos M. F., Stevens C. F., Brose N., Rizo J., Rosenmund C. and Südhof T. C. (2001) Synaptotagmin I functions as a calcium regulator of release probability. *Nature* 410, 41–49.

Franks D. J., Tunnicliff G. and Ngo T. T. (1980) Inactivation of adenylate cyclase by phenylglyoxal and other dicarbonyls: evidence for existence of essential arginyl residues. *Biochim. Biophys. Acta* 611, 358–62.

Gallimberti L., Spella M. R., Soncini C. A. and Gessa G. L. (2000) Gamma-hydroxybutyric acid (GHB) in the treatment of alcohol and heroin. *Alcohol* 20, 257–62.

Godschalk M., Dzoljic M. R. and Bonta I. L. (1976) Antagonism of gamma-hydroxybutyrate-induced hypersynchronization in the ECoG of the rat by anti-petit mal drugs. *Neurosci. Lett.* 3, 145–50.

Godschalk M., Dzoljic M. R. and Bonta I. L. (1977) Slow wave sleep and a state resembling absence epilepsy induced in the rat by γ-hydroxybutyrate. *Eur. J. Pharmacol.* 44, 105–11.

Hanson P. I., Heuser J. E. and Jahn R. (1997) Neurotransmitter release – four years of SNARE complexes. *Curr. Opin. Neurobiol.* 7, 310–15.

Hechler V., Bourguignon J. J., Wermuth C. G., Mandel P. and Maitre M. (1985) γ-Hydroxybutyrate uptake by rat brain striatal slices. *Neurochem. Res.* 10, 387–96.

Kanner B. I. (1993) Glutamate transporters from brain – a novel neurotransmitter transporter family. *FEBS Lett.* 325, 95–99.

Kaufman E. E. and Nelson T. (1991) An overview of γ-hydroxybutyrate catabolism: the role of the cytosolic NADP⁺-dependent oxidoreductase EC 1.1.1.19 and of a mitochondrial hydroxyacid–oxoacid transhydrogenase in the initial, rate-limiting step in this pathway. *Neurochem. Res.* 16, 965–74.

Kemmel V., Taleb O., Perard A., Andriamampandry C., Siffert J. C., Mark J. and Maitre M. (1998) Neurochemical and electrophysiological evidence for the existence of a functional γ-hydroxybutyrate system in NCB-20 neurons. *Neuroscience* 86, 989–1000.

Knight D. E. and Baker P. F. (1982) Calcium-dependence of catecholamine release from bovine adrenal medulla cells after exposure to intense electric fields. *J. Membr. Biol.* 68, 107–40.

Laborit H. (1964) Sodium 4-hydroxybutyrate. *Int. J. Neuropharmacol.* 3, 433–52.

Lapierre O., Montplaisir J., Lamarre M. and Bedard M. A. (1990) The effect of gamma-hydroxybutyrate on nocturnal and diurnal sleep of normal subjects: further considerations on REM sleep-triggering mechanisms. *Sleep* 13, 24–30.

Levi G. and Raiteri M. (1993) Carrier-mediated release of neurotransmitters. *Trends Neurosci.* 16, 415–19.

MacDermot J., Higashida H., Wilson S. P., Matsuzawa H., Minna J. and Nirenberg M. (1979) Adenylate cyclase and acetylcholine release regulated by separate serotonin receptors of somatic cell hybrids. *Proc. Natl Acad. Sci. USA* 76, 1135–39.

McCormick S. J. and Tunnicliff G. (1998) Inhibitors of synaptosomal γ-hydroxybutyrate transport. *Pharmacology* 57, 124–31.

McGivan J. D. and Pastor-Anglada M. (1994) Regulatory and molecular aspects of mammalian amino acid transport. *Biochem. J.* 299, 321–34.

Maitre M., Cash C., Weissmann-Nanopoulos D. and Mandel P. (1983) Depolarization-evoked release of γ-hydroxybutyrate from rat brain slices. *J. Neurochem.* 41, 287–90.

Muller C., Viry S., Miehe M., Andriamampandry C., Aunis D. and Maitre M. (2002) Evidence for a γ-hydroxybutyrate (GHB) uptake by rat brain synaptic vesicles. *J. Neurochem.* 80, 899–904.

Nelson N. (1998) The family of Na⁺/Cl⁻ neurotransmitter transporters. *J. Neurochem.* 71, 1785–803.

Roberts E. and Frankel S. (1950) γ-Aminobutyric acid in brain: its formation from glutamic acid. *J. Biol. Chem.* 187, 55–63.

Roth R. H. and Giarman N. J. (1969) Conversion *in vivo* of γ-aminobutyric acid to γ-hydroxybutyric acid in mammalian brain. *Biochem. Pharmacol.* 18, 247–50.

Scharf M. B., Brown D., Woods M., Brown L. and Hirschowitz J. (1985) The effects and effectiveness of γ-hydroxybutyrate in patients with narcolepsy. *J. Clin. Psychiatr.* 46, 222–25.

Scharf M. B., Hauck M., Stover R., McDannold M. and Berkowitz D. (1998) Effects of gamma-hydroxybutyrate on pain, fatigue, and alpha sleep anomaly in patients with fibromyalgia. Preliminary report. *J. Rheumatol.* 25, 1986–90.

Schuldiner S. (1994) A molecular glimpse of vesicular monoamine transporters. *J. Neurochem.* 62, 2067–78.

Scrima L., Hartman P. G., Johnson F. H., Thomas E. E. and Hiller F. C. (1990) The effects of γ-hydroxybutyrate on the sleep of narcolepsy patients: a double-blind study. *Sleep* 3, 479–90.

Snead O. C. (2000) Evidence for a G protein-coupled γ-hydroxybutyric acid receptor. *J. Neurochem.* 75, 1986–96.

Snead O. C. and Morley B. J. (1981) Ontogeny of γ-hydroxybutyric acid. I. Regional concentration in developing rat, monkey and human brain. *Brain Res.* 227, 579–89.

Solway J. and Sadove M. S. (1965) 4-Hydroxybutyrate: a clinical study. *Anesth. Analg. Curr. Res.* 44, 532–39.

Stein W. D. (1990) *Channels, Carriers, and Pumps: An Introduction to Membrane Transport.* Academic Press, San Diego, CA.

Takahashi K. (1968) The reaction of phenylglyoxal with arginine residues in proteins. *J. Biol. Chem.* 243, 6171–79.

Thalmann R. H. (1988) Evidence that guanosine triphosphate (GTP)-binding proteins control a synaptic response in brain: effect of pertussis toxin and GTPγS on the late inhibitory postsynaptic potential of hippocampal CA3 neurons. *J. Neurosci.* 8, 4589–602.

Tunnicliff G. (1979) Inhibition by pyridoxal 5'-phosphate of γ-aminobutyric acid receptor binding to synaptic membranes of cat cerebellum. *Biochem. Biophys. Res. Commun.* 87, 712–18.

Tunnicliff G. (1980a) Inhibition of mouse brain synaptosomal γ-aminobutyric acid transport by pyridoxal 5'-phosphate. *Biochem. Biophys. Res. Commun.* 97, 1024–30.

Tunnicliff G. (1980b) Influence of pyridoxal phosphate on the binding of GABA to its transporter. *Brain Res. Bull.* 5 (Suppl. 2), 101–03.

Tunnicliff G. (1994) Stimulation of mouse brain γ-hydroxybutyric acid transport by pyridoxal 5'-phosphate. *Pharmacology* 49, 271–77.

Tunnicliff G. and Ngo T. T. (1978) Functional role of arginine residues in glutamic acid decarboxylase from brain and bacteria. *Experientia* 34, 989–90.

Vayer P. and Maitre M. (1988) Regional differences in depolarization-induced release of γ-hydroxybutyrate from rat brain slices. *Neurosci. Lett.* 87, 99–103.

Vayer P., Charlier B., Mandel P. and Maitre M. (1987) Effect of anticonvulsant drugs on gamma-hydroxybutyrate release from hippocampal slices: inhibition by valproate and ethosuximide. *J. Neurochem.* 49, 1022–24.

Vayer P., Ehrhardt J.-D., Gobaille S., Mandel P. and Maitre M. (1988) Gamma-hydroxybutyrate distribution and turnover rates in discrete brain regions of the rat. *Neurochem. Int.* 12, 53–59.

Vickers M. D. (1969) Gammahydroxybutyric acid. *Int. Anaesth. Clin.* 7, 75–89.

White R. P. and Samson F. E. (1956) Effects of fatty acid anions on electroencephalogram of unanesthetized rabbits. *Am. J. Physiol.* 186, 271–78.

Winters W. D. and Spooner C. E. (1965) Various seizure activities following gamma-hydroxybutyrate. *Int. J. Neuropharmacol.* 4, 197–200.

Young J. D., Mason D. K. and Fincham D. A. (1988) Topographical similarities between harmaline inhibition sites on Na$^+$-dependent amino acid transport system ASC in human erythrocytes and Na$^+$-independent system ASC in horse erythrocytes. *J. Biol. Chem.* 263, 140–43.

5 Action of γ-hydroxybutyrate on neuronal excitability and underlying membrane conductances

V. Crunelli and N. Leresche

1. Introduction

Despite the long-standing, though very limited, clinical use of γ-hydroxybutyric acid (GHB) (Mamelak and Webster 1981; Mamelak *et al.* 1986; Lapierre *et al.* 1990; Scrima *et al.* 1990; Kleinschmidt *et al.* 1998; Gallimberti *et al.* 1992; Poldrugo and Addolorato 1999) and its recently recognized abuse potential (Dyer 1991; Chin *et al.* 1992; Friedman *et al.* 1995; Stell and Ryan 1996; Galloway *et al.* 1997; Tunnicliff 1997), and in contrast to the large body of biochemical and behavioural data (reviewed in Bernasconi *et al.* (1999) and Maitre (1997)), there is only a surprisingly small number of electrophysiological studies that have investigated the cellular actions of GHB on neuronal excitability, and even fewer that have characterized the underlying changes in membrane conductances.

This chapter starts with a brief overview of the pharmacological (mainly EEG and behavioural) effects of GHB in man and experimental animals, with emphasis on the relationship between injected dose and plasma/brain levels of GHB (for a comprehensive analysis of the biochemical and clinical aspects of GHB action, the reader is referred to the relevant chapters in this book). The main sections (3–10) of this chapter present a critical review of the GHB-elicited changes in neuronal excitability, which are discussed by considering each neuronal class separately. Both *in vitro* intracellular/patch electrode studies as well as *in vivo* and *in vitro* extracellular recordings are summarized, as only a combined analysis of these data would help highlight the similarities and discrepancies between the results of electrophysiological versus biochemical and behavioural findings. This systematic approach also allows linking the cellular actions of GHB within a given neuronal population to the particular (dys)functions to which that neuronal population is known to contribute (e.g. sleep and absence seizures for thalamic neurons).

The chapter ends with a summary (Section 11) of the major points that emerge from this detailed analysis of GHB action on different neuronal classes, in an attempt to focus future clinical and experimental investigations on the still unresolved issues concerning the (patho)physiological and pharmacological actions of this substance. The emerging picture is one that clearly indicates an involvement of GABA$_B$, but not of GHB, receptors in all but one of the GHB effects recorded in *in vitro* electrophysiological studies: this is in sharp contrast with the large number of GHB receptor-mediated actions of GHB that can be measured either biochemically *in vitro/in vivo* or electrophysiologically *in vivo*. The paucity of GHB receptor agonists and antagonists, their very limited use and (for some) their questionable selectivity have not helped resolve the existing controversies on the pharmacological actions of this substance. Nevertheless, a cellular GABA$_B$ receptor-mediated mechanism that provides the scientific rationale for the use of GHB in anaesthesia and in the treatment

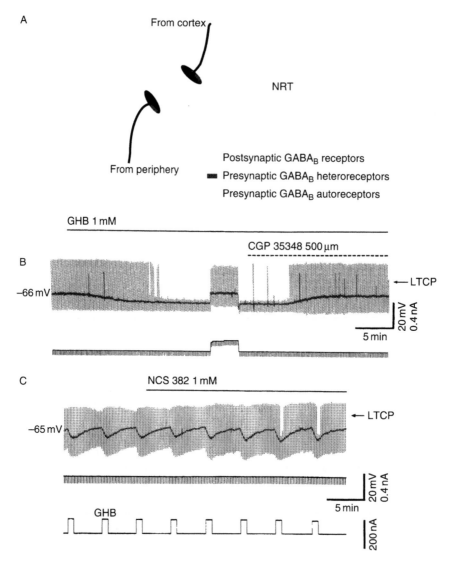

Figure 5.1 Postsynaptic effect of GHB on TC neurons in brain slices. (A) Schematic diagram of sensory, cortical and nucleus reticularis (NRT) synapses on a TC neuron showing the location of presynaptic and postsynaptic $GABA_B$ receptors on which GHB has been shown to act. Note that NRT neurons are GABAergic. (B) Intracellular voltage record (top trace) shows the hyperpolarization elicited by GHB to be abolished by the subsequent application of CGP 35348. Upward deflections in the trace represent low threshold Ca^{2+} potentials (LTCPs). Downward deflections represent the voltage response to hyperpolarizing current pulses (bottom trace) that were used to measure the input resistance of the neuron. The reduction in input resistance associated with the response to GHB persists during repolarization of the membrane potential (by steady dc injection) to the pre-GHB value. (C) Repeated iontophoretic applications of GHB elicit a rapid-onset hyperpolarization (uppermost trace) that is not blocked by bath application of NCS 382. Note how during continuous perfusion of NCS 382 there is a clear, though small, increase in the response to GHB, which in turn leads to a failure of evoking LTCPs, as indicated by the absence of upward voltage deflections accompanying the peak of the last two responses to GHB. The middle trace is the current injected via the intracellular electrode, whereas the bottom trace represents the iontophoretic current used to eject GHB. (B) and (C) reproduced with permission from Williams *et al.* (1995). (See Color Plate IV.)

of narcoleptic patients has been identified, as has a role for these receptors in the GHB model of absence epilepsy. On the other hand, the precise sequence of the neuronal processes involved in the potential GHB abuse or its use as an anti-craving agent in alcohol/ heroin withdrawal remains to be fully elucidated, since essential cellular electrophysiological investigations in the majority of neurons within the brain reward system are lacking.

2. GHB dose versus plasma/brain levels

In healthy volunteers, low doses of GHB (10 mg/kg p.o.) elicit sedative, anxiolytic and myorelaxant actions, while intermediate doses of GHB (2.0–3.0 g/subject p.o.) increase stage 4 (and stage 3, in some studies) of non-REM sleep without a major effect on total sleep time (Lapierre *et al.* 1990; Van Cauter *et al.* 1997). It is therefore not surprising that treatment of narcoleptic patients with 20–40 mg/kg p.o. has been shown to lead to an improvement of nocturnal sleep quality by decreasing stage 1 and increasing δ sleep (stage 3–4) with fewer awakenings (Scrima *et al.* 1990; Nishino and Mignot 1997). At higher doses (50–60 mg/kg p.o. or 20–30 mg/kg i.v.), GHB can bring about unconsciousness, and has been successfully used alone, or as an adjuvant, in anaesthesia (Laborit *et al.* 1960; Solway and Sadove 1965; Vickers 1969; Kleinschmidt *et al.* 1998). Finally, doses above 60 mg/kg p.o. are known to induce (within 30–40 min of the administration) tonic-clonic seizures, respiratory depression and profound coma (Metcalf *et al.* 1966; Dyer 1991). As a GHB dose of 100 mg/kg p.o. brings about a peak serum concentration of 100 μg/ml, it has been suggested that the effects on sleep that occur with doses of 30 mg/kg p.o. are elicited by serum levels of 50 μg/ml or less (Hoes *et al.* 1980; Mamelak 1989). On the other hand, 100 mg/kg i.v. gives an initial plasma level of 200–300 μg/ml (Helrich *et al.* 1964).

In rats,[1] GHB doses of 50–100 mg/kg i.p. increase (around 20%) slow wave sleep duration (measured over a 4 h period), while 25 mg/kg has no effect (Godschalk *et al.* 1977). Higher doses (200 mg/kg i.p.) induce a synchronous, bilateral and symmetrical EEG pattern (i.e. spike and wave discharges, SWDs) associated with sudden behavioural arrest (but no convulsions), that is reminiscent of human and experimental absence seizures (Godschalk *et al.* 1977; Snead 1991a). Administration of 3.5 mM/kg i.p. of GHB results in brain GHB levels of 7 and 33 μg/g at 6 and 30 min, respectively, after injection, while the threshold concentration necessary to elicit SWDs, the hallmark of experimental and clinical absence seizures, is around 30 μg/g (240 μM) (Snead 1991a).[2]

3. Thalamus

We begin our systematic analysis of neuronal GHB responses with the thalamus as this brain area plays an important role both in the sleep–waking cycle and absence epilepsy, a physiological function and a neurological condition, respectively, where putative actions of (endogenous and exogenous) GHB have been implicated on the basis of both clinical and experimental results (Scrima *et al.* 1990; Lapierre *et al.* 1990; Snead 1995; Van Cauter *et al.* 1997). At present, the available electrophysiological data on the effect of GHB in the thalamus are restricted to *in vitro* studies in brain slices and to thalamocortical (TC) neurons, the cells that receive sensory information from the periphery and provide the thalamic output to the cortex (Fig. 5.1A).

3.1. Postsynaptic action

Bath application of GHB produces a dose-dependent hyperpolarization of rat and cat TC neurons (Williams *et al.* 1995) (Fig. 5.1B). At the smallest concentration tested (100 μM), GHB elicits a hyperpolarization of 2–7 mV, while the maximum effect (13 mV) is observed with 3 mM (i.e. 10 mM hyperpolarizes TC neurons by the same amount). The iontophoretic application of GHB elicits an identical response to the one seen during bath applications, though with a faster onset (around 400 ms) and offset (Figs 5.1C and 5.2E). The hyperpolarization persists in the presence of tetrodotoxin (TTX) and in a low Ca^{2+}/high Mg^{2+} medium, is associated with a small decrease in input resistance (10 and 38% with 100 μM and 3 mM, respectively) (Fig. 5.1B), and has a reversal potential of -94 mV (when $[K^+]_o = 6.25$ mM). Thus, GHB has a direct postsynaptic action on TC neurons that is mediated by the opening of K^+ channels.

As the electroresponsiveness and firing patterns of rat and cat TC neurons are highly voltage-dependent (Crunelli *et al.* 1987a,b; Turner *et al.* 1997), the changes in TC neuron excitability produced by the GHB-elicited hyperpolarization are strongly concentration-dependent, and include not only a straightforward decrease in firing (as one would expect by the increased membrane polarization) but also a change in firing pattern with associated initiation of burst firing (Fig. 5.2) (Williams *et al.* 1995). Low concentrations of GHB (100 μM), in fact, hyperpolarize TC neurons into the voltage region (-65 to -75 mV) where action potential burst firing is generated via the intrinsic δ sleep oscillation (Figs 5.2A,C and E). These are rhythmic oscillations (0.5–4 Hz) of the membrane potential that are elicited as a pacemaker activity by single TC neurons (Figs 5.2B,D) during the deep stages (EEG δ rhythm) of non-REM sleep (for a full description of the ionic mechanisms underlying the generation of δ sleep oscillations, see McCormick and Pape (1990), Leresche *et al.* (1991) and Steriade *et al.* (1991)). Higher concentrations of GHB (0.5–3 mM) bring the membrane potential to more negative values than those where δ oscillations occur (Fig. 5.1B), leading to a full block of the intrinsically generated firing and to a progressive reduction in the ability of synaptic inputs to reach action potential threshold. Note, however, that in TC neurons with the largest input resistance, a GHB concentration of 100 μM can elicit such a large hyperpolarization that the δ sleep oscillations first start and then terminate as the neuron is hyperpolarized through the oscillation voltage range during the drug wash-in (Fig. 5.2C). The latter finding, therefore, clearly indicates that concentrations of GHB smaller than 100 μM would be able to elicit δ oscillations in TC neurons.[3]

The GHB-elicited hyperpolarization and associated changes in TC neuron excitability are blocked by the selective $GABA_B$ receptor antagonist CGP 35348 (0.4–1.0 mM) (Fig. 5.1B) (Williams *et al.* 1995), indicating that they occur via activation of the postsynaptic $GABA_B$ receptors (Fig. 5.1A) that are present on these neurons (Crunelli *et al.* 1988; Soltesz *et al.* 1989; Crunelli and Leresche 1991). In contrast, the selective GHB receptor antagonist NCS 382 (0.1–5.0 mM) (Maitre *et al.* 1990) has no effect on its own, does not block the GHB-induced hyperpolarization (Figs 5.1C and 5.2E), but appears to potentiate slightly the action of GHB (Fig. 5.1C).

3.2. Presynaptic action

Together with a direct $GABA_B$ receptor-mediated hyperpolarizing action, GHB also affects the synaptic potentials that are recorded in TC neurons following stimulation of sensory, cortical and intrathalamic afferents (Fig. 5.1A).

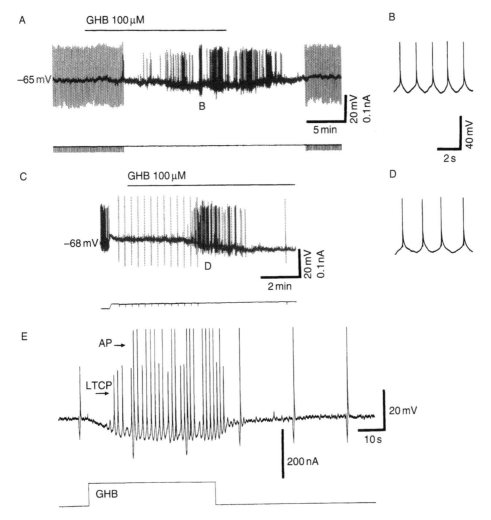

Figure 5.2 The postsynaptic GABA$_B$ receptor-mediated action of GHB elicits intrinsic δ sleep oscillations in TC neurons. (A) Bath application of a relatively small concentration of GHB (100 μM) elicits a clear hyperpolarization (top trace). Although the peak of the hyperpolarization is only about 4 mV, this change in voltage is sufficient to bring the membrane potential into the voltage region where the intrinsic δ sleep oscillations occur. The bottom trace is the current injected via the intracellular electrode. (B) Enlargement of a portion of the GHB-elicited δ oscillation, shown in (A). (C) A similar result is obtained in another TC neuron, where, however, the δ oscillations are only transiently expressed as the GHB-elicited hyperpolarization is larger, and thus the voltage region critical for the oscillation is only traversed during the drug wash-in. The initial portion of the trace show the δ sleep oscillations evoked by steady dc injection (bottom trace) prior to the GHB application. (D) Enlargement of a portion of the GHB-elicited δ oscillation, shown in (C). (E) In another TC neuron, the δ sleep oscillation could be induced by the iontophoretic application of GHB. Note the short latency of the GHB-elicited hyperpolarization (top trace) elicited by the iontophoretic application of GHB (bottom trace). This record was obtained in the presence of 1 mM NCS 382. (A)–(E): Reproduced with permission from Williams *et al.* (1995).

GHB reversibly decreases the amplitude of the sensory glutamatergic EPSPs evoked in rat TC neurons of the dorsal lateral geniculate nucleus and the ventrobasal thalamic complex by low-frequency electrical stimulation of the optic tract (Fig. 5.3) and the medial lemniscus, respectively (Emri *et al.* 1996a). This effect is observed in every cell and is dose-dependent, with the smallest GHB concentration tested (100 μM) eliciting a 36% reduction (Fig. 5.3C) and 5 mM producing a 80% decrease in amplitude. As QX 314-filled intracellular electrodes were used to block the K^+ channel responsible for the postsynaptic $GABA_B$ receptor-mediated response elicited by GHB (see the previous section), it is reasonable to conclude that the effect of GHB on these synaptic potentials occurs via an action on presynaptic receptors located on the afferent sensory fibres (Fig. 5.1A).

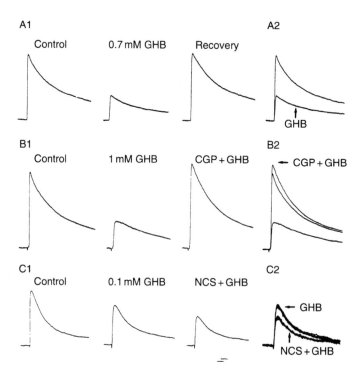

Figure 5.3 Presynaptic $GABA_B$ receptor-mediated effect of GHB on sensory EPSPs in TC neurons in brain slices. (A1) EPSPs evoked by electrical stimulation of the optic tract in a TC neuron of the dorsal lateral geniculate nucleus are reversibly reduced (59%) by GHB. Each trace is the average of ten EPSPs. (A2) Superimposition of the control and GHB traces shown in (A1). (B1) The GHB-elicited reduction (67%) in the EPSP is reversibly abolished by CGP 35348 (1 mM), which restores the amplitude of the synaptic response to a value greater than the control. The latter effect indicates the presence of a tonic activation of $GABA_B$ receptors by ambient GABA (see Emri *et al.* 1996b). Each trace is the average of ten EPSPs. (B2) Superimposition of the traces shown in (B1). (C1) Application of 1 mM NCS 382 does not block the GHB-elicited reduction (26%) in the EPSP but produces a further reduction (20%). Each trace is the average of ten EPSPs. (C2) Superimposition of five EPSPs recorded in the presence of GHB alone or in the presence of GHB and NCS 382. (A1)–(C2): Reproduced with permission from Emri *et al.* (1996a).

This presynaptic action of GHB is reversibly blocked by CGP 35348 (0.4–1.0 mM) (Fig. 5.3B) (Emri *et al.* 1996a). On the other hand, 0.1–5 mM NCS 382 has no effect on its own and does not antagonize the EPSP reduction elicited by GHB (0.5–1.0 mM). When a low concentration (100 μM) of GHB is used, however, NCS 382 (1 mM) evokes a further decrease (20%) of the EPSP amplitude (Fig. 5.3C). Interestingly, a similar effect of NCS 382 is observed on the EPSP reduction (36%) elicited by a low concentration (0.1 μM) of the selective GABA$_B$ receptor agonist baclofen, so that, in the presence of baclofen and NCS 382 (1 mM), the EPSP is reduced by a further 14% (Emri *et al.* 1996a). Thus, as is the case for its postsynaptic action, the presynaptic action of GHB on the sensory EPSP in TC neurons is mediated by GABA$_B$ receptors (Fig. 5.1A). Another point worth stressing from the studies of Emri *et al.* (1996a,b) is that comparison of equi-effective concentrations of GHB and baclofen shows the former to be a 1000-fold less potent than the latter in reducing the sensory EPSP: this is true both for the lowest (100 μM) and maximal (5 mM) GHB concentration used.

We have recently started to investigate the action of GHB on the EPSC evoked by electrical stimulation of the corticofugal afferents, that is the other major glutamatergic input to TC neurons, and on GABA$_A$ IPSCs elicited by stimulation of the GABAergic neurons of the nucleus reticularis thalami (Fig. 5.1A). The amplitude of cortical EPSCs is decreased (30%) by 500 μM GHB in each cell tested, and this effect is blocked by CGP 56999 (100 nM) (Fig. 5.4A). Interestingly, GHB (500 μM) decreases (30%) the GABA$_A$ IPSC only in 75% of TC neurons (Fig. 5.4B), in a dose-dependent manner (30% and 17% reduction at 500 and 250 μM, respectively). In neurons where up to 1 mM GHB is ineffective against GABA$_A$ IPSCs, subsequent application of 10 μM baclofen reduces their amplitude by 95% (Fig. 5.4C). The effects of GHB on GABA$_A$ IPSCs are antagonized by 100 nM CGP 56999, while NCS 382 (1 mM) has either no effect on the GHB-evoked reduction (Fig. 5.4D) or produces a further decrease (20%).

In summary, the action of GHB on TC neurons *in vitro* consists of (i) a postsynaptic GABA$_B$ receptor-mediated hyperpolarization that, depending on the concentration, brings about either an increase or a decrease in excitability, and (ii) a reduction of sensory EPSPs, cortical EPSCs and intrathalamic IPSCs, which is mediated by GABA$_B$ receptors located presynaptically on the respective afferent fibres. NCS 382, the 'selective' (but see Section 3.4.2) GHB receptor antagonist, is unable to block these presynaptic and postsynaptic actions of GHB, but shows effects that are consistent with it being a partial agonist. It is unlikely that this partial agonist action is the result of a direct interaction of NCS 382 with presynaptic and postsynaptic GABA$_B$ receptors as NCS 382 does not bind to GABA$_B$ receptors (Maitre *et al.* 1990; Snead 1996; Bernasconi *et al.*, Chapter 3 of this volume). Nevertheless, no electrophysiological study has attempted to block the partial agonist actions of NCS 382 with a GABA$_B$ receptor antagonist in the thalamus (or in other brain areas). Finally, the thalamic investigations reviewed here clearly indicate that the notion of 'micromolar' versus 'millimolar' concentrations of GHB acting on GHB and GABA$_B$ receptors, respectively (Snead 1995, 1996; Maitre 1997; Hu *et al.* 2000), is totally unfounded (see endnote 1). In fact, (i) a GHB concentration of 100 μM (or lower; see Section 3.1 and endnote 3) can elicit physiologically relevant effects mediated both by presynaptic and postsynaptic GABA$_B$ receptors, and (ii) these 'micromolar' GHB concentrations are similar to those required to observe the only two examples of GHB receptor-mediated electrophysiological responses to GHB observed so far *in vitro*, that is 25–100 μM in NCB-20 cell line (Kemmel *et al.* 1998) and 100–600 μM in CA1 hippocampal neurons (Berton *et al.* 1999) (see endnote 1, and Sections 4.2 and 9).

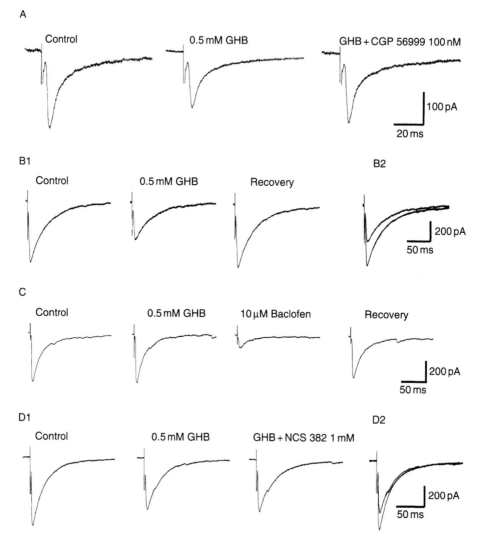

Figure 5.4 Effect of GHB on cortical EPSCs and IPSCs in TC neurons in brain slices. (A) GHB (500 μM) reduces the amplitude of the electrically evoked corticothalamic EPSC (see Fig. 5.1A), and this effect is fully antagonized by the GABA$_B$ antagonist CGP 56999 (100 nM). EPSCs were recorded in the presence of gabazine (1 μM). (B1) The GABA$_A$ IPSC evoked by stimulation of the nucleus reticularis thalami (see Fig. 5.1A) is reversibly decreased by 500 μM GHB. (B2) Superimposition of the traces shown in (B1). (C) In another TC neuron, the amplitude of the GABA$_A$ IPSC evoked by electrical stimulation of the nucleus reticularis thalami is unaffected by 500 μM GHB. Following wash-out of GHB, application of 10 μM baclofen markedly and reversibly reduces the IPSC amplitude. (D1) Lack of effect of 1 mM NCS 382 on the GHB-induced reduction of the IPSC amplitude. (D2) Superimposition of the traces shown in (D1). All records in (B), (C) and (D) were obtained in the presence of APV (50 μM) and CNQX (10 μM) with $E_{Cl} = 0$ mV, as described in Le Feuvre *et al.* (1997).

The cellular actions of GHB on TC neurons will now be discussed in the context of its ability to induce anaesthesia in man and absence seizures in experimental animals, and to instate a normal sleep pattern in narcoleptic patients.

3.3. Thalamic GHB action and sleep

A large number of the cellular mechanisms occurring in cortical, thalamic and brainstem neurons during different stages of sleep have been characterized by the extensive experimental work carried out *in vitro* and *in vivo* in the last twenty years (Steriade *et al.* 1994). In particular, the identity of, and the interplay between, the ionic conductances of different types of cortical and thalamic neurons that are responsible for the EEG patterns of sleep spindles and δ waves have been almost fully elucidated. For the purpose of this section, a brief summary of the mechanisms underlying the firing patterns of TC neurons is now presented, as they are essential for an understanding of TC neuron electroresponsiveness (and thus of the overall thalamic output) in different behavioural states (Leresche *et al.* 1991; Steriade *et al.* 1994). Depending on the membrane potential, which *in vivo* is set by the behavioural state-dependent activity of (mainly cholinergic) brainstem afferents (Steriade *et al.* 1994), a TC neuron can either (i) tonically fire single action potentials (during the awake/attentive state) when its membrane potential is greater than -65 mV, or (ii) intrinsically generate δ oscillations and associated burst firing as a neuronal pacemaker (during EEG δ rhythm, 0.5–4 Hz, of stage 3–4 non-REM sleep) when its membrane potential is in the range between -65 and -75 mV (see the legend of Fig. 5.2 for further details). Note that the δ oscillation originating from single TC neurons is unable *per se* to sustain the EEG δ rhythm, as it requires cortical synchronizing inputs (Steriade *et al.* 1994).

Within this range of behavioural state-dependent firing patterns of TC neurons, it is not difficult to envisage how the action of increasing doses of GHB on these neurons will progressively engender a condition of sedation and sleep/anaesthesia. Small GHB concentrations/doses, acting on postsynaptic receptors, will hyperpolarize TC neurons into the voltage region where δ oscillations occur (see Fig. 5.2A), thus promoting an increase in δ sleep. Intermediate concentrations/doses will take the membrane potential to more negative voltages than those where δ oscillations occur, and into a membrane potential region where TC neurons may be brought to firing only by large sensory and cortical synaptic inputs. Larger (toxic) concentrations/doses will take the membrane potential to even more negative values, so that TC neurons will become, for all intents and purposes, unresponsive to all synaptic potentials (i.e. external and internal stimuli), a cellular electrical condition reminiscent of coma. The dose-dependent decrease of sensory and cortical EPSPs as well as of intrathalamic IPSPs, which is elicited by the presynaptic action of GHB, would act side by side with the postsynaptically mediated GHB hyperpolarization, effectively contributing to an increasing electrical isolation of TC neurons from intrathalamic and extrathalamic influences.

This scenario provides a solid cellular thalamic mechanism to explain the sedative/anaesthetic action of GHB and the selective increase in δ sleep that accompanies the overall clinical improvements obtained with GHB therapy in narcoleptic patients (Lapierre *et al.* 1990; Scrima *et al.* 1990; Nishino and Mignot 1997). The increase in REM sleep observed in narcoleptic patients undergoing GHB treatment might not result from a direct action of GHB on REM sleep mechanisms, but could solely be the consequence of the enhanced non-REM sleep (Moruzzi 1972; Steriade and Llinas 1988). The thalamic processes described here, however, should not be taken as the sole contributor to the sedation/sleep/anaesthesia/(coma) induced by progressively larger (and toxic) doses of GHB, as they are undoubtedly re-enforced by the action of GHB on other neuronal types, in particular cortical neurons (Olpe and Koella 1979; Godbout *et al.* 1995). Unfortunately, it is not possible at present to draw a detailed cellular scenario of the action of GHB in the cerebral cortex, as the GHB effects on the ionic conductances and the behavioural state-dependent firing patterns of different cortical cell types have not been investigated (see Section 7).

Finally, we need to ask the question as to whether the aforementioned sequence of cellular thalamic processes that occur following *in vivo* administration of GHB is mediated by GHB and/or $GABA_B$ receptors. An involvement of GHB receptors in the GHB-induced sedation is suggested by the ability of systemically injected NCS 382 (100 mg/kg i.p.) to abolish this pharmacological action of GHB (Schmidt *et al.* 1991) (but see problems relating to NCS 382 in Section 3.4.2). At present, however, it would be premature to suggest a role for thalamic GHB receptors in the GHB-elicited increase in stage 3–4 slow wave sleep, as no GHB receptor-mediated cellular response to GHB has so far been detected electrophysiologically in the thalamus *in vitro* (Williams *et al.* 1995; Emri *et al.* 1996a). On the other hand, a role for thalamic $GABA_B$ receptors in the sleep–wake cycle is suggested by the decreased duration of stage 3–4 slow wave sleep following intrathalamic injection of $GABA_B$ antagonists in freely moving cats (Juhasz *et al.* 1994). Therefore, although no study has directly investigated the effect of $GABA_B$ antagonists on the sedation/sleep/anaesthesia elicited by systemic administration of GHB, it is reasonable to predict that thalamic $GABA_B$ receptors are involved in this pharmacological/clinical action of GHB as $GABA_B$ antagonists abolish the GHB-induced δ sleep oscillations and decrease in synaptic potential amplitude in TC neurons *in vitro* (Williams *et al.* 1995; Emri *et al.* 1996a).

3.4. The GHB model of absence epilepsy

Among the various effects elicited by intermediate doses of GHB (150–250 mg/kg i.p. in the rat), there is a behavioural and EEG pattern that is highly reminiscent of human absence epilepsy, including sudden behavioural arrest, some myoclonic phenomena and the abrupt appearance of SWDs on an otherwise normal background EEG (Godschalk *et al.* 1976, 1977; Snead *et al.* 1976). The properties and mechanisms of GHB-induced absence seizures have been extensively investigated by Snead and his group (see Snead, Chapter 8 of this volume), and the GHB model is now considered one of the most robust and reliable pharmacological models of absence epilepsy in rodents. Nowadays, γ-butyrolactone (GBL) (100 mg/kg i.p.), a biologically inactive compound that is rapidly metabolized in GHB by a peripheral lactonase, is used instead of GHB, as it produces behavioural and EEG correlates of absence seizures 'similar to GHB but with a faster onset and a more predictable dose–response' (Snead 1992). Note that the doses of GHB that are used for the GHB absence epilepsy model in rodents and felines only infrequently induce minor convulsive patterns, while much higher doses are required to evoke full convulsive seizures (Snead *et al.* 1976). In view of this finding, and as the processes involved in convulsive epilepsy and generalized absence seizures are markedly different (Avoli *et al.* 1990; Niedermeyer 1996; Duncan 1997; Delgado-Escueta *et al.* 1999), the mechanisms underlying the GHB-elicited absence seizures should not be used to draw any parallel with the pathophysiological events involved in the expression of the tonic–clonic episodes that accompany the intake of high doses of GHB in humans (Dyer 1991; Chin *et al.* 1992; Friedman *et al.* 1995).

Extensive experimental work, corroborated by some human studies (Williams 1953), has shown that the generation/expression of SWDs, the characteristic feature of absence seizure, requires the interaction of cortex and thalamus and does not appear to involve limbic structures. On the basis of electrophysiological evidence (Crunelli *et al.* 1988; Soltesz *et al.* 1989), it had been suggested that rhythmic sequences of a $GABA_B$ IPSP and a low threshold Ca^{2+} potential might underlie the expression of SWDs in TC neurons (Crunelli and Leresche 1991). In a ferret thalamic slice preparation, it has been possible to observe this series of cellular events following application of a $GABA_A$ antagonist (Bal *et al.* 1995). However, *in vivo* intracellular recordings in a genetic absence model (the Genetic Absence Epilepsy Rats from Strasbourg, GAERS) and in the penicillin/bicuculline model have indicated that during SWDs the majority of TC neurons do not express low threshold Ca^{2+} potentials but rhythmic (8–10 Hz) sequences of $GABA_A$ IPSPs superimposed on a long-lasting hyperpolarization (Steriade and Contreras 1995; Pinault *et al.* 1998; Charpier *et al.* 1999), which might represent summating $GABA_B$ IPSPs. Unfortunately, a similarly detailed intracellular analysis of the membrane conductances underlying the activity of thalamic (and cortical) neurons during GHB-induced SWDs is not available. Nor would it be appropriate to assume a close similarity of cellular thalamic (and cortical) electrical events between the GAERS and GHB models, as the higher SWD frequency in the former (8–10 Hz) compared to the latter (4–6 Hz) model (Vergnes and Marescaux 1994; Snead 1995; Hu *et al.* 2000) suggests differences in their underlying mechanisms.

3.4.1. *$GABA_B$ receptors and the GHB model of absence epilepsy*

Notwithstanding the difficulties in identifying the precise $GABA_B$ receptor component involved in the activity of thalamic and cortical neurons during SWDs, one of the most interesting findings originating from the thalamic $GABA_B$ receptor hypothesis (Crunelli and Leresche 1991) has undoubtedly been the demonstration that absence seizures (i.e. SWDs and their behavioural correlates) are blocked by systemic and intrathalamic injection of $GABA_B$ antagonists in the GHB (and every other experimental) model of this generalized epilepsy (Liu *et al.* 1992; Snead 1992; Hosford *et al.* 1992, 1995; Noebels 1994). On the other hand, the systemic and intrathalamic injections of $GABA_B$ agonists exacerbate absence seizures, again in the GHB and other models of absence epilepsy (Liu *et al.* 1992; Snead 1992; Hosford *et al.* 1992, 1995; Noebels 1994).

A presynaptic $GABA_B$ receptor mechanism in the GHB model has been favoured on the basis of biochemical and electrophysiological findings (Snead 1995, 1996; Maitre 1997; Hu *et al.* 2000). As outlined below, however, the rejection of a mechanism involving postsynaptic $GABA_B$ receptors is undermined by a number of inaccuracies.

(i) The biochemical results that are used to support a presynaptic mechanism in GHB-induced SWDs are those that show systemic and thalamic injections of GHB to decrease basal GABA (but not glutamate) release in the ventrobasal thalamus (Banerjee and Snead 1995), while systemic injection of GBL reduces both basal and K^+-evoked GABA (but not glutamate) release in cortical layers I–III (Hu *et al.* 2000). However, it is difficult to understand how basal release could play a major role in the strong synaptic interactions occurring within the thalamocortical loop during SWDs (Avoli *et al.* 1990; Pinault *et al.* 1998; Steriade *et al.* 1994). Furthermore, the *in vivo* results on cortical evoked release (Hu *et al.* 2000) cannot be reproduced in cortical slices (see Bernasconi *et al.*, Chapter 3 of this volume), and the thalamic release data (Banerjee and Snead 1995) cannot be supported by the *in vitro* electrophysiological data in TC neurons (see Section 3.2 and Fig. 5.4).

(ii) It has been suggested (Banerjee and Snead 1995; Snead 1996; Maitre 1997) that the 'millimolar' GHB concentrations required to elicit a postsynaptic effect in TC neurons *in vitro* (Williams *et al.* 1995) can account for the anaesthetic/toxic effects of systemic GHB but not for its ability to elicit SWDs, as the brain concentrations necessary to induce SWDs *in vivo* are in the 'micromolar' range (threshold: 240 μM) (Snead 1991a). First, as explained earlier (see Section 3.1), it is not correct to correlate the 'millimolar' ED_{50} of an electro-physiological response (i.e. hyperpolarization) with a particular change in excitability for neurons that possess highly voltage-dependent firing patterns. Second, the thalamic processes generating SWDs and sleep spindles have been suggested to be tightly linked (Niedermeyer 1996), and the latter activity is generated in TC neurons by similar or smaller levels of hyperpolarization than δ sleep oscillations (Steriade *et al.* 1994). Therefore, GHB concentrations of 100 μM or lower (i.e. capable of eliciting $GABA_B$ receptor-mediated δ sleep oscillations; see Fig. 5.2) (Williams *et al.* 1995) are sufficient to engender in TC neurons a postsynaptic receptor-mediated hyperpolarization inductive of SWDs.

On the basis of these arguments and until new evidence becomes available, therefore, it is more appropriate to consider that both the *presynaptic* (i.e. decrease in synaptic potentials) and the *postsynaptic* (i.e. small hyperpolarization) $GABA_B$ *receptor*-mediated actions of 'micromolar' concentrations of GHB are those that are blocked by intrathalamically and systemically injected $GABA_B$ antagonists in their prevention of GHB-induced SWDs.

3.4.2. *GHB receptors and the GHB model of absence epilepsy*

Systemic injection of NCS 382 blocks the behavioural and EEG expressions of absence seizures in the GHB (Snead 1996) and other models of absence epilepsy (Maitre *et al.* 1990; Snead 1996; Aizawa *et al.* 1997),[4] though the effect of intrathalamic injections of NCS 382 has not been investigated. In contrast to the action of $GABA_B$ antagonists, however, the block of GHB-elicited SWDs by systemically injected NCS 382 cannot at present be easily related to any electrophysiological action. In the thalamus, in fact, no NCS 382-sensitive electrophysiological action of GHB at the single cell level *in vitro* has so far been disclosed (Williams *et al.* 1995; Emri *et al.* 1996a). It may also be possible that systemically injected NCS 382 decreases GHB-elicited SWDs not via thalamic but via cortical GHB receptors, as application of this antagonist in the upper cortical layers abolishes the GHB-evoked decrease in GABA release (Hu *et al.* 2000).

Notwithstanding the significance of these results, a straightforward acceptance of an involvement of GHB receptors in GHB-elicited SWDs is questioned by the following find-ings/arguments. First, the interpretation of data concerning NCS 382 is confounded by its ability to inhibit competitively GHB-dehydrogenase (Maitre 1997). As the involvement of a GHB-derived GABA pool in the pharmacological actions of GHB cannot be excluded,[5] the possibility remains that NCS 382 blocks GHB-elicited SWDs (and other GHB effects) via inhibition of GHB-dehydrogenase and not via its presumed selective antagonism of GHB receptors. Second, trans-4-hydroxycrotonic acid, a potent GHB receptor agonist (Maitre *et al.* 1983a; Hechler *et al.* 1993) that is not a substrate for the GHB-dehydrogenase (Maitre 1997) and does not bind to $GABA_B$ receptors (Bernasconi *et al.*, Chapter 3 of this volume), is unable to evoke SWDs in control animals or to aggravate them in GAERS when injected i.p. at doses even fourfold higher than GHB (Depaulis *et al.* 1988; Snead 1991b). This would clearly exclude a role for GHB receptors in GHB-elicited SWDs, unless it is assumed that different GHB receptor subtypes (some with higher affinities for GHB, others for *trans*-4-hydroxycrotonic acid) exist in the brain (the evidence for which is still weak and controversial)

(Hechler *et al.* 1990; Maitre 1997). Third, the correlation between the developmental profile of GHB receptors and GHB-elicited SWDs might simply represent a coincidental occurrence and not be the result of a cause–effect relationship (see the discussion in Snead (1994)). Fourth, the up-regulation of GHB binding sites at the onset of, and during, GHB-elicited SWDs (Banerjee *et al.* 1993) could be a consequence of the epileptic activity as it occurs just at, and not before, the onset of GHB-induced SWDs (Banerjee *et al.* 1993), and it is also present in adult GAERS exhibiting SWDs (Snead *et al.* 1990).

In summary, from all the data reviewed in the absence epilepsy sections, it is apparent that 'micromolar' GHB concentrations, identical to the brain GHB levels necessary for the induction of SWDs, act via thalamic (and possibly cortical) presynaptic and postsynaptic GABA$_B$ receptors to elicit SWDs in rats. While the block by systemic and intracortically injected NCS 382 might suggest a putative role for GHB receptors in GHB-elicited SWDs, confirmation should await results showing (i) potent GHB receptor agonists able to evoke SWDs, (ii) truly selective GHB receptor antagonists able to abolish SWDs elicited by GHB or by selective and potent GHB agonists, and (iii) electrophysiological responses mediated by GHB receptors in *in vitro* studies on single neurons.

4. Hippocampus

The information available on the electrophysiological action of GHB in the hippocampus is restricted to CA1 pyramidal neurons, and, while a number of investigators have used the hippocampal slice, only one study has focused on the effects of GHB *in vivo*.

4.1. In vivo *study*

Aizawa *et al.* (1997) have tested the action of baclofen and GHB on long-term potentiation (LTP) by recording extracellularly the population spike (PS) in the CA1 region following electrical stimulation of the CA3 area in urethane-anaesthetized mice. They find that baclofen (5 mg/kg i.p.) transiently (i.e. 10 min) decreases the PS by 50%. Tetanic stimulation (applied 15 min after baclofen injection) brings about the same potentiation as in saline pre-treated mice, but while LTP remains stable in the control group for up to 150 min, in the baclofen-treated animals it starts decaying almost immediately after the tetanus and continues to do so up to 90 min after the tetanus. CGP 35348 (200 mg/kg i.p.) has no effect of its own on the PS either before or after tetanic stimulation (in contrast to the findings of Davies *et al.* (1991) in the rat *in vitro*), but blocks the baclofen-induced decay of LTP. GBL (50 mg/kg i.p.) significantly decreases the PS before tetanic stimulation (as baclofen does), but increases it immediately after the tetanus and up to a level that reaches statistical significance 90 min after the tetanic stimulation. NCS 382 (50 mg/kg i.p.) appears to potentiate slightly the GBL-induced decrease in the PS but partially reverses the GBL-induced increase in LTP. Unfortunately, Aizawa *et al.* (1997) did not test the action of NCS 382 alone or against the baclofen-elicited changes in the PS and LTP, nor the action of CGP 35348 against the GBL-elicited changes in these parameters.

4.2. In vitro *studies*

In CA1 pyramidal cells of the rat hippocampus *in vitro*, the postsynaptic action of GHB is very similar to the one in TC neurons. Thus, GHB (0.25–10 mM) produces a concentration-dependent and reversible hyperpolarization that is mediated by switching on an outward current (Xie and Smart 1992a). Although no information concerning the threshold concentration is given, the illustrated examples indicate a clear, though small, hyperpolarization with 250–500 μM GHB. Xie and Smart (1992a) show that this action of GHB is insensitive to TTX (1 μM), bicuculline and picrotoxin (both 50 μM), persists when neurons are impaled with KCl-filled electrodes, but is blocked by Ba^{2+} (0.5–1 mM) and Zn^{2+} (300 μM). The hyperpolarization is also abolished by CGP 35348 (20–100 μM) and CGP 36742 (20–100 μM), indicating that GHB directly activates postsynaptic $GABA_B$ receptors linked to a K^+ current.

Xie and Smart (1992b) were the first to show a reduction of synaptic potentials by GHB. Thus, at concentrations of 1–10 mM, GHB depresses (20–80%) the monosynaptic EPSP evoked in CA1 pyramidal neurons by stimulation of the Schaffer collateral/commissural pathway in a concentration-dependent manner. Pure $GABA_A$ IPSPs, evoked by localized electrical stimulation in the presence of NMDA and AMPA receptor antagonists, are also reversibly reduced (20–80%) (Fig. 5.5A) by concentrations of GHB similar to those that are effective on the EPSPs. All the effects of GHB on synaptic potentials are antagonized by CGP 36742 (20–100 μM), indicating an action mediated by activation of presynaptic $GABA_B$ receptors (Fig. 5.5A).

A recent study (Berton *et al.* 1999) has re-investigated the effect of low doses of GHB on the Schaffer collaterals EPSP in rat CA1 pyramidal neurons *in vitro*. The NMDA and AMPA

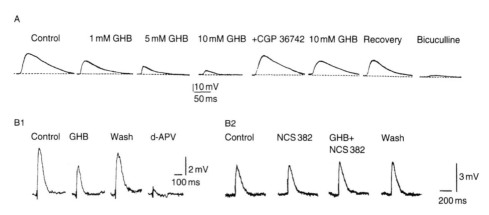

Figure 5.5 Presynaptic GHB receptor- and $GABA_B$ receptor-mediated effects of GHB on synaptic potentials in CA1 hippocampal pyramidal neurons in brain slices. (A) Pharmacologically isolated $GABA_A$ IPSPs evoked by localized electrical stimulation of stratum radiatum, and recorded using KCl- (3 M) filled intracellular electrodes, in the presence of CNQX (20 μM), APV (40 μM) and high Mg^{2+} (4 mM). Application of low millimolar concentrations of GHB dose-dependently reduces the amplitude of the IPSP. Ten minutes after removal of 10 mM GHB, application of CGP 36742 (300 μM) has no effect on its own but markedly reduces the response of 10 mM GHB. After wash-out of these two drugs, application of bicuculline (20 μM) abolishes the evoked IPSP. Membrane potential: −90 mV. Each trace is composed of three superimposed, successive IPSPs. (B1) Pharmacologically isolated NMDA EPSPs evoked by electrical stimulation of the Schaffer collaterals (DNQX, 10 μM; bicuculline, 30 μM). In the continuous presence of CGP 35348 (500 μM), application of 600 μM GHB reversibly reduces the amplitude of the EPSP. Following wash-out of GHB, d-APV (30 μM) blocks the EPSP. (B2) The GHB-elicited reduction of the NMDA EPSP is mediated by GHB receptors, as it does not occur in the presence of NCS 382 (500 μM). Resting membrane potential in (B1) and (B2) −65 mV. Reproduced with permission from Xie and Smart (1992b) (A) and Berton *et al.* (1999) (B1, B2).

components of this EPSP were isolated by adding CNQX and APV, respectively, to the perfusion medium, while $GABA_A$ and $GABA_B$ receptors were blocked by the presence of bicuculline (30 μM) and CGP 35348 (500 μM), respectively. GHB (100, 300 and 600 μM) decreases the NMDA EPSP (by 17%, 39% and 34%, respectively) (Fig. 5.5B1), and this reduction is fully antagonized by NCS 382 (500 μM), which has no effect of its own (Fig. 5.5B2). In addition, GHB (600 μM) decreases the AMPA EPSP and its paired pulse facilitation, with both effects, again, being antagonized by NCS 382 (500 μM). These results represent the only example of a GHB receptor-mediated action of GHB on single neurons *in vitro* (i.e. an NCS 382-sensitive electrophysiological response), and indicate that in *in vitro* electrophysiological experiments GHB acts on GHB receptors when it is applied *after* a $GABA_B$ antagonist (see below for a detailed discussion of this point).

The effect of GHB at the rat Schaffer collateral–CA1 neuron synapse has also been investigated by recording extracellularly the synaptic field potential (FP) *in vitro* (King *et al.* 1997). GHB dose-dependently depresses the FP, with a threshold concentration of about 1 mM and an IC_{50} of about 11 mM. The action of GHB is inhibited (but never abolished) by large concentrations of the $GABA_B$ antagonist, 2-OH-saclofen (800 μM), a result that is in line with previous data showing agonist-like properties of 2-OH-saclofen on $GABA_B$ receptors (Emri *et al.* 1996b). NCS 382, at a concentration of up to 5 mM, has no effect on the GHB-induced changes in FP, but produces a concentration-dependent increase in the FP slope when applied alone (with 100 μM and 2 mM producing 11% and 65% enhancement, respectively). In addition, removal of NCS 382 brings about a relatively fast, though only partial, recovery of this effect, as even after 60 min wash-out of NCS 382 the FP remains elevated to about 40% of the control value. From these results, King *et al.* (1997) conclude that either NCS 382 is an inverse agonist at hippocampal GHB receptors or that there is tonic activation of GHB receptors in this brain area.

In summary, the postsynaptic action of GHB on CA1 pyramidal neurons *in vitro* consists of a K^+-mediated hyperpolarization that occurs via activation of $GABA_B$ receptors, though any potential sensitivity of this effect to a GHB receptor antagonist remains to be investigated. At the presynaptic level, the action of GHB is to reduce both EPSPs and $GABA_A$ IPSPs. There is a discrepancy, however, between the results of Xie and Smart (1992b) and King *et al.* (1997) on the one hand, and those of Berton *et al.* (1999) on the other, as in the latter study the action of GHB (100–600 μM) on the Schaffer collateral EPSP is shown to be mediated by GHB receptors. Berton *et al.* (1999) claim that their ability to detect an NCS 382-sensitive presynaptic action of GHB is due to the use of 'smaller GHB concentrations than those used by Xie and Smart (1992b)' and to 'the continuous block of $GABA_B$ receptors in their standard recording medium'. The first suggestion appears unlikely as King *et al.* (1997) have observed a $GABA_B$ receptor-mediated effect of 100–500 μM GHB at the same hippocampal synapses, and in other brain areas the presynaptic action of 100–500 μM GHB *in vitro* has been found to be insensitive to NCS 382 (Emri *et al.* 1996a; Fig. 5.3). Although Berton *et al.* (1999) correctly point out that saturation and desensitization of GHB receptors might occur when using a high concentration of GHB (Maitre 1997), it is difficult to see how there could be a large difference in saturation/desensitization of GHB receptors between their GHB concentration of 600 μM and the 1.0 mM GHB used by Xie and Smart (1992b). On the other hand, it is obvious that in the experiments of Berton *et al.* (1999) the *continuous* presence of a $GABA_B$ antagonist in the perfusion medium would have made any pre- and postsynaptic $GABA_B$ receptor-mediated action of GHB undetectable. The results of Berton *et al.* (1999), therefore, raise the important question as to why unmasking a presynaptic GHB receptor-mediated electrophysiological effect *in vitro* requires *previous* block of

GABA$_B$ receptors. It would clearly be interesting to test whether the same is true for the presynaptic GHB action on other neuronal types as well as for the GHB effects mediated by postsynaptic GHB receptors.

5. Substantia nigra

The electrophysiological evidence on the action of GHB in neurons of the substantia nigra is mainly restricted to the dopaminergic neurons of the pars compacta, and is based upon a number of *in vivo* studies and one *in vitro* investigation.

5.1. In vitro *study*

Using intracellular microelectrodes, Harris *et al.* (1989) have investigated the effect of GHB in presumed dopamine-containing neurons of the guinea pig substantia nigra (pars compacta). As previously observed by Olpe and Koella (1979) *in vivo* (see below), not all neurons in this brain area respond to GHB. Thus, Harris *et al.* (1989) observe an effect of GHB only in a subset (63%) of neurons, with the responders having similar passive membrane properties to the non-responders except for the presence of inward and outward rectification in the majority of cells in the former group. GHB elicits a dose-dependent hyperpolarization, with the smallest response (0.5–3.0 mV) observed at 0.3–1.0 mM and a maximum effect (14 mV) at 30 mM. As for the GHB-induced hyperpolarization already described in TC and hippocampal neurons, this GHB action in the pars compacta neurons *in vitro* is associated with a small decrease in input resistance, and is abolished by Ba^{2+} (2.4 mM). In addition, it appears to be insensitive to TTX (0.5–0.9 μM), TEA (10 mM), Cd^{2+} (1 mM) and 4-AP (5 mM). The only illustrated example of a GHB response in the presence of TTX, however, is recorded at a membrane potential of -50 mV while the corresponding control trace is at -55 mV (cf. Fig. 3 in Harris *et al.* 1989), indicating that a small TTX-sensitive response to GHB might have gone undetected due to the different membrane potential. Surprisingly, the reversal potential of the GHB-elicited hyperpolarization is between -60 and -65 mV in five cells, and between -65 and -100 mV in the remaining twelve cells. The GHB-evoked hyperpolarization is slightly reduced (27%) by bicuculline (200 μM, $n=3$), a result that would explain why, in a number of neurons, the reversal potential (-65 mV) of the GHB-induced hyperpolarization was closer to the Cl$^-$ (around -65 mV) equilibrium potential. Taken together, these data would support the notion that part of the GHB action described in this study occurs via GABA$_A$ receptors, activated either directly by GHB or indirectly by GHB increasing GABA release from the neighbouring GABAergic neurons of the pars compacta/reticulata. As GHB does not bind to the GABA$_A$ receptor (maximal concentration tested: 100 μM) (Bernasconi *et al.* 1992), one should favour the latter possibility. This explanation might appear to be in contrast with the reported TTX insensitivity of the GHB-elicited hyperpolarization. However, we have already explained how an experimental inaccuracy might have prevented Harris *et al.* (1989) from detecting a TTX-sensitive component in their GHB response. Moreover, the TTX concentration used by Harris *et al.* (1989) might have been insufficient for a full block of the Na$^+$ current in substantia nigra neurons (cf. Haüsser *et al.* 1995).

In addition to a hyperpolarizing action, Harris *et al.* (1989) report that GHB facilitates high and low threshold Ca^{2+} potentials, though the number of neurons tested (or showing this response?) is very small ($n=3$). In addition, as different membrane potentials are used in the various experimental protocols employed for this analysis (see Figs 5 and 6 in

Harris *et al.* (1989)), it is difficult to ascertain how much of this putative action of GHB on Ca^{2+} potentials is due to the presence of additional voltage-dependent currents at different membrane potentials and/or to the difficulty of controlling, under current-clamp conditions, the effect of the steep voltage dependence of the conductance(s) underlying these Ca^{2+} potentials.

5.2. In vivo *studies*

The first report on the effect of GHB on the firing of dopamine-containing neurons of the pars compacta is by Roth *et al.* (1973). Using chloral hydrate anaesthetized rats, these authors show that GHB (100–350 mg/kg i.v.) and GBL (200–400 mg/kg i.p.) decrease the firing rate of these neurons, while having little effect on locus coeruleus and raphe neurons. Using iontophoretic application of GHB in similarly anaesthetized rats, Olpe and Koella (1979) found that only 50% of the pars compacta neurons are sensitive to GHB, and that this action is insensitive to iontophoretically applied bicuculline.

In chloral hydrate-anaesthetized rats, Diana *et al.* (1991) have confirmed that a reduction in the firing rate of pars compacta neurons occurs at 400 (but not 200) mg/kg i.v. of GHB. In unanaesthetized, paralysed rats, however, the same authors find that doses of 50–400 mg/kg i.v. increase (10–56%) the firing rate of dopamine neurons, while higher (subtoxic) doses (1–1.5 g/kg i.v.) of GHB abolish spontaneous firing. Interestingly, an i.p. injection of GHB (750 mg/kg) produces an initial increase (23%) in firing at 15 min post-injection: this is followed by a decrease (29%) at 30 min, which is preceded and accompanied by a reduction in burst firing.

The changes in firing rate/pattern elicited by GHB in neurons of the substantia nigra pars compacta have been carefully analysed by Engberg and Nissbrandt (1993) in chloral hydrate-anaesthetized rats. Low doses of GHB (<200 mg/kg i.v.) induce a small increase in the overall firing rate, but undoubtedly the most striking effect highlighted by these authors is a regularization of the firing pattern and a marked suppression of burst firing. At higher doses (200–400 mg/kg i.v.), the burst firing is fully blocked and an overall decrease in firing is then observed. Interestingly, baclofen (8–16 mg/kg i.v.) mimics in every aspect the action of GHB. All these actions of GHB and baclofen are blocked by CGP 35348 (200 mg/kg i.v.). Note also that during the regularization of firing and block of burst firing elicited by low doses of GHB (200 mg/kg i.v.), there is a reduction in striatal dopamine levels measured by microdialysis (Nissbrandt *et al.* 1994). This effect, however, appears not to be due to a direct action of GHB in the striatum as intrastriatal application of GHB increases dopamine levels in this area (Nissbrandt *et al.* 1994).

To clarify further the receptors involved in the GHB-elicited changes in the firing properties of pars compacta neurons, Erhardt *et al.* (1998) have compared the effect of the $GABA_B$ antagonist SCH 50911 (75 mg/kg i.v.) with the action of NCS 382 (100 mg/kg i.v.). As was the case with CGP 35348 (200 mg/kg i.v.) (Engberg and Nissbrandt 1993), SCH 50911 blocks all actions of systemically injected baclofen and GHB. NCS 382, on the other hand, does not counteract any of the effects of the two agonists, except for blocking the regularization of firing induced by low doses of GHB ($\leq$100 mg/kg).

In summary, there is a good agreement among the *in vivo* studies on the dopamine neurons of the pars compacta, as they show an increase in firing rate and a marked reduction/suppression of burst firing following low doses of GHB (Diana *et al.* 1993; Engberg and Nissbrandt 1993). It is unlikely that the increase in firing of the dopaminergic pars compacta neurons is mediated via the GABAergic neurons of the substantia nigra pars reticulata as the latter cells are mainly unaffected (61% of neurons) or slightly inhibited by i.v. injections

(400 mg/kg) of GHB (Diana *et al.* 1993), though any action of GHB on the GABAergic cells of the pars compacta has not been investigated. Assuming that the increase in firing and associated burst suppression produced by i.v. injection of low doses of GHB, therefore, are generated by a direct (postsynaptic) action of this drug on the recorded dopaminergic neuron, it is then difficult to see how the GHB-induced hyperpolarization described *in vitro* by Harris *et al.* (1989) might bring about a suppression of burst firing (since this firing pattern is elicited/facilitated at potentials more negative than the regular firing pattern). Alternatively, one might speculate that GHB regularizes the firing pattern by a direct action on somatic Ca^{2+} currents, in particular the L type, as this current has been shown to be critical for the burst firing ability of dopaminergic pars compacta neurons (Ping and Shepard 1999; Amini *et al.* 1999). Unfortunately, the only data on the effect of GHB on high threshold Ca^{2+} spikes of pars compacta neurons show a putative facilitatory action of GHB on these potentials (Harris *et al.* 1989), but we have already mentioned that difficulties exist in the interpretation of these results (see Section 5.1). On the other hand, in dopaminergic neurons of the ventral tegmental area and in the NCB-20 cell line GHB produces a clear reduction of high threshold Ca^{2+} spikes and I_{Ba}, respectively (Kemmel *et al.* 1998; Madden and Johnson 1998) (see Sections 6 and 9). Indeed, as the GHB-mediated burst suppression is reduced by $GABA_B$ antagonists, it may be that this action of GHB is mediated by the 'classical' reduction of Ca^{2+} currents that follows $GABA_B$ receptor activation (Wu and Saggau 1995). It is then reasonable to suggest that low doses of GHB *in vivo* by decreasing high threshold Ca^{2+} current(s) reduce the size and number of the high threshold Ca^{2+} current-mediated plateau potentials and after-depolarizations that underlie the generation of burst firing in dopaminergic pars compacta neurons (Ping and Shepard 1999; Amini *et al.* 1999). At higher doses of GHB *in vivo*, a larger reduction of Ca^{2+} currents, together with a postsynaptic hyperpolarization, may bring the membrane potential to more negative potentials, thus creating a condition for marked decrease in all firing patterns.

The ability of $GABA_B$ antagonists and NCS 382 to antagonize the burst suppression observed at the lowest GHB doses (25–100 mg/kg i.v.) represents a clear indication that this action is mediated via GHB *and* $GABA_B$ receptors, while only $GABA_B$ receptors are responsible for the full inhibition of firing elicited by higher doses of GHB (>100 mg/kg).

6. Ventral tegmental area

With the growing problem of GHB abuse and its potential anti-craving properties in alcohol withdrawal syndrome, it is surprising that only one study (Madden and Johnson 1998) has investigated the changes in membrane conductances underlying the action of GHB in the brain regions (prefrontal cortex, nucleus accumbens, ventral tegmental area) associated with drug-reward mechanisms. Thus, in the dopamine-containing neurons of the ventral tegmental area, GHB elicits a concentration-dependent, bicuculline (30 μM) and strychnine (10 μM) insensitive hyperpolarization (ED_{50}: 0.88 mM) (Fig. 5.6A) (Madden and Johnson 1998): this is associated with a decrease in input resistance and a full inhibition of the spontaneous (1–4 Hz) firing present in these neurons at their resting membrane potential *in vitro*. The threshold concentration of GHB (100 μM) elicits a 2–3 mV hyperpolarization and a maximum effect (20 mV) is obtained with 10 mM. Under voltage-clamp conditions, GHB produces an outward current (associated with an increase in chord conductance) (Fig. 5.6B1), whose reversal potential follows changes in the extracellular K^+ concentration similar to those predicted by the Nernst equation for a pure K^+ conductance (Fig. 5.6B2). The hyperpolarization (Fig. 5.6A) and the outward current elicited by GHB (1 and 10 mM) are blocked

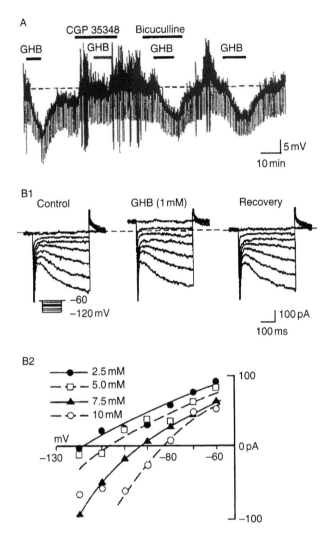

Figure 5.6 Effect of GHB on dopaminergic neurons of the ventral tegmental area in brain slices. (A) Continuous intracellular record showing that the hyperpolarization elicited by 1 mM GHB is abolished by CGP 35348 (300 μM) but not by bicuculline (30 μM). Downward deflections represent the voltage response to brief current pulses (70 pA). Membrane potential: -60 mV (dashed line). (B1) Application of GHB evokes an outward current at -60 mV (dashed line). Currents evoked by the illustrated voltage steps show an increase in membrane conductance. (B2) Net current–voltage plot of the effect of GHB in a single neuron. The reversal potential of the GHB-elicited current is progressively shifted to the right by higher extracellular K^+ concentrations.

Reproduced with permission from Madden and Johnson (1998).

by CGP 35348 (300 μM), indicating an action mediated by $GABA_B$ receptors. On the other hand, NCS 382 (10 μM) has no effect on the outward current evoked by 10 mM GHB. In addition, Madden and Johnson (1998) observed that the number of high threshold Ca^{2+} spikes evoked by depolarizing voltage steps in the presence of TTX (0.3 μM) and TEA (10 mM) were reversibly reduced by 10 mM GHB. Unfortunately, the pharmacological antagonism of this effect of GHB either by CGP 35348 or NCS 382 was not investigated.

In summary, one of the two postsynaptic effects of GHB on the dopaminergic neurons of the ventral tegmental area is similar to the one described in all other neuronal types examined *in vitro*: it consists of a $GABA_B$ receptor-mediated hyperpolarization that is elicited by the opening of K^+ channels, is insensitive to a GHB receptor antagonist and is observed at GHB concentrations as low as 100–300 μM. The other action of GHB on ventral tegmental area cells, that is the reduction in the number of high threshold Ca^{2+} current transients, would appear to be in contrast with the results of Harris *et al.* (1989) obtained in the dopaminergic neurons of the substantia nigra (pars compacta). The latter data, however, are not devoid of technical/interpretation biases (as explained in previous sections). Furthermore, (i) as presynaptic $GABA_B$ receptor activation decreases hippocampal EPSPs and IPSPs via a reduction of high threshold Ca^{2+} current (Wu and Saggau 1995), and (ii) if one assumes that the pharmacology of high threshold Ca^{2+} currents in the soma and at nerve terminals is not very dissimilar, then the GHB-evoked decrease in the number of high threshold Ca^{2+} spikes in neurons of the ventral tegmental area (Madden and Johnson 1998) is in agreement with the $GABA_B$ receptor-mediated reduction of thalamic and hippocampal EPSPs and IPSPs elicited by GHB (Xie and Smart 1992b; Emri *et al.* 1996a).

From a functional perspective, both GHB effects on ventral tegmental area neurons (i.e. the hyperpolarization and the reduction in Ca^{2+} spikes) should contribute to a decrease in firing. As discussed by Madden and Johnson (1998), however, this is in contrast to the increased dopamine release that is known to mediate the rewarding properties of a drug (but see MacDonald and Barker (1979)). On the other hand, the electrophysiological correlates of these biochemical findings are not available: except for the study of Madden and Johnson (1998), in fact, there has been no attempt to analyse in detail the pre- and postsynaptic actions of GHB on the monoamine- and amino acid-mediated synaptic transmission (as well as on the intrinsic properties of the different neurons) in the nucleus accumbens and ventral tegmental area, in the GABAergic cells of the pars compacta/reticulata and in the different types of prefrontal cortex neurons. We strongly feel that all this information is essential before any meaningful attempt can be made at understanding the cellular effects of GHB on the overall electrical output of single neurons and networks of neurons in each of these regions, and the role that each of these effects play in establishing GHB reward/abuse features (Colombo *et al.* 1998). Clearly, the anti-craving properties of baclofen, which have recently been reported in alcohol-, cocaine-, heroin- and nicotine-dependent rats (Brebner *et al.* 1999; Colombo *et al.* 2000; Fattore *et al.* 2000) and in man (Poldrugo and Addolorato 1999; Addolorato *et al.* 2000), suggest a strong potential for $GABA_B$ agonist-based treatments of drug withdrawal, and provide additional justification for a systematic, cellular electrophysiological investigation of the $GABA_B$ and GHB receptor-dependent mechanisms operating in the brain regions involved in drug addiction.

7. Cerebral cortex

Kozhechkin (1980) has investigated the action of iontophoretically applied GHB on unidentified cortical neurons in the sensorimotor cortex of paralysed, artificially ventilated rabbits. Of the eighty-six extracellularly recorded units, 55% and 12% show a reduction and an increase in firing, respectively, while the remaining 33% are unaffected by GHB. Interestingly, the decrease in firing is abolished by bicuculline (four out of six neurons), while the increase in firing is mainly observed in neurons that discharge in bursts. As half of these burst firing neurons were recorded in animals that had developed a poorly identified 'traumatic epilepsy', the GHB-elicited increase in firing should be interpreted with caution. It is also worth pointing out that in the illustrated examples (Fig. 1 in Kozhechkin (1980)) the spontaneous firing rate of neurons showing a GHB-mediated decrease in firing is higher than those showing an increase in firing during application of GHB.

Olpe and Koella (1979) also found that iontophoretic application of GHB decreases the firing rate of the majority (85%) of unidentified cortical cells in chloral hydrate-anaesthetized rats, but, in contrast to the results of Kozhechkin (1980), this effect is insensitive to concomitant iontophoretic application of bicuculline. In addition, Olpe and Koella (1979) observed that the systemic injection of a relatively low dose of GHB (100 mg/kg i.p.) increases the firing rate in two (out of five) cells, while the remaining neurons are unaffected. Large doses of GHB (300–600 mg/kg i.p.) inhibit the majority of cortical cells ($n = 9$) located between 400 and 2,000 μM below the cortical surface, with the remaining showing either no effect ($n = 3$) or a small increase in firing ($n = 2$). As the iontophoretic technique tends to deliver high concentrations of a drug in the proximity of the recorded neuron, it may not be surprising that only a decrease in firing rate is observed when employing this route of drug application.

Using urethane-anaesthetized rats, Godbout *et al.* (1995) have confirmed and enlarged the observations of Olpe and Koella (1979), though the definition of low versus high doses of systemically applied GHB is somewhat different in the two studies. Thus, an i.p. injection of low doses (5–10 mg/kg) of GHB increases the firing in 66% of layer II–VI prefrontal cortex neurons ($n = 12$ out of 18), with the remaining neurons showing either a decrease ($n = 3$) or no effect ($n = 3$). High doses (160–320 mg/kg) decrease the firing rate in 80% of neurons ($n = 12$ out of 15), with the remaining neurons showing either an increase ($n = 1$) or no effect ($n = 2$). Just over one-third of the twenty units tested with intermediate doses of GHB (20, 40 and 80 mg/kg) showed an increase in firing, with the remaining being either unaffected (33% of units) or showing a decrease (33% of units). Some of the neurons in the groups tested with different doses were antidromically activated from the striatum, and no difference in the type of response to GHB was observed between antidromically and non-antidromically activated neurons. No major difference in maximum potency, latency and duration of the GHB effect is observed between cells excited and those inhibited by low and/or high doses of GHB. Unfortunately, the pre-drug firing rate of neurons showing either a decrease or an increase in firing is not reported. Thus, as the only illustrated examples of the GHB-induced increase and decrease in firing are from two cells showing a low (40 spikes/s) and a high (280 spikes/s) firing rate, respectively (cf. Fig. 1 in Godbout *et al.* (1995)), one wonders whether some of the differences in the effect of GHB observed by Godbout *et al.* (1995) and Kozhechkin (1980) might also depend on the firing rate of the neuron prior to drug application. It is also worth stressing that in most cases the inhibition elicited by high doses of GHB was followed by a strong and long-lasting excitation (100% increase of baseline firing and 1 h, respectively, as judged solely from the illustrated example), and thus similar to the increase in firing induced by low doses of GHB. Administration of NCS 382 (10 and 320 mg/kg i.p.), either 1 or 30 min prior to the injection of an equivalent dose of GHB, has no action of its own, fails to affect the GHB-evoked decrease in firing, but blocks the GHB-induced increase in firing.

In summary, although some discrepancy exists in the definitions of low versus high doses, from these three electrophysiological studies on unidentified (except in Godbout *et al.* 1995) cortical cells in anaesthetized animals it is possible to conclude that (i) the increase in firing produced by low doses (5–20 mg/kg in Godbout *et al.* 1995) of systemically injected GHB occurs via an action on GHB receptors, (ii) the decrease in firing observed with high doses

(>100 mg/kg) of GHB does not involve GHB receptors, and (iii) intermediate doses (40–100 mg/kg) would clearly show both types of responses. The ability of iontophoretically applied GHB to decrease firing allows us to speculate that GHB probably has a direct inhibitory action on the recorded cortical neurons or on cells lying in close proximity. On the other hand, a similar conclusion cannot be drawn from the excitation observed with iontophoretically applied GHB (Kozhechkin 1980), as these data may contain some uncontrolled bias due to a developing pathology during the experimental session (see above). Thus, the known problems associated with *in vivo* studies and systemic injections, together with the lack of *in vitro* studies investigating the *direct* action of GHB on different types of cortical neurons, make it difficult to ascertain whether the GHB receptor-mediated increase in cortical cell firing (and its block by NCS 382) represents a direct cortical response or is mediated by GHB receptors located at sites distant from the cortex. The decrease in basal and K^+-evoked GABA release observed in cortical layer I–III following systemic injection of GBL (100 mg/kg i.p.) (Hu *et al.* 2000) might be responsible for the excitation of cortical cells in lower layers. As these changes in GABA release are partly antagonized by intracortically injected NCS 382 (100 μM) (Hu *et al.* 2000), it is plausible that the NCS 382-sensitive excitation of cortical neurons by systemic GHB is occurring at the cortical level. As the GHB-induced cortical release of GABA is also antagonized (in part) by intracortically administered CGP 35348 (100 μM) (Hu *et al.* 2000), it would be interesting to check whether the excitation of cortical neurons by systemic GHB is also sensitive to $GABA_B$ antagonists. Finally, whether the GHB receptor-mediated increase in cortical cell firing is mediated by a pre- or postsynaptic mechanism will need electrophysiological evaluation in identified cortical neurons *in vitro*.

8. Other mammalian neuronal types

Hösli *et al.* (1983) have investigated the action of GHB on organotypic cultures of spinal, brainstem and cerebellar neurons prepared from foetal or newborn rats. GHB (0.1–1.0 mM) causes a hyperpolarization (2–12 mV) that, at times, is even larger than the one produced by GABA (at least on spinal and brainstem neurons). The hyperpolarization is associated with a small decrease (27%) in input resistance (though this is not apparent in the illustrated traces), is abolished by bicuculline (10 μM), is transformed into a depolarizing response when recording with KCl electrodes, and appears to be due to a direct action of GHB on the recorded neuron, as experiments were conducted in a medium that abolishes synaptic transmission. Thus, the properties of the GHB-mediated hyperpolarization of spinal, brainstem and cerebellar neurons are highly reminiscent of those observed following activation of $GABA_A$ receptors, and are very different from those resulting from the direct postsynaptic hyperpolarization elicited by GHB in all other mammalian neuronal types so far examined using brain slices (but see Harris *et al.* (1989)). Whether this discrepancy reflects (a) some unique characteristics of spinal, brainstem and cerebellar neurons, and/or (b) the inability of GHB receptors to physiologically develop during the 11–53 days in organotypic culture conditions, remains to be determined.

9. GHB action in NCB-20 cell line

Maitre's group has recently investigated the presence of a functional GHB system in differentiated, clonal neurohybridoma NCB-20 cells (Kemmel *et al.* 1998; Maitre *et al.*, Chapter 14 of this volume). Together with the presence of binding sites and an uptake system specific for GHB, these authors describe a GHB-mediated inhibition of Ca^{2+} currents (using Ba^{2+} as the charge carrier) in seventy (out of ninety-two) cells (Figs 5.7A and B). The strongest effect is observed at the maximal activation of I_{Ba} (about -20 mV): at this voltage, a 27% reduction is seen with 25–50 μM GHB and a maximum inhibition (value not reported) occurred with 100 μM GHB. The GHB inhibition of I_{Ba} is blocked by application of 25–50 μM NCS 382 (Fig. 5.7A), but is insensitive even to high concentrations (100–150 nM) of the $GABA_B$ antagonist CGP 55845 (Fig. 5.7B). Though a low and a high threshold component of I_{Ba} were identified in these cells, any differential effect of GHB on these two components was not investigated. Interestingly, in some cells, I_{Ba} is reduced (40%) by 50 μM baclofen but not by GHB, while in other cells this current is sensitive to GHB but not to baclofen.

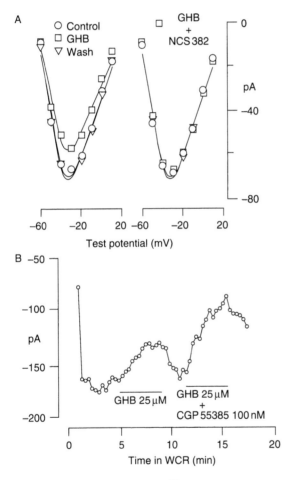

Figure 5.7 Effect of GHB on Ca^{2+} currents of NCB-20 cell. (A) Current–voltage plot of Ca^{2+} current (measured with Ba^{2+} as the charge carrier, I_{Ba}) shows the reversible reduction of this current by 25 μM GHB (left plot). Co-application of GHB and NCS 382 (25 μM) fully abolishes the effect of GHB (right plot). (B) Time course of peak I_{Ba} (test potential: -10 mV). Application of CGP 55385 has no effect on the GHB- (25 μM) elicited reduction of I_{Ba}.

Reproduced with permission from Kemmel *et al.* (1998).

In summary, this study on a differentiated neuronal cell line provides direct evidence of a conductance change (i.e. Ca^{2+} current decrease) that results from the selective activation of GHB receptors. If confirmed on neuronal membranes, this result would support the notion that GHB activation of presynaptic GHB and $GABA_B$ receptors brings about similar conductance changes and functional responses (i.e. decrease in Ca^{2+} current leading to reduced transmitter release and synaptic potential amplitude). In addition, this study provides electrophysiological support for the biochemical findings indicating that GHB and $GABA_B$ receptors might represent separate molecular entities (Cash *et al.* 1996; Mathivet *et al.* 1997).

10. GHB action on recombinant $GABA_B$ receptors

Following the recent cloning of $GABA_B$ receptors (Kaupmann *et al.* 1997), Lingenhoehl *et al.* (1999) have investigated the action of GHB on recombinant $GABA_B$ receptors co-expressed with inwardly rectifying K^+ channels (Kir3) in Xenopus oocytes (Bernasconi *et al.*, Chapter 3 of this volume). GHB produces a concentration-dependent increase in I_{Kir}, with a threshold concentration that, as can be seen from their illustrated examples, is close to 300 μM (Fig. 5.8A).

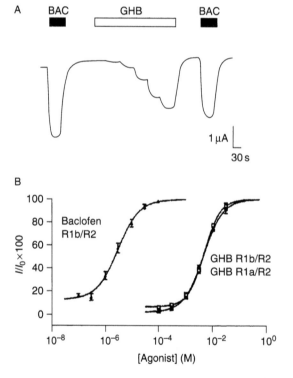

Figure 5.8 Effect of GHB on recombinant $GABA_B$ receptors in oocytes. (A) Continuous current trace recorded from an oocyte expressing $GABA_B$R1a/R2 and Kir3.1/2/4 channels. Continuous open line above the current indicates the successive application of 0.1, 0.3, 1.0, 3.0, 10.0 and 30.0 mM GHB. Filled lines indicate the time of application of 10 μM l-baclofen, before and after GHB applications. (B) Normalized dose–response curves of GHB- and l-baclofen-induced currents show the similarity in the response of GHB in oocytes expressing either the $GABA_B$R1a or $GABA_B$R1b isoform (right curves), and the weaker potency of this substance when compared to l-baclofen (left curve).

Reproduced with permission from Lingenhoehl *et al.* (1999).

The EC_{50} of the GHB response is about 4.5 mM (both for the $GABA_BR1a$ and $GABA_BR1b$ isoforms) (Fig. 5.8B), and the maximal response is about 70% of the effect produced by 10 μM L-baclofen. The GHB-induced current is insensitive to NCS 382 (300 μM), but is fully antagonized by a number of competitive $GABA_B$ antagonists, including CGP 35348 and CGP 54626A. It is worth stressing that the equi-effective concentrations of GHB at recombinant and native $GABA_B$ receptors in binding studies differ by a factor of 100, as is the case for other $GABA_B$ agonists. More importantly, the threshold concentration and the EC_{50} of GHB in these functional experiments on recombinant $GABA_B$ channels (Lingenhoehl *et al.* 1999) are similar to the threshold concentration and the EC_{50} of the GHB effect on native $GABA_B$ receptors in the majority of *in vitro* electrophysiological studies (Harris *et al.* 1989; Xie and Smart 1992a,b; Williams *et al.* 1995; Emri *et al.* 1996a; King *et al.* 1997; Madden and Johnson 1998).

In summary, these results strengthen the findings of the majority of electrophysiological *in vitro* studies on native $GABA_B$ receptors by demonstrating a direct, though weak, agonist action of GHB at recombinant $GABA_B$ receptors.

11. Conclusions

From this review of the electrophysiological responses elicited by endogenous and exogenous GHB in different neuronal types, the following points emerge.

1   The only direct evidence of a GHB receptor-mediated conductance change is the reduction of high threshold Ca^{2+} currents that has been observed in a differentiated neurohybridoma (NCB-20) cell line (Kemmel *et al.* 1998).

2   The only GHB receptor-mediated electrophysiological action of GHB (i.e. the decrease of the Schaffer collateral EPSPs) observed so far in neurons *in vitro* requires the previous block of $GABA_B$ receptors by a $GABA_B$ receptor antagonist (Berton *et al. 1999)*. Although the conductance underlying this EPSP reduction has not been investigated, by analogy with the effect of GHB in NCB-20 cells (Kemmel *et al.* 1998) it is reasonable to suggest that activation of *presynaptic GHB receptors* in the hippocampus might elicit an inhibition of Ca^{2+} currents leading to the observed reduction in EPSP amplitude. In this respect, therefore, presynaptic GHB receptors might possess similar effector systems (but see Ratomponirina *et al.* (1995) and Snead (2000)) to presynaptic $GABA_B$ auto- and heteroreceptors (Bernasconi *et al.*, Chapter 3 of this volume). The possible reasons underlying the ability to demonstrate a GHB receptor-mediated electrophysiological effect only after $GABA_B$ receptor blockade are addressed in point 5 below.

3   There is at present no direct or indirect evidence of a GHB receptor-mediated electrophysiological response that involves activation of putative postsynaptic GHB receptors. Whether this is the result of 'inappropriate' *in vitro* conditions (i.e. no study has investigated the presence of a postsynaptic GHB receptor-mediated effect following the block of $GABA_B$ receptors; see point 2 above) and/or is a reflection of the relative paucity of postsynaptic GHB receptors, as has been suggested by biochemical investigations (Maitre *et al.* 1983a,b; Maitre 1997), remain to be determined.

4   GHB acts as a weak agonist at pre- and postsynaptic $GABA_B$ receptors in all neuronal classes where it has been electrophysiologically tested so far (i.e. thalamus, hippocampus, substantia nigra pars compacta and ventral tegmental area). This finding is supported by similar results obtained in binding studies on native $GABA_B$ receptors (Mathivet *et al.* 1997), and has recently been directly demonstrated by electrophysiological experiments on reconstituted $GABA_BR1$–R2 receptors co-expressed with Kir3 channels in oocytes (Lingenhoehl *et al.* 1999).

a   *At the postsynaptic level*, the action of GHB on native $GABA_B$ receptors involves the opening of inwardly rectifying K^+ channels (Xie and Smart 1992a; Madden and Johnson 1998), resulting in membrane hyperpolarization (Harris *et al.* 1989; Williams *et al.* 1995). Whether this hyperpolarization brings about either inhibition (as one would expect from neuronal hyperpolarization) or excitation closely depends not only on the GHB concentration but also on the intrinsic membrane properties of each neuronal type. As has been demonstrated for TC neurons (Williams *et al.* 1995), it is erroneous to describe highly voltage-dependent physiological (and pathological) effects of an endogenous or exogenous substance in terms of the unifying notion of a simple change in membrane potential (e.g. hyperpolarization) and its overall ED_{50}. Similar detailed analysis of the effects of GHB on the firing patterns of dopaminergic and GABAergic cells of the substantia nigra, as well as of different types of cortical neurons, is essential to clarify the mechanisms underlying the GHB-elicited excitation and inhibition observed *in vivo* in these brain regions (Engberg and Nissbrandt 1993; Godbout *et al.* 1995).

b   *At the presynaptic level*, the action of GHB on native $GABA_B$ receptors brings about a reduction in the amplitude of EPSPs and IPSPs in hippocampal CA1 pyramidal cells and in TC neurons (Xie and Smart 1992b; Emri *et al.* 1996a; Figs 5.3 and 5.4A). Unfortunately, there is no direct evidence on the conductance(s) involved in this $GABA_B$ receptor-mediated action of GHB. By analogy with the action of baclofen on hippocampal and thalamic presynaptic $GABA_B$ auto- and heteroreceptors (Wu and Saggau 1995; Guyon and Leresche 1995), however, it is reasonable to suggest that the presynaptic $GABA_B$ receptor-mediated action of GHB involves a decrease in high voltage-activated Ca^{2+} current(s).

5   NCS 382-sensitive electrophysiological effects of endogenous and exogenous GHB observed *in vivo* have not been completely replicated *in vitro*, as NCS 382 is able to block GHB-elicited responses in the latter experimental condition only following the block of $GABA_B$ receptors. These findings raise a number of questions.

a   How long do $GABA_B$ receptors need to be blocked before a GHB response becomes detectable? Does the block of $GABA_B$ receptors *in vitro* change the conformational properties of GHB receptors, thus allowing the binding of the agonist GHB? Is this an indication of an as yet undiscovered GHB–$GABA_B$ receptor complex (Snead 1995)? Is there an 'endogenous blocker' of $GABA_B$ receptors that is washed away in the standard conditions used for electrophysiological *in vitro* recordings and that needs to be replaced by the 'exogenously applied $GABA_B$ antagonist' before an electrophysiological effect of GHB on GHB receptors becomes evident? If so, why is this 'endogenous blocker' not washed away during the standard conditions used for *in vitro* biochemical analysis? Is the previous block of $GABA_B$ receptors also essential to disclose an *in vitro* electrophysiological effect of GHB that is mediated by postsynaptic GHB receptors?

b   Although NCS 382 does not bind to $GABA_B$ receptors (Maitre *et al.* 1990; Snead 1996), a number of unresolved issues relate to the 'selectivity' of this (the only available) GHB receptor antagonist. (i) Since the contribution of a GHB-derived GABA pool to the action of GHB cannot at present be fully excluded (see Section 3.4.2), the competitive block of GHB-dehydrogenase by NCS 382 should always be kept in mind when interpreting *in vivo* and *in vitro* results, in particular when the antagonist is used at relatively high concentrations (Maitre 1997). (ii) The inability of the GHB agonist *trans*-4-hydroxycrotonic acid to elicit SWDs raises doubts on the involvement of GHB receptors in experimental absence epilepsy. This result, then, indirectly questions whether the block by NCS 382 of GHB-elicited SWDs (and other *in vivo* effects of GHB) is indeed due to its antagonism of GHB receptors. (iii) In the cerebellum, where no GHB receptors are present (Maitre 1997), doses of NCS 382 similar to those that block other *in vivo* pharmacological actions of GHB markedly decrease cGMP level by an action additive to that of GBL (Bernasconi *et al.*, Chapter 3 of this volume). Interestingly, careful examination of a recent biochemical study on GABA release in upper cortical layers shows a clear partial agonist action of NCS 382, 10 min after GHB injection (cf. Fig. 5 in Hu *et al.* (2000)). (iv) Why have most of the biochemical investigations (except the two mentioned in (iii) above) failed to detect the partial/inverse agonist properties of NCS 382 that are observed in electrophysiological studies (Williams *et al.* 1995; Emri *et al.* 1996a; King *et al.* 1997)? Alternatively, are these partial/inverse agonist properties of NCS 382 an artefact of the *in vitro* electrophysiological approach? All these controversial findings clearly suggest a pressing need for the development of highly selective GHB antagonists and for a more widespread use of (the already available) potent GHB agonists (Maitre *et al.* 1983b; Hechler *et al.* 1993; Hédou *et al.* 2000; Schmidt-Mutter *et al.* 1999).

6   The electrophysiological effects of exogenous GHB in single TC neurons disclose the cellular thalamic mechanism underlying its therapeutic use in anaesthesia and in the treatment of narcolepsy. Thus, the hyperpolarization-mediated induction of δ sleep oscillations (Williams *et al.* 1995) and the reduction of sensory (Emri *et al.* 1996a), cortical and intrathalamic (see Fig. 5.4) synaptic potentials would, together with a similar (though yet undisclosed) cortical action, contribute to setting the TC loop into a state inductive of deep (stage 3–4) sleep. These thalamic actions of exogenous GHB occur via activation of pre- and postsynaptic GABA$_B$ receptors, as indicated (i) by the results of *in vitro* studies in TC neurons (Williams *et al.* 1995; Emri *et al.* 1996a) and (ii) by the ability of intrathalamically injected GABA$_B$ antagonists to decrease the deep stages of non-REM sleep (Juhasz *et al.* 1994). Although a key genetic abnormality present in narcoleptic patients has recently been discovered (Peyron *et al.* 2000; Thannickal *et al.* 2000), the thalamic mechanism for the GHB-elicited increase in δ sleep oscillations described here provides a solid cellular rationale for the GHB therapy for narcoleptic patients until an appropriate gene therapy becomes available for this disease.

7   The involvement of thalamic GABA$_B$ receptors in the mechanisms leading to the generation/expression of GHB-elicited SWDs has been demonstrated. In contrast, the inability of *trans*-4-hydroxycrotonic acid to elicit SWDs and the inhibition of the GHB-dehydrogenase by NCS 382 make it difficult to draw any solid conclusion on a putative role for GHB receptors in GHB-induced SWDs. This uncertainty is not helped by the fact that, to the best of our knowledge, none of the other potent GHB agonists (e.g. NCS 356) (Maitre *et al.* 1983a; Hechler *et al.* 1993; Hédou *et al.* 2000) have been tested for the ability to induce absence seizures in normal animals or to aggravate them in genetic rat and mouse models of this disease.

8   No informed description of the electrical events contributing to the cellular mechanisms involved in GHB abuse and its therapeutic use in drug dependence/withdrawal is possible at present. This stems from the lack of a systematic (*in vitro*) analysis of the action of GHB in different cortical, nucleus accumbens and substantia nigra neurons. As so many effects

of GHB are mediated via $GABA_B$ receptors, it is interesting to note that recent data have highlighted the anti-craving properties of this $GABA_B$ receptor agonist against a number of substances of abuse (Brebner *et al.* 1999; Addolorato *et al.* 2000; Colombo *et al.* 2000). Unfortunately, as is the case for GHB receptors, the precise cellular contribution of $GABA_B$ receptors in the different neurons of the brain reward system is at present largely unknown.

9   The pre- and postsynaptic $GABA_B$ receptor-mediated electrophysiological responses of different neuronal types to high/toxic concentrations of GHB indicate a clear contribution of these receptors to GHB intoxication. This conclusion does not preclude an additional, though as yet undisclosed, role for putative GHB receptors in this pathology.

10   Due to the poor selectivity of NCS 382 (the only available GHB receptor antagonist, see point 5b above), any attempt to unveil a (patho)physiological role for a putative endogenous GHB system seems at present unrealistic.

In summary, no electrophysiological *in vitro* investigation has so far been able to detect a GHB receptor-mediated electrophysiological response to endogenous or exogenous GHB (defined as a response antagonized by NCS 382) in any type of CNS neuron under normal conditions. A GHB receptor-mediated effect of GHB has been observed in one study, but only after $GABA_B$ receptors had been blocked with a $GABA_B$ antagonist. This scenario is in sharp contrast to the relative ease with which a GHB receptor-mediated response to GHB can be electrically recorded *in vivo* or biochemically measured *in vivo* and *in vitro*.

The $GABA_B$ receptor-mediated induction of δ sleep oscillations and decrease of synaptic potential amplitude elicited by GHB in TC neurons provides a cellular basis for the therapeutic use of GHB in narcoleptic patients and in anaesthesia.

The involvement of $GABA_B$ receptors in GHB-elicited SWDs has been conclusively demonstrated, whereas that of GHB receptors still awaits further experimentation. Similarly, the precise sequence of electrical events underlying the generation/expression of GHB-induced SWDs in thalamic and cortical neurons remains to be investigated.

Notwithstanding the impact and significance of the last thirty years of research to our knowledge of the neuronal actions of exogenous and endogenous GHB, essential electrophysiological investigations at the single cell level are still needed to discover the precise site and mechanism of action of GHB in a number of brain regions. This is particularly true for those areas that play a key role in drug addiction and withdrawal, as only one such study is available in the ventral tegmental area, with none in the prefrontal cortex, nucleus accumbens and substantia nigra. This lack of information severely limits our current understanding of the cellular mechanisms involved in GHB abuse and leaves the development of any therapeutic approach without a clearly defined molecular/cellular target (but see Nestler (2001)). The recent promising results obtained with baclofen in human studies of drug dependence/withdrawal suggest the need for appropriate double blind controlled trials to compare the therapeutic potential of this $GABA_B$ agonist and GHB in the treatment of drug addiction/withdrawal.

Full molecular and functional characterization of cloned GHB receptor(s) (Cash *et al.* 1996), development of potent and selective GHB receptor antagonists (devoid of any action on GHB-dehydrogenase), and the use of potent GHB receptor agonists (that, unlike GHB, are not a substrate for GABA synthesis) are each seen as an essential milestone in clarifying the still unresolved issues of potential physiological roles for the neuronal GHB system, the mechanisms underlying some of GHB pharmacological actions and the pathological consequences of genetically determined or abuse-related increases in the brain levels of GHB.[6]

Acknowledgements

We are indebted to Drs Zs. Emri, Z. Monnier, J. P. Turner and S. R. Williams, who were involved in the experiments on TC neurons, and to Prof. D. I. Wallis, Dr A. Guyon, Dr S. W. Hughes, Dr H. R. Parri and Ms K. L. Blethyn for critical comments on this chapter. The work in our laboratories is supported by the Wellcome Trust (V. C.) and the CNRS (N. L.).

Notes

1 In rat brain preparations, GHB binds to high- (K_d: 30–580 nM) and low- (K_d: 2.3–16 μM) affinity GHB sites (Maitre *et al.* 1997), as well as to $GABA_B$ sites with a K_i of 125 and 80–120 μM when using [^{3}H]baclofen and [^{3}H]GABA, respectively (Benavides *et al.* 1982; Bernasconi *et al.* 1992; Mathivet *et al.* 1997). GHB shows no activity in $GABA_A$ (tested only up to 100 μM) and NMDA binding assays (Bernasconi *et al.* 1992). In addition, GHB binding sites differ from $GABA_B$ sites in regional brain distribution (Hechler *et al.* 1992; Turgeon and Albin 1993) and development profile (Knott *et al.* 1993; Snead 1994).

2 Data on plasma/brain levels in other species can be found in Jouvet *et al.* (1961), Snead (1978), Snead and Morley (1981) and Snead *et al.* (1976, 1980).

3 It is worth highlighting here that when recorded in current-clamp conditions the size of a substance-induced response (and therefore the ability to detect it) depends not only on the applied concentration but also on the input resistance of the recorded neuron. For a given concentration of GHB, therefore, neurons with a higher input resistance will produce a larger hyperpolarization than those that have a smaller resistance (either as their intrinsic feature or as a result of a poor seal with, or damage by, the recording electrode). It follows that in studies using current-clamp recording techniques the threshold concentration of GHB and the amplitude of its response might have been over- and underestimated, respectively. In addition, as discussed by Madden and Johnson (1998), because of the GHB uptake system the 'true' GHB concentration at the receptor sites is smaller than the exogenous concentration applied to the *in vitro* preparations through the perfusion medium. Indeed, low millimolar concentrations of GABA and glutamate are often required to elicit an electrophysiological response in similar *in vitro* conditions (Crunelli *et al.* 1988).

4 Although the sensitivity of experimental SWDs to GHB (and $GABA_B$; see the previous section) receptor antagonists may be interpreted as indicating that alterations of these brain systems are one predisposing factor for the generation of SWDs, it may solely result from the ability of these drugs to alter thalamocortical transmission. Indeed, the only reported abnormalities in the GHB and GABA systems in any of the genetic models of absence epilepsy relate to post-seizures animals (Snead *et al.* 1990; Spreafico *et al.* 1993; Richards *et al.* 1995; Sutch *et al.* 1999), and might therefore be a consequence, more than a cause, of the seizures.

5 In contrast to initial observations (Möhler *et al.* 1976; Doherty and Roth 1978), it has been shown that GHB can be metabolized into GABA by a GHB-dehydrogenase (Vayer *et al.* 1985). As the conflicting evidence on this issue continues to accumulate (Kaufman and Nelson 1991; Mathivet *et al.* 1997; Hechler *et al.* 1997; Gobaille *et al.* 1999; Bernasconi *et al.*, Chapter 3 of this volume), the possibility of a GHB-derived GABA pool should always be kept in mind when interpreting the pharmacological actions of GHB and, in particular, their antagonism by NCS 382 which has been shown to inhibit competitively GHB-dehydrogenase (Maitre 1997).

6 Experiments conducted after completion of this chapter have highlighted a GHB receptor-mediated electrophysiological response of GHB in thalamic slices. In the presence of a $GABA_B$ receptor antagonist (CGP 56999, 100 nM), GHB (500 μM) decreased (30%) the corticothalamic EPSC recorded in TC neurons, and this effect was fully antagonized by NCS 382 (1 mM). While further studies will be required to elucidate why a prior blockade of $GABA_B$ receptors is essential to disclose GHB receptor-mediated electrophysiological effects of GHB, this result will allow a novel line of interpretation of the cellular thalamic actions of GHB in relation to sleep and absence epilepsy mechanisms.

References

Addolorato G., Caputo F., Capristo E., Colombo G., Gessa G. L. and Gasbarrini G. (2000) Ability of baclofen in reducing alcohol craving and intake: II – Preliminary clinical evidence. *Alcoholism: Clin. Exp. Res.* 24, 67–71.

Aizawa M., Ito Y. and Fukuda H. (1997) Roles of γ-aminobutyric acid$_B$ (GABA$_B$) and γ-hydroxybutyric acid receptors in hippocampal long-term potentiation and pathogenesis of absence seizures. *Biol. Pharm. Bull.* 20, 1066–70.

Amini B., Clark J. W. Jr and Canavier C. C. (1999) Calcium dynamics underlying pacemaker-like and burst firing oscillations in midbrain dopaminergic neurons: a computational study. *J. Neurophysiol.* 82, 2249–61.

Avoli M., Gloor P., Kostopoulos G. and Naquet R. (1990) *Generalized Epilepsy: Neurobiological Approaches.* Birkhäuser, Boston.

Bal T., von Krosigk M. and McCormick D. A. (1995) Synaptic and membrane mechanisms underlying synchronized oscillations in the ferret LGNd *in vitro. J. Physiol.* 483, 641–63.

Banerjee P. K. and Snead O. C. (1995) Presynaptic gamma-hydroxybutyric acid (GHB) and gamma-aminobutyric acid$_B$ (GABA$_B$) receptor-mediated release of GABA and glutamate (GLU) in rat thalamic ventrobasal nucleus (VB): a possible mechanism for the generation of absence-like seizures induced by GHB. *J. Pharmacol. Exp. Ther.* 273, 1534–43.

Banerjee P. K., Hirsche E. and Snead O. C. (1993) Gamma-hydroxybutyric acid induced spike and wave discharges in rats: relation to high-affinity [3H]gamma-hydroxybutyric acid binding sites in the thalamus and cortex. *Neuroscience* 56 (1), 11–21.

Benavides J., Rumigny J. F., Bourguignon J. J., Cash, C. D., Wermuth C. G., Mandel P., Vincendon G. and Maitre, M. (1982) High-affinity binding site for gamma-hydroxybutyrate in rat brain. *Life Sci.* 30, 953–61.

Bernasconi R., Lauber J., Marescaux C., Vergnes M., Martin P., Rubio V., Leonhardt T., Reymann N. and Bittiger H. (1992) Experimental absence seizures: potential role of γ-hydroxybutyric acid and GABA$_B$ receptors. In Generalized non-convulsive epilepsy: focus on GABA$_B$ receptors. *J. Neural Transm.* (Suppl. 35) 155–77.

Bernasconi R., Mathivet P., Bischoff S. and Marescaux C. (1999) Gamma-hydroxybutyric acid: an endogenous neuromodulator with abuse potential? *Trends Neurosci.* 20, 135–41.

Bernasconi R., Mathivet P., Otten U., Bettler B., Bischoff S. and Marescaux C. (2002) Part of γ-hydroxybutyrate pharmacological actions are mediated by GABA$_B$ receptors. In *Gamma-Hydroxybutyrate: Molecular, Functional and Clinical Aspects.* (Tunnicliff G. and Cash C. D., eds), pp. 28–63. Taylor and Francis, London.

Berton F., Brancucci A., Beghè F., Cammalleri M., Demuro A., Francesconi W. and Gessa G. (1999) γ-Hydroxybutyrate inhibits excitatory postsynaptic potentials in rat hippocampal slices. *Eur. J. Pharmacol.* 380, 109–16.

Brebner K., Froestl W., Andrews M., Phelan R. and Roberts D. C. S. (1999) The GABA$_B$ agonist CGP 44532 decreases cocaine administration in rats: demonstration using a progressive ratio and a discrete trials procedure. *Neuropharmacology* 38, 1797–804.

Cash C. D., Hechler V., Mersel M. and Maitre M. (1996) Kinetic characterisation and solubilisation of γ-hydroxybutyrate receptors from rat brain. *Neurosci. Lett.* 209, 25–28.

Charpier S., Leresche N., Deniau J.-M., Mahon S., Hughes S. W. and Crunelli V. (1999) On the putative contribution of GABA$_B$ receptors to the electrical events occurring during spontaneous spike and wave discharges. *Neuropharmacology* 38, 1699–706.

Chin M. Y., Kreutzer R. A. and Dyer J. E. (1992) Acute poisoning from γ-hydroxybutyrate in California. *West J. Med.* 156, 380–84.

Colombo G., Agabio R., Lobina C., Reali R. and Gessa G. L. (1998) Involvement of GABA$_A$ and GABA$_B$ receptors in the mediation of discriminative stimulus effects of gamma-hydroxybutyric acid. *Physiol. Behav.* 64, 293–302.

Colombo G., Agabio R., Carai M. A., Lobina C., Pani M., Reali R., Addolorato G. and Gessa G. L. (2000) Ability of baclofen in reducing alcohol intake and withdrawal severity: I – Preclinical evidence. *Alcoholism: Clin. Exp. Res.* 24, 58–66.

Crunelli V. and Leresche N. (1991) A role for GABA$_B$ receptors in excitation and inhibition of thalamocortical cells. *Trends Neurosci.* 14 (1), 16–21.

Crunelli V., Kelly J. S., Leresche N. and Pirchio M. (1987a) The ventral and dorsal lateral geniculate nucleus of the rat: intracellular recordings *in vitro. J. Physiol.* 384, 587–601.

Crunelli V., Leresche N. and Parnavelas J. G. (1987b) Membrane properties of morphologically identified X and Y cells in the lateral geniculate nucleus of the cat *in vitro. J. Physiol.* 390, 243–56.

Crunelli V., Haby M., Jassik-Gerschenfeld D., Leresche N. and Pirchio M. (1988) Cl$^-$ and K$^+$-dependent inhibitory postsynaptic potentials evoked by interneurones of the rat lateral geniculate nucleus. *J. Physiol.* 399, 153–76.

Davies C. H., Starkey S. H., Pozza M. F. and Collingridge G. L. (1991) GABA$_B$ autoreceptors regulate the induction of LTP. *Nature* 349, 609–11.

Delgado-Escueta A. V., Wilson W. A., Olsen R. W. and Porter R. J. (1999) *Jasper's Basic Mechanisms of the Epilepsies. Adv. Neurol.*, Vol. 79. Lippincott Williams and Wilkins, Philadelphia.

Depaulis A., Bourguignon J.-J., Marescaux C., Vergnes M., Schmitt M., Micheletti G. and Warter J.-M. (1988) Effects of gamma-hydroxybutyrate and gamma-butyrolactone derivatives on spontaneous generalized non-convulsive seizures in the rat. *Neuropharmacology* 27 (7), 683–89.

Diana M., Mereu G., Mura A., Fadda F., Passino N. and Gessa G. (1991) Low doses of γ-hydroxybutyric acid stimulate the firing rate of dopaminergic neurons in unanaesthetized rats. *Brain Res.* 566, 208–11.

Diana M., Pistis M., Muntoni A. and Gessa G. (1993) Heterogeneous responses of substantia nigra pars reticulata neurones to γ-hydroxybutyric acid administration. *Eur. J. Pharmacol.* 230, 363–65.

Doherty J. D. and Roth R. H. (1978) Metabolism of γ-hydroxy-[1-14C] butyrate by rat brain: relationship to the Krebs cycle and metabolic compartmentation of amino acids. *J. Neurochem.* 30, 1305–09.

Duncan J. S. (1997) Idiopathic generalized epilepsies with typical absences. *J. Neurol.* 244, 403–11.

Dyer J. E. (1991) γ-hydroxybutyrate: a health-food product producing coma and seizure like activity. *Am. J. Emerg. Med.* 9, 321–24.

Emri Z., Antal K. and Crunelli V. (1996a) Gamma-hydroxybutyric acid decreases thalamic sensory excitatory postsynaptic potentials by an action on presynaptic GABA$_B$ receptors. *Neurosci. Lett.* 216, 121–24.

Emri Z., Turner J. P. and Crunelli V. (1996b) Tonic activation of presynaptic GABA$_B$ receptors on thalamic sensory afferents. *Neuroscience* 72, 689–98.

Engberg G. and Nissbrandt H. (1993) γ-Hydroxybutyric acid (GHBA) induces pacemaker activity and inhibition of substantia nigra dopamine neurons by activating GABA$_B$-receptors. *Naunyn-Schmied. Arch. Pharmacol.* 348, 491–97.

Erhardt S., Andersson B., Nissbrandt H. and Engberg G. (1998) Inhibition of firing rate and changes in the firing pattern of nigral dopamine neurons by γ-hydroxybutyric acid (GHBA) are specifically induced by activation of GABA$_B$ receptors. *Naunyn-Schmied. Arch. Pharmacol.* 357, 611–19.

Fattore L., Cossu G., Martellotta M. C. and Fratta W. (2000) Effect of baclofen on intravenous self-administration in mice and rats. *Eur. J. Neurosci.* 12 (Suppl. 11), 204.

Friedman J., Westlake R. and Furman M. (1995) 'Grievous bodily harm': gamma hydroxybutyrate abuse leading to a Weernicke–Korsakoff syndrome. *Neurology* 46, 469–71.

Gallimberti L., Ferri M., Ferrara S. D., Fadda F. and Gessa G. L. (1992) Gamma-hydroxybutyrate acid for treatment of alcohol dependence: a double-blind study. *Alcoholism: Clin. Exp. Res.* 16, 673–76.

Galloway G. P., Frederick S. L., Staggers F. E. Jr, Gonzales M., Stalcup S. A. and Smith D. E. (1997) Gamma-hydroxybutyrate: an emerging drug of abuse that causes physical dependence. *Addiction* 92, 89–96.

Gobaille S., Hechler V., Andriamampandry C., Kemmel V. and Maitre M. (1999) γ-hydroxybutyrate modulates synthesis and extracellular concentration of γ-aminobutyric acid in discrete rat brain regions *in vivo. J. Pharmacol. Exp. Ther.* 290, 303–09.

Godbout R., Jelenic P., Labrie C., Schmitt M. and Bourguignon J.-J. (1995) Effect of gamma-hydroxybutyrate and its antagonist NCS-382 on spontaneous cell firing in the prefrontal cortex of the rat. *Brain Res.* 673, 157–60.

Godschalk M., Dzoljic M. R. and Bonta I. L. (1976) Antagonism of gamma-hydroxybutyrate-induced hypersynchronization in the ECoG of rat by anti petit mal drugs. *Neurosci. Lett.* 3, 1173–78.

Godschalk M., Dzoljic M. R. and Bonta I. L. (1977) Slow wave sleep and a state resembling absence epilepsy induced in the rat by γ-hydroxybutyrate. *Eur. J. Pharmacol.* 44, 105–11.

Guyon A. and Leresche N. (1995) Modulation by different GABA$_B$ receptor types of voltage-activated calcium currents in rat thalamocortical neurones. *J. Physiol.* 485, 29–42.

Harris N. C., Webb C. and Greenfield S. A. (1989) The effects of gamma-hydroxybutyrate on the membrane properties of guinea-pig pars compacta neurons in the substantia nigra *in vitro*. *Neuroscience* 31, 363–70.

Haüsser M., Stuart G., Racca C. and Sakmann B. (1995) Axonal initiation and active dendritic propagation of action potentials in substantia nigra neurons. *Neuron* 15, 637–47.

Hechler V., Schmitt M., Bourguignon J. J. and Maitre M. (1990) trans-Hydroxycrotonic acid binding sites in brain: evidence for a subpopulation of γ-hydroxybutyrate sites. *Neurosci. Lett.* 110, 204–09.

Hechler V., Gobaille S. and Maitre M. (1992) Selective distribution pattern of gamma-hydroxybutyrate receptors in rat forebrain and midbrain as revealed by quantitative autoradiography. *Brain Res.* 572, 345–48.

Hechler V., Peter P., Gobaille S., Bourguignon J. J., Schmitt M., Ehrhardt J. D., Mark J. and Maitre M. (1993) Gamma-hydroxybutyrate ligands possess antidopaminergic and neuroleptic-like activities. *J. Pharmacol. Exp. Ther.* 264, 1406–14.

Hechler V., Ratomponirina C. and Maitre M. (1997) γ-hydroxybutyrate conversion into GABA induces displacement of GABA$_B$ binding that is blocked by valproate and ethosuximide. *J. Pharmacol. Exp. Ther.* 281, 753–60.

Hédou G., Chasserot-Golaz S., Kemmel V., Gobaile S., Roussel G., Artault J.-C., Andriamampandry C., Aunis D. and Maitre M. (2000) Immunohistochemical studies of the localization of neurons containing the enzyme that synthesizes dopamine, GABA, or γ-hydroxybutyrate in the rat substantia nigra and striatum. *J. Comp. Neurol.* 426, 549–60.

Helrich M., McAslan T. and Skolnik S. (1964) Correlation of blood levels of 4-hydroxybutyrate with state of consciousness. *Anaesthesiology* 25, 771–75.

Hoes M., Vree T. and Guelen P. (1980) Gammahydroxybutyric acid as hypnotic. Clinical and pharmacological evaluation of gammahydroxybutyric acid as hypnotic in man. *Encephale* 69, 93–99.

Hosford D. W., Clark S., Cao Z., Wilson W. A. Jr, Lin F.-H., Morrisett R. A. and Huin A. (1992) The role of GABA$_B$ receptor activation in absence seizures of lethargic (*Ih/Ih*) mice. *Science* 257, 398–401.

Hosford D. A., Lin F.-H., Kraemer D. L., Cao Z., Wang Y. and Wilson J. T. Jr (1995) Neural network of structures in which GABA$_B$ receptors regulate absence seizures in the lethargic (*Ih/Ih*) mouse model. *J. Neurosci.* 15 (11), 7367–76.

Hösli L., Hösli E., Lehmann R., Schneider J. and Borner M. (1983) Action of γ-hydroxybutyrate and GABA on neurones of cultured rat central nervous system. *Neurosci. Lett.* 37, 257–60.

Hu R. Q., Banerjee P. K. and Snead O. C. (2000) Regulation of γ-aminobutyric acid (GABA) release in cerebral cortex in the γ-hydroxybutyric acid (GHB) model of absence seizures in rat. *Neuropharmacology* 39, 427–39.

Jouvet M., Cier A., Mounier D. and Valatx J. L. (1961) Effets du 4-hydroxybutyrate de sodium sur l'E.E.G. et le comportement du Chat. *C. R. Soc. Biol.* 6, 1313–16.

Juhasz G., Emri Z., Kékesi K. A., Salfay O. and Crunelli V. (1994) Blockade of thalamic GABA$_B$ receptors decreases EEG synchronization. *Neurosci. Lett.* 172, 155–58.

Kaufman E. E. and Nelson T. (1991) An overview of γ-hydroxybutyrate catabolism: the role of the cytosolic NAPD+-dependent oxidoreductase EC 1.1.1.19 and of a mitochondrial hydroxyacid–oxoacid transhydrogenase in the initial, rate-limiting step in this pathway. *Neurochem. Res.* 16, 965–74.

Kaupmann K., Huggel K., Heid J., Flo P. J., Bischoff S., Mickel S. J., McVaster G., Angst C., Bittiger H., Froestl W. and Bettler B. (1997) Expression cloning of GABA$_B$ receptors uncovers similarity to metabotropic glutamate receptors. *Nature* 386, 239–46.

Kemmel V., Taleb O., Perard A., Andriamampandry C., Siffert J. C., Mark J. and Maitre M. (1998) Neurochemical and electrophysiological evidence for the existence of a functional γ-hydroxybutyrate system in NCB-20 neurons. *Neuroscience* 86, 989–1000.

King M. A., Thinschmidt J. S. and Walker D. W. (1997) γ-Hydroxybutyrate (GHB) receptor ligand effects on evoked synaptic field potentials in CA1 of the rat hippocampal slice. *J. Neural Transm.* 104, 1177–93.

Kleinschmidt S., Grundmann U., Knocke T., Silomon M., Bach F. and Larsen R. (1998) Total intravenous anaesthesia with gamma-hydroxybutyrate (GHB) and sufentanil in patients undergoing coronary artery bypass graft surgery: a comparison in patients with unimpaired and impaired left ventricular function. *Eur. J. Anaesthesiol.* 15 (5), 559–64.

Knott C., Maguire J. J. and Bowery N. G. (1993) Age-related regional sensitivity to pertussin toxin-mediated reduction in GABA$_B$ receptor binding in rat brain. *Mol. Brain Res.* 18, 353–57.

Kozhechkin S. X. (1980) Microiontophoretic study of the mechanism of action of gamma-hydroxybutyric acid. *Bull. Exp. Biol. Med.* 88, 1293–96.

Laborit H., Jovany J., Gerard J. and Fabiani F. (1960) Résumé d'une étude experimentale et clinique sur un substrat métabolique à action centrale inhibitrice; le 4-hydroxy-butyrate de Na. *Presse Med.* 50, 1867–69.

Lapierre O., Montplaisir M., Lamarre M. and Bedard, M. A. (1990) The effect of gamma-hydroxybutyrate on nocturnal and diurnal sleep of normal subjects: further considerations on REM sleep-triggering mechanisms. *Sleep* 13, 24–30.

Le Feuvre Y., Fricker D. and Leresche N. (1997) GABA$_A$ receptor-mediated IPSCs in rat thalamic sensory nuclei: patterns of discharge and tonic modulation by GABA$_B$ autoreceptors. *J. Physiol.* 502 (1), 91–104.

Leresche N., Lightowler S., Soltesz I., Jassik-Gerschenfield D. and Crunelli V. (1991) Low-frequency oscillatory activities intrinsic to rat and cat thalamocortical cells. *J. Physiol.* 441, 155–74.

Lingenhoehl K., Brom R., Heid J., Beck P., Froestl W., Kaupmann K., Bettler B. and Mosbacher J. (1999) γ-Hydroxybutyrate is a weak agonist at recombinant GABA$_B$ receptors. *Neuropharmacology* 38, 1667–73.

Liu Z., Vergnes M., Depaulis A. and Marescaux C. (1992) Involvement of intrathalamic GABA$_B$ neurotransmission in the control of absence seizures in the rat. *Neuroscience* 48, 87–93.

McCormick D. A. and Pape H.-C. (1990) Properties of a hyperpolarization-activated cation current and its role in rhythmic oscillation in thalamic relay neurones. *J. Physiol.* 431, 291–318.

MacDonald R. L. and Barker J. L. (1979) Anticonvulsant and anesthetic barbiturates: different postsynaptic actions in cultured mammalian neurons. *Neurology* 29, 432–47.

Madden T. E. and Johnson S. W. (1998) Gamma-hydroxybutyrate is a GABA$_B$ receptor agonist that increases a potassium conductance in rat ventral tegmental dopamine neurons. *J. Pharmacol. Exp. Ther.* 287, 261–65.

Maitre M. (1997) The γ-hydroxybutyrate signalling system in brain: organization and functional implications. *Prog. Neurobiol.* 51, 337–61.

Maitre M., Hechler V., Vayer P., Gobaille S., Cash C. D., Schmitt M. and Bourguignon J.-J. (1990) A specific γ-hydroxybutyrate receptor ligand possesses both antagonistic and anticonvulsant properties. *J. Pharm. Exp. Ther.* 255, 657–63.

Maitre M., Rumigny J. F., Benavides J., Bourguignon J. J. Wermuth C. G., Cash C. and Mandel P. (1983a) High affinity binding site for gamma-hydroxybutyric acid in rat brain. *Adv. Biochem. Psychopharmacol.* 37, 441–53.

Maitre M., Rumigny J. F., Cast C. and Mandel P. (1983b) Subcellular distribution of γ-hydroxybutyrate binding sites in rat brain. Principal localization in the synaptosomal fraction. *Biochem. Biophys. Res. Commun.* 110, 262–65.

Maitre M., Kemmel V., Andriamampandry C., Gobaille S. and Aunis D. (2002) The role of γ-hydroxybutyrate in brain function. In *Gamma-Hydroxybutyrate: Molecular, Functional and Clinical Aspects.* (Tunnicliff G. and Cash C. D., eds), pp. 236–47. Taylor and Francis, London.

Mamelak M. (1989) Gammahydroxybutyrate: an endogenous regulator of energy metabolism. *Neurosci. Biobehav. Rev.* 13, 187–98.

Mamelak M. and Webster P. (1981) Treatment of narcolepsy and sleep apnea with gamma-hydroxybutyrate: a clinical and polysomnographic case study. *Sleep* 4, 105–11.

Mamelak M., Scharf M. B. and Wood M. (1986) Treatment of narcolepsy with gamma-hydroxybutyrate. A review of clinical and sleep laboratory findings. *Sleep* 9, 285–89.

Mathivet P., Bernasconi R., De Barry J., Marescaux C. and Bittiger H. (1997) Binding characteristics of γ-hydroxybutyric acid as a weak but selective $GABA_B$ receptor agonist. *Eur. J. Pharmacol.* 321, 67–75.

Metcalf D. R., Emde R. N. and Stripe J. T. (1966) An EEG-behavioral study of sodium hydroxybutyrate in humans. *Electroencephalogr. Clin. Neurophysiol.* 20, 506–12.

Möhler H., Patel A. J. and Balazs R. (1976) Gamma-hydroxybutyrate degradation in the brain *in vivo*: negligible direct conversion to GABA. *J. Neurochem.* 27, 253–58.

Moruzzi G. (1972) The sleep-walking cycle. *Ergebn. Phusiol.* 64, 1–165.

Nestler E. J. (2001) Molecular basis of long-term plasticity underlying addiction. *Nature Rev. Neurosci.* 2, 119–28.

Niedermeyer E. (1996) Primary (idiopathic) generalized epilepsy and underlying mechanisms. *Clin. Electroencephalogy* 27, 1–21.

Nishino S. and Mignot E. (1997) Pharmacological aspects of human and canine narcolepsy. *Prog. Neurobiol.* 52 (1), 27–78.

Nissbrandt H., Elverfors A. and Engberg G. (1994) Pharmacologically induced cessation of burst firing in nigral dopamine neurons: significance for the terminal dopamine efflux. *Synapse* 17, 217–24.

Noebels J. L. (1994) Genetic and phenotypic heterogeneity of inherited spike-and-wave epilepsies. In *Idiopathic Generalized Epilepsies* (Malafosse A., Genton P., Hirsch E., Marescaux C., Broglin D. and Bernasconi R., eds), pp. 215–25. John Libbey, London.

Olpe H.-R. and Koella W. P. (1979) Inhibition of nigral and neocortical cells by γ-hydroxybutyrate: a microiontophoretic investigation. *Eur. J. Pharmacol.* 53, 359–64.

Peyron C., Faraco J., Rogers W., Ripley B., Overeem S., Charnay Y., Nevsimalova S., Aldrich M., Reynolds D., Albin R., Li R., Hungs M., Pedrazzoli M., Padigaru M. *et al.* (2000) A mutation in a case of early onset narcolepsy and a generalized absence of hypocretin peptides in human narcoleptic brains. *Nature Med.* 6, 991–97.

Pinault D., Leresche N., Charpier S., Deniau J.-M., Marescaux C., Vergnes M. and Crunelli V. (1998) Intracellular recordings in thalamic neurones during spontaneous spike and wave discharges in rats with absence epilepsy. *J. Physiol.* 509, 449–56.

Ping H. X. and Shepard P. D. (1999) Blockade of SK-type Ca^{2+}-activated K^+ channels uncovers a Ca^{2+}-dependent slow afterdepolarization in nigral dopamine neurons. *J. Neurophysiol.* 81, 977–84.

Poldrugo F. and Addolorato G. (1999) The role of γ-hydroxybutyric acid in the treatment of alcoholism: from animal to clinical studies. *Alcohol Alcoholism* 34, 15–24.

Ratomponirina C., Hodé Y., Hechler V. and Maitre M. (1995) γ-Hydroxybutyrate receptor binding in rat brain is inhibited by guanyl nucleotides and pertussis toxin. *Neurosci. Lett.* 189, 51–53.

Richards D. A., Lemos T., Whitton P. S. and Bowery N. G. (1995) Extracellular GABA in the ventrobasal thalamus of rats exhibiting spontaneous absence epilepsy: a microdialysis study. *J. Neurochem.* 65 (4), 1674–80.

Roth R. H., Walters J. R. and Aghajanian G. K. (1973) Effect of impulse flow on the release and synthesis of dopamine in the rat striatum. In *Frontiers in Catecholamine Research*, pp. 567–74.

Schmidt C., Gobaille S., Hechler V., Schmitt M., Bourguignon J. J. and Maitre M. (1991) Anti-sedative and anti-cataleptic properties of NCS-382, a γ-hydroxybutyrate receptor antagonist. *Eur. J. Pharmacol.* 203, 393–97.

Schmidt-Mutter C., Muller C., Zwiller J., Gobaille S. and Maitre M. (1999) Gamma-hydroxybutyrate and cocaine administration increases mRNA expression of dopamine D_1 and D_2 receptors in rat brain. *Neuropsychopharmacology* 21, 662–69.

Scrima L., Hartman P. G., Johnson F. H. Jr, Thomas E. E. and Hiller C. (1990) The effects of γ-hydroxybutyrate on the sleep of narcolepsy patients: a double blind study. *Sleep* 13, 479–90.

Snead O. C. (1978) Gamma-hydroxybutyrate in the monkey. I. Electroencephalographic, behavioural and pharmacokinetic studies. *Neurology* 28, 636–42.

Snead O. C. (1991a) The γ-hydroxybutyrate model of absence seizures: correlation of regional brain levels of γ-hydroxybutyric acid and γ-butyrolactone with spike wave discharges. *Neuropharmacology* 30, 161–67.

Snead O. C. (1991b) Effect of γ-crotonolactone in experimental models of generalized absence seizures. *Brain Res.* 543, 10–14.

Snead O. C. (1992) Evidence for GABA$_B$-mediated mechanisms in experimental generalized absence seizures. *Eur. J. Pharmacol.* 213, 343–49.

Snead O. C. (1994) The ontogeny of [^{3}H]γ-hydroxybutyrate and [^{3}H]GABA$_B$ binding sites: relation to the development of experimental absence seizures. *Brain Res.* 659, 147–56.

Snead O. C. (1995) Basic mechanisms of generalized absence seizures. *Ann. Neurol.* 37, 146–57.

Snead O. C. (1996) Antiabsense seizure activity of specific GABA$_B$ and γ-hydroxybutyric acid receptor antagonists. *Pharmacol. Biochem. Behav.* 53, 73–79.

Snead O. C. (2000) Evidence for a G protein-coupled γ-hydroxybutyric acid receptor. *J. Neurochem.* 75, 1986–96.

Snead O. C., Bearden L. J. and Pegram V. (1980) Effect of acute and chronic anticonvulsant administration on endogenous γ-hydroxybutyrate in rat brain. *Neuropharmacology* 19, 47–52.

Snead O. C. and Morley B. J. (1981) Ontogeny of γ-hydroxybutyric acid. I. Regional concentration in developing rat, monkey and human brain. *Dev. Brain Res.* 1, 579–89.

Snead O. C., Yu R. K. and Huttenlocker P. R. (1976) Gamma hydroxybutyrate: correlation of serum and cerebrospinal fluid levels with electroencephalographic and behavioral effects. *Neurology.* 26, 51–56.

Snead O. C., Hechler V., Vergnes M., Marescaux C. and Maitre M. (1990) Increased γ-hydroxybutyric acid receptors in thalamus of a genetic animal model of petit mal epilepsy. *Epilepsy Res.* 7, 121–28.

Snead O. C. (2002) γ-Hydroxybutyrate and absence seizure activity. In *Gamma-Hydroxybutyrate: Molecular, Functional and Clinical Aspects* (Tunnicliff G. and Cash C. D., eds), pp. 150–68. Taylor and Francis, London.

Soltesz I., Lightowler S., Leresche N. and Crunelli V. (1989) On the properties and origin of the GABA$_B$ inhibitory postsynaptic potential recorded in morphologically identified projection cells of the cat dorsal lateral geniculate nucleus. *Neuroscience* 33, 23–33.

Solway J. and Sadove M. (1965) 4-Hydroxybutyrate. A clinical study. *Anesth. Analg.* 44, 532–41.

Spreafico R., Mennini T., Danober L., Cagnotto A., Regondi M. C., Miari A., De Blas A., Vergnes M. and Avanzini G. (1993) GABA$_A$ receptor impairment in the genetic absence epilepsy rats from Strasbourg (GAERS): an immunocytochemical and receptor binding autoradiographic study. *Epilepsy Res.* 15, 229–38.

Stell I. M. and Ryan J. M. (1996) γ-Hydroxybutyrate is a new recreational drug that may lead to loss of consciousness. *Br. Med J.* 313, 424.

Steriade M. and Contreras D. (1995) Relations between cortical and thalamic cellular events during transition from sleep patterns to paroxysmal activity. *J. Neurosci.* 15, 623–42.

Steriade M., Contreras D. and Amzica F. (1994) Synchronized sleep oscillations and their paroxysmal developments. *Trends Neurosci.* 17, 199–208.

Steriade M., Curró Dossi R. and Nuñez A. (1991) Network modulation of a slow intrinsic oscillation of cat thalamocortical neurons implicated in sleep delta waves: cortically induced synchronization and brainstem cholinergic suppression. *J. Neurosci.* 11, 3200–17.

Steriade M. and Llinas R. (1988) The functional state of the thalamus and the associated neuronal interplay. *Physiol. Rev.* 68, 649–742.

Sutch R. J., Davis C. C. and Bowery N. G. (1999) GABA release and uptake measured in crude synaptosomes from Genetic Absence Epilepsy Rats from Strasbourg (GAERS). *Neurochem. Int.* 34 (5), 415–25.

Thannickal T. C., Moore R. Y., Nienhuis R., Ramanathan L., Gulyani S., Aldrich M., Cornford M. and Siegel J. M. (2000) Reduced number of hypocretin neurons in human narcolepsy. *Neuron* 27, 469–74.

Tunnicliff, G. (1997) Sites of action of gamma-hydroxybutyrate (GHB) – a neuroactive drug with abuse potential. *Clin. Toxicol.* 35, 581–90.

Turgeon S. M. and Albin R. L. (1993) Pharmacology, distribution, cellular localization and development of $GABA_B$ binding in rodent cerebellum. *Neuroscience* 55, 311–23.

Turner J. P., Anderson C. M., Williams S. R. and Crunelli V. (1997) Morphology and membrane properties of neurones in the cat ventrobasal thalamus *in vitro*. *J. Physiol.* 505, 707–26.

Van Cauter E., Plat L., Scharf M. B., Leproult R., Cespedes S., L'Hermite-Baleriaux M. and Copinschi G. (1997) Simultaneous stimulation of slow-wave sleep and growth hormone secretion by gamma-hydroxybutyrate in normal young men. *J. Clin. Invest.* 100 (3), 745–53.

Vayer P., Mandel P. and Maitre M. (1985) Conversion of γ-hydroxybutyrate to γ-aminobutyrate *in vivo*. *J. Neurochem.* 45, 810–14.

Vergnes M. and Marescaux C. (1994) Pathophysiological mechanisms underlying genetic absence epilepsy in rats. In *Idiopathic Generalized Epilepsies* (Malafosse A., Genton P., Hirsch E., Marescaux C., Broglin D. and Bernasconi R., eds), pp. 151–68. John Libbey, London.

Vickers M. D. (1969) γ-Hydroxybutryic acid. *Intern. Anesthesiol. Clin.* 7, 75.

Williams D. A. (1953) A study of thalamic and cortical rhythms in petit mal. *Brain* 76, 56–69.

Williams S. R., Turner J. P. and Crunelli V. (1995) Gamma-hydroxybutyrate promotes oscillatory activity of rat and cat thalamocortical neurones by a tonic $GABA_B$ receptor-mediated hyperpolarization. *Neuroscience* 66, 133–41.

Wu L.-G. and Saggau P. (1995) $GABA_B$ receptor-mediated presynaptic inhibition in guinea-pig hippocampus is caused by reduction of presynaptic Ca^{2+} influx. *J. Physiol.* 485.3, 649–57.

Xie X. and Smart T. G. (1992a) γ-Hydroxybutyrate hyperpolarizes hippocampal neurones by activating $GABA_B$ receptors. *Eur. J. Pharmacol.* 212, 291–94.

Xie X. and Smart T. G. (1992b) γ-Hydroxybutyrate depresses monosynaptic excitatory and inhibitory postsynaptic potentials in rat hippocampal slices. *Eur. J. Pharmacol.* 223, 193–96.

6 Regulation of central dopamine by γ-hydroxybutyrate

S. G. Howard and P. K. Banerjee

Introduction

γ-Hydroxybutyrate (GHB) is a naturally occurring compound in the brain which exhibits distinct neurophysiological (Maitre 1997) and neuropharmacological actions (Howard and Feigenbaum 1997). The mechanism of action of GHB is of interest not only as a pharmacological tool, but also because it is a drug of abuse, and its notoriety is increasing. By many accounts GHB has been used recreationally and has been associated with drug-facilitated sexual assault (Schwartz *et al.* 2000; Galloway *et al.* 2000). This use of the drug amongst teenagers and young adults is not only a cause for concern because of what the drug is being used for, but also because of the possibility of an accidental overdose when combined with alcohol or barbiturates (Ledray 1996; *Morbidity and Mortality Weekly Report* 1997).

GHB is a four-carbon fatty acid derivative which has been used for many years as an adjuvant to anesthetic treatment (Blumenfeld *et al.* 1962; Solway and Sadove 1965). It has a close structural resemblance to γ-aminobutyric acid (GABA), a major inhibitory neurotransmitter, and it has been considered both a precursor and a metabolite of GABA, as well as a GABA agonist. Over the past ten years, it has become apparent that not only does GHB possess neuromodulating properties, but also that considerably more research is necessary if we are to understand the mechanism of action of this drug. A unique property of GHB is that systemic administration can increase the concentration of GHB in the brain by 100 fold, a property not normally associated with any compound that possesses signaling capacity (Maitre 1997). Certain of the central actions of GHB might be explainable with reference to its effects on central dopaminergic neurons (Feigenbaum and Howard 1996a; Hechler *et al.* 1993; Menon *et al.* 1976). This review explores the evidence.

Distribution of GHB in the central nervous system

While GHB appears to be found in most brain regions, Snead (1991) has demonstrated that there is a variation between regions and nuclei in the brain. Levels of GHB appear to be highest in the substantia nigra and the ventral tegmental area. These regions comprise the area of the brain richest in dopamine (DA) cell bodies (Dahlstrom and Fuxe 1964). This is consistent with the hypothesis that GHB regulates central dopaminergic activity (Maitre 1997). In the developing brain, however, the picture is not as clear. During development, the highest concentrations of GHB are found in the hypothalamus and cortex. High-affinity binding sites do not appear until postnatal day 17, and then they appear first in the hippocampus, followed on postnatal day 18 in the frontal cortex (Snead 1994). One can only assume that if GHB has any regulatory role with respect to dopamine, then it is not so in the developing brain, since dopamine and the dopamine receptors are present early in gestation and reach adult levels by the third postnatal week (Dobbing and Sands 1979).

Benavides *et al.* (1982) demonstrated that GHB has an affinity for specific binding sites in the brain. Additional evidence supports the idea of receptors specific for GHB (Maitre 1997), and we now believe these receptors are linked to G protein activation (Snead 2000). Using a series of lesions and chemical denervation, Hechler *et al.* (1989) demonstrated that GHB receptors were not located on dopamine terminals, but rather on intrinsic neurons in the caudate nucleus and in the hippocampus. In the neostriatum, the interneurons are either cholinergic (Butcher and Butcher 1974) or GABAergic (Kawaguchi *et al.* 1995). It was later determined that the GABAergic interneurons also expressed immunoreactivity for enkephalin (Kawaguchi *et al.* 1995). This finding provides some insight into our data demonstrating that naloxone blocks the effect of GHB on dopamine release (Howard and Feigenbaum 1997).

Effect of GHB on levels and synthesis of dopamine

Dopamine acts as a neurotransmitter in certain well-defined pathways (Molinoff and Axelrod 1971; Dahlstrom and Fuxe 1964). Like several other drugs of abuse, GHB interferes with dopaminergic neuronal activity. More than thirty years ago, Gessa *et al.* (1966, 1968) examined the effect of GHB on dopamine in the brain and observed increases in the concentration of this monoamine. This was confirmed by Walters and Roth (1972). Since then, a great deal of work has been done to study the action of GHB on dopamine synthesis and release, and to determine the mechanism of action of GHB in this regard. The first series of investigations described the effects of GHB on the levels and synthesis of dopamine (Anden and Stock 1973; Kehr *et al.* 1972; Anden *et al.* 1971, 1973). These studies were followed by histofluorescent analysis of the different catecholamine systems (Arluison *et al.* 1982), and it was concluded that the dopamine fibers appeared swollen, with an abnormal appearance, suggesting an altered metabolism from the effects of GHB. The increases in dopamine after GHB could be reduced or prevented by treatment with d-amphetamine (Anden *et al.* 1973; Roth *et al.* 1973). These increases in brain dopamine were not only reversed by d-amphetamine, but also by other sympathomimetics (Roth *et al.* 1973) and by K^+ ions (Stock *et al.* 1973). Taken together, all of the early studies demonstrated an inhibition of dopamine release; however, the mechanism through which this inhibition occurred was not yet clear.

Effect of GHB on the release of dopamine

The early work on GHB was followed by studies to determine the mechanism of action of GHB, as well as the effect that GHB has on other transmitter systems. In fact, these experiments have shown that GHB evokes a variety of effects on various neurotransmitter systems, especially that of dopamine. The studies demonstrated that determining the mechanism of action of GHB would not be easy and some controversy resulted owing to differing experimental designs, conditions and controls. In particular, controversy arose when push–pull cannula techniques or *in vivo* dialysis techniques were employed. Cheremy *et al.* (1977) were one of the first groups to report an increase in dopamine release in the cat caudate nucleus. Maitre *et al.* (1990); Hechler *et al.* (1991) and Gobaille *et al.* (1994), using *in vivo* microdialysis, also reported a brief increase in dopamine release following GHB administration. In an attempt to resolve the discrepancy between the early work and the more recent work of Maitre and co-workers, two papers – one research (Howard and Feigenbaum 1997) and one review (Feigenbaum and Howard 1996a) – examined the data and possible differences in procedures and experimental design, as well as the resulting pharmacological data.

The results from our *in vivo* dialysis experiments (Howard and Feigenbaum 1997) clearly demonstrated two important points. First, GHB clearly inhibited striatal dopamine release in the awake, freely moving animal, but, depending on the anesthetic used, might not inhibit release in the anesthetized animal. The drug interaction which occurs between an anesthetic and any drug should always be taken into consideration. However, this important interaction was not initially considered in the first microdialysis studies examining the effect of GHB on dopamine release. GHB is itself an anesthetic, and the likelihood is high that an interaction would occur when used in combination with other commonly used anesthetics such as urethane or chloral hydrate. These studies compared the effect of systemically administered GHB on dopamine release in the awake or in the anesthetized animal. A comparison was made between chloral hydrate, which produced an increase in release of dopamine, and urethane, which produced a variable effect in dopamine release after GHB treatment. Several investigators have found that chloral hydrate reverses the effects of dopamine agonists and antagonists on the release of dopamine (Nielson and Moore 1982; Bunney *et al.* 1973; Carlson *et al.* 1987). However, Kelland *et al.* (1989) have shown that chloral hydrate inhibits the dopamine neuronal firing pattern independent of other drugs. These data demonstrated unequivocally that GHB does produce a decrease in the release of dopamine in the awake, unanesthetized animal.

The second point was discovered while trying to determine how the differences found in several other microdialysis studies were obtained. The importance of the concentration of Ca^{2+} in the dialysate was again confirmed in this study (Howard and Feigenbaum 1997). The concentration of Ca^{2+} in the dialysate is critical. Concentrations of Ca^{2+} higher than 2.3 mM have been shown to increase dopamine release in the striatal dopamine system. The effect of Ca^{2+} on dopamine release was not only demonstrated by these authors (Howard and Feigenbaum 1997), but also by numerous other investigators (Westerink *et al.* 1988; Moghaddam and Bunney 1989; DeBoer *et al.* 1990; Timmerman and Westerink 1991). These studies revealed that by altering the concentration of Ca^{2+}, release of dopamine (and other neurotransmitters) could be increased or decreased. The recent controversy regarding the inhibitory versus stimulatory effect of GHB on dopamine release in the *in vivo* dialysis experiments may have been resolved by taking into account the concentration of Ca^{2+} in the dialysate. GHB did produce a decrease in release of dopamine in the awake animal when the concentration of Ca^{2+} in the dialysate was 2.3 mM or less. However, GHB could produce varying effects when combined with other anesthetics, such as urethane or chloral hydrate, and these are reviewed in the article by Howard and Feigenbaum (1997).

While it has long been accepted that GHB decreases the release of dopamine (for a review, see Feigenbaum and Howard (1996)), the significance or functional relevance of this GHB-mediated decrease in dopamine release is yet to be elucidated. Since it is generally believed that the rewarding or the reinforcing properties of the psychotropic drugs are mediated by increased dopamine release, the potentially rewarding properties of GHB must be mediated by a non-dopaminergic, possibly a GABAergic, mechanism (as in the case of reinforcing properties of general sedative-hypnotics which do not alter dopamine release). GHB decreases ethanol and opiate withdrawal syndromes in humans (Gallimberti *et al.* 1992, 1993) and thus has a potential role in alleviating ethanol withdrawal symptoms. This could well be dependent on the effect of GHB on the GABAergic system, since GHB is known to decrease GABA release in many brain areas (Banerjee and Snead 1995; Hu *et al.* 2000). Ethanol is also known to act on $GABA_A$ receptors by modulating brain neurosteroid levels (Van Doren *et al.* 2000), while $GABA_A$ active neurosteroids have been shown to possess affinity for GHB binding sites in the brain (Banerjee *et al.* 1998).

Apart from its ability to reduce dopamine release, GHB also possesses several anti-dopaminergic (neuroleptic) activities. For example, GHB produces short-acting catalepsy in both rats and mice, which is reversed by the specific GHB receptor antagonist NCS-382 (Navarro *et al.* 1998; Ratomponirina *et al.* 1998). In addition, GHB produces hyperthermia and effectively blocks apomorphine-induced hypothermia and stereotyped behavior (Hechler *et al.* 1993). These responses to GHB along with its effect on dopamine release may prove to be of interest in drug development.

Mechanisms through which GHB alters neurochemical transmission

Despite the fact that GHB is, in part, metabolically derived from GABA, and is structurally similar to GABA, it fails to compete for $GABA_A$ sites. Moreover, $GABA_A$ ligands do not show appreciable affinity for the GHB binding sites, with the exception of neuroactive steroids which have been shown to possess low affinity for GHB binding sites (Banerjee *et al.* 1998).

Though the evidence supports a lack of effect on $GABA_A$ mechanisms, several recent studies have suggested that GHB may act as a 'selective' partial agonist at central $GABA_B$ receptors. For example, GHB decreases GABA release in many rat brain regions in a manner similar to that of (−)-baclofen, which is equally effective in reducing glutamate release. GHB does not modulate glutamate release, suggesting that GHB lacks affinity for the pre-synaptic $GABA_B$ receptors present on glutamatergic terminals (Banerjee and Snead 1995; Snead *et al.* 2000).

The most obvious candidate to mediate the effects of GHB on dopamine neurons would seem to be the GHB receptor. Indeed, the GHB receptor antagonist NCS-382 has been reported to prevent increases in striatal dopamine following GHB treatment (Maitre *et al.* 1990). However, this too may prove controversial. If GHB is acting solely as a GHB receptor agonist, then the postsynaptic effects of GHB should be blocked or at least attenuated by the NCS-382, but they are not (Madden and Johnson 1998; Williams *et al.* 1995; Xie and Smart 1992). Indeed, several studies indicate that activation of $GABA_B$ receptors by GHB actually is responsible for its action on dopaminergic neurotransmission.

Not long ago, a report by Madden and Johnson (1998) concluded that a mechanism through which GHB produced its inhibitory actions on dopaminergic activity was one involving K^+-dependent membrane hyperpolarization, after the activation of $GABA_B$ receptors. This mechanism would in fact agree with much of the data found in the literature. As an example, Engberg and Nissbrandt (1993) observed that $GABA_B$ receptors mediated the GHB-induced inhibition of dopamine neuronal activity and that $GABA_B$ antagonists reduced the effect. Consistent with these results are the findings of Godbout *et al.* (1995) which derived from measuring the spontaneous firing rate of prefrontal cortical neurons. The authors demonstrated that NCS-382 blocked only the excitation produced by low doses of GHB, while having no effect on high doses of GHB. Similarly, a study by Lingenhoehl *et al.* (1999) demonstrated that in recombinant $GABA_B R1/R2$ receptors, expressed in *Xenopus oocytes*, GHB only demonstrated weak agonistic properties, which were neither potentiated by (−)-baclofen nor blocked by NCS-382. Moreover, several radioligand binding studies have reported that (−)-baclofen does not displace [^{3}H]GHB from high-affinity GHB binding sites, certainly a requirement for a potential agonist. More recently, however, Lorente *et al.* (2000) have found that GABA$_B$ receptor subtypes respond to GHB and the effect can be mimicked by baclofen and blocked by GABA$_B$ antagonists.

Let us consider all of the data in a slightly different way. If we assume that the GHB receptors are located on the intrinsic neurons of the neostriatum and hippocampus as the data of Hechler *et al.* (1999) suggest, and if we also accept that these interneurons express both GABA and enkephalin as demonstrated by Kawaguchi *et al.* (1995), then stimulation of these GHB receptors by GHB might produce several different responses, for example:

1   GHB acting at its receptor, or at an isoform of the GABA$_B$ receptor, might inhibit dopamine release by producing the K^+-dependent membrane hyperpolarization reported by Madden and Johnson (1998), or perhaps GHB could decrease cell excitability by inducing hyperpolarization through activation of inwardly rectifying K^+ conductance, an effect mimicked by baclofen (Lorente *et al.* 2000).

2   The levels of dopamine would increase, consistent with hyperpolarization of the membrane and existing reports (for a review, see Roth *et al.* 1980 and Feigenbaum and Howard 1996a).

3   Dopamine-dependent behaviors would be interrupted (for a review, see Feigenbaum and Howard 1996b).

4   Naloxone acting on the same neuron (which expresses both GABA and enkephalin) could reverse the effect of GHB, consistent with the results of Howard and Feigenbaum (1997) and Feigenbaum and Howard (1996a,b), thus altering the effect of GHB on dopamine release.

5   The D2 receptor 'knockout' animals would demonstrate no opiate reward response, consistent with the data of Maldonado *et al.* (1997).

6   Amphetamine, or other dopaminergic releasing drugs, would be able to reverse the effect of GHB on dopamine release, as reported by numerous investigators (e.g. Roth *et al.* 1980).

If all of the evidence is taken together, what mechanism could we hypothesize to explain how GHB exerts its effect by altering the response of several transmitter systems, not just dopamine? The data is consistent with one compelling hypothesis, which would suggest a complex mechanism of action. GHB stimulates GHB receptors, or an isoform of the GABA$_B$ receptor, producing a decrease in the release of GABA. The resultant decrease in extracellular GABA would produce an inhibition of dopamine release by hyperpolarization of K^+-dependent membranes. Since the GHB receptors are on interneurons which contain both GABA and enkephalin, the interaction of these transmitters will also play a role in the final outcome. Recent data by Lorente *et al.* (2000) have demonstrated that GABA$_B$ receptors will form heteromeric complexes from GABA$_B$R1 and GABA$_B$R2 subtypes. These assemblies of GABA$_B$R1 and GABA$_B$R2 are activated by GHB, producing an inwardly rectifying K^+ current. This effect is mimicked by baclofen and is inhibited by GABA$_B$ antagonists. These data would suggest that GHB is not acting at a specific GABA$_B$ receptor, but instead at a site near the receptor, or at a complex of subtypes of the GABA$_B$ receptor. If GHB is acting at a complex of subtypes of the GABA$_B$ receptor, then the sites are separate from one another, yet they are linked to each other pre- and/or postsynaptically. This hypothesis would then be consistent with the above data, and suggests a new, albeit complex mechanism of action for GHB. Consequences of GHB action on dopaminergic neurons can be summarized as:

$$\text{GHB} \;\rightarrow\; \begin{array}{c}\text{K}^+\text{-dependent}\\ \text{hyperpolarization}\end{array} \;\rightarrow\; \begin{array}{c}\text{Inhibition of}\\ \text{impulse flow}\end{array} \;\rightarrow\; \begin{array}{c}\text{Decrease in DA}\\ \text{release}\end{array}$$

This produces:

$$\begin{array}{c}\text{Increase}\\ \text{in intra-}\\ \text{neuronal DA}\end{array} \rightarrow \begin{array}{c}\text{Upregulation}\\ \text{of DA}\\ \text{receptors}\end{array} \rightarrow \begin{array}{c}\text{Increase}\\ \text{in DA}\\ \text{synthesis}\end{array} \rightarrow \begin{array}{c}\text{Altered GABA}\\ \text{and opioid}\\ \text{response}\end{array} \rightarrow \begin{array}{c}\text{Altered}\\ \text{behavior}\end{array}$$

Conclusion

While a potential mechanism of action for GHB has been hypothesized above, until further research has been done, great caution should be exercised in considering this compound as a substitute for another drug of abuse such as cocaine (Schmidt-Mutter *et al.* 1999), alcohol abuse (Gallimberti *et al.* 1992), opiate withdrawal (Gallimberti *et al.* 1993) or in any other clinical situation. Such cautions have been voiced in the past by Roth *et al.* (1980).

References

Anden N.-E. and Stock G. (1973) Inhibitory effect of gamma hydroxybutyric acid and gamma-aminobutyric acid on the dopamine cells in the substantia nigra. *Naunyn-Schmied. Arch. Pharmacol.* 279, 89–92.

Anden N.-E., Corrodi H., Fuxe K. and Ungerstedt U. (1971) Importance of nervous impulse flow for the neuroleptic induced increase in amine turnover in central dopamine neurons. *Eur. J. Pharmacol.* 15, 193–99.

Anden N.-E., Magnussen T. and Stock G. (1973) Effect of drugs influencing monoamine mechanisms on the increase in brain dopamine produced by axotomy or treatment with gamma hydroxybutyric acid. *Naunyn-Schmied. Arch. Pharmacol.* 15, 193–99.

Arluison M., Javoy-Agid F., Feuerstein C., Tauc M., Conrath-Verrier M. and Mailly P. (1982) Histofluorescence analysis of several systems of catecholaminergic nerve fibres within the rat neostriatum revealed by either restricted lesions of the substantia nigra or gamma-hydroxybutyrate. *Brain Res. Bull.* 9, 355–65.

Banerjee P. K. and Snead O. C. (1995) Presynaptic gamma-hydroxybutyric acid (GHB) and gamma-aminobutyric acid B (GABA$_B$) receptor-mediated release of GABA and glutamate (GLU) in rat thalamic ventrobasal nucleus (VB): a possible mechanism for the generation of absence-like seizures induced by GHB. *J. Pharmacol. Exp. Ther.* 273, 1534–43.

Banerjee P. K., Liu C. C. and Snead O. C. (1998) Steroid-inhibition of [^{3}H] γ-hydroxybutyric acid (GHB) binding in thalamic relay nuclei increases during absence seizures. *Brain Res.* 813, 343–50.

Benavides L., Rumigny J. F., Bourguignon J. J., Cash C. D., Wermuth C. G., Mandel P., Vincendon G. and Maitre M. (1982) High affinity binding site for γ-hydroxybutyric acid in rat brain. *Life Sci.* 30, 953–61.

Blumenfeld M., Suntay R. G. and Harmel M. H. (1962) Sodium gamma-hydroxybutyric acid: a new anesthetic adjuvant. *Anesth. Analg. Curr. Res.* 41, 721–26.

Bunney B. S., Walters J. R. and Aghajanian G. K. (1973) Dopaminergic neurons: effect of antipsychotic drugs and amphetamine on single cell activity. *J. Pharmacol. Exp. Ther.* 185, 560–71.

Butcher S. G. and Butcher L. L. (1974) Origin and modulation of acetylcholine activity in the neostriatum. *Brain Res.* 71, 167–71.

Carlson J. H., Bergstrom D. A., Weick B. G. and Walters J. R. (1987) Neurophysiological investigation of the effects of the D-1 agonist SKF 38393 on the tonic activity of substantia nigra dopamine neurons. *Synapse* 1, 411–16.

Cheramy A., Nieoullon A. and Glowinski J. (1977) Stimulating effects of gamma-hydroxybutyrate on dopamine release from the caudate nucleus and the substantia nigra of the cat. *J. Pharmacol. Exp. Ther.* 203, 283–93.

Dahlstrom A. and Fuxe L. (1964) Evidence for the existence of monoamine-containing neurons in the central nervous system. I. Demonstration of monoamines in the cell bodies of brain stem neurons. *Acta Physiol. Scand.* 62 (Suppl. 232), 1–55.

DeBoer P., Damsma G., Fibiger H. C., Timmerman W., deVries J. B. and Westerink B. H. (1990) Dopaminergic-cholinergic interactions in the striatum: the critical significance of calcium concentrations in brain microdialysis. *Naunyn-Schmied. Arch. Pharmacol.* 342, 528–34.

Dobbing J. and Sands J. (1979) Comparative aspects of the brain growth spurt. *Early Human Dev.* 3, 79–83.

Engberg G. and Nissbrandt H. (1993) Gamma-hydroxybutyric acid (GHBA) induces pacemaker activity and inhibition of substantia nigra dopamine neurons by activating GABA$_B$ receptors. *Naunyn-Schmied. Arch. Pharmacol.* 348, 491–97.

Feigenbaum J. J. and Howard S. G. (1996a) Does gamma-hydroxybutyrate inhibit or stimulate central DA release? *Int. J. Neurosci.* 88, 53–69.

Feigenbaum J. J. and Howard S. G. (1996b) Naloxone reverses the inhibitory effect of gamma-hydroxybutyrate on central DA release *in vivo* in awake animals: a microdialysis study. *Neurosci. Lett.* 218, 55–58.

Gallimberti L., Ferri M., Ferrara S. D., Fadda F. and Gessa G. L. (1992) Gamma-hydroxybutyric acid for treatment of alcohol dependence: a double blind study. *Alcoholism: Clin. Exp. Res.* 16, 77–81.

Gallimberti L., Cibin M., Pagnin P., Sabbion R., Pnai P. P., Priatsu R., Ferrara S. D. and Gessa G. L. (1993) Gamma-hydroxybutyric acid for the treatment of opiate withdrawal syndrome. *Neuropsychopharmacology* 9, 77–81.

Galloway G. P., Frederick-Osborne S. L., Seymour R., Contini S. E. and Smith D. E. (2000) Abuse and therapeutic potential of gamma-hydroxybutyric acid. *Alcohol* 20, 263–69.

Gessa G. L., Vargiu L., Crabai F., Boero G., Caboni F. and Camba R. (1966) Selective increase of brain dopamine induced by gamma-hydroxybutyrate. *Life Sci.* 5, 1921–30.

Gessa G. L., Crabai F., Vargiu L. and Spano P. F. (1968) Selective increase of brain dopamine induced by γ-hydroxybutyrate: study of the mechanism of action. *J. Neurochem.* 15, 377–81.

Gobaille S., Schmidt C., Cupo A., Herbrecht F. and Maitre M. (1994) Characterization of methionine-enkephalin release in the rat striatum by *in vivo* dialysis: effects of gamma-hydroxybutyrate on cellular and extracellular methionine-enkephalin levels. *Neuroscience* 60, 637–48.

Godbout R., Jelenic P., Labrie C., Schmitt M. and Bourguignon J. J. (1995) Effect of gamma-hydroxybutyrate and its antagonist NCS-382 on spontaneous cell firing in the prefrontal cortex of the rat. *Brain Res.* 673, 157–60.

Hechler V., Gobaille S. and Maitre M. (1989) Localization studies of gamma-hydroxybutyrate receptors in rat striatum and hippocampus. *Brain Res. Bull.* 23, 129–35.

Hechler V., Gobaille S., Bourguignon J. J. and Maitre M. (1991) Extracellular events induced by gamma-hydroxybutyrate in striatum: a microdialysis study. *J. Neurochem.* 56, 938–44.

Hechler V., Peter P., Gobaille S., Bourguignon J., Schmitt M., Ehrhardt J., Mark J. and Maitre M. (1993) γ-Hydroxybutyrate ligands possess antidopaminergic and neuroleptic activities. *J. Pharmacol. Exp. Ther.* 264, 1406–14.

Howard S. G. and Feigenbaum J. J. (1997) Effect of gamma-hydroxybutyrate on central dopamine release *in vivo*. A microdialysis study in awake and anesthetized animals. *Biochem. Pharmacol.* 53, 103–10.

Hu R. Q., Banerjee P. K. and Snead O. C. (2000) Regulation of GABA release in the cerebral cortex in the gamma-hydroxybutyrate (GHB) model of absence seizures in rats. *Neuropharmacology* 39, 427–39.

Kawaguchi Y., Wilson C. J., Augood S. J. and Emson P. C. (1995) Striatal interneurons: chemical, physiological and morphological characterization. *Trends Neurosci.* 18, 527–35.

Kehr W., Carlsson A., Lindqvist M., Magnusson T. and Atack C. (1972) Evidence for a receptor-mediated feedback control of striatal tyrosine hydroxylase activity. *J. Pharm. Pharmacol.* 24, 744–47.

Kelland M. D., Freeman A. S. and Chiodo L. A. (1989) Chloral hydrate anaesthesia alters the responsiveness of identified midbrain neurons to dopamine agonist administration. *Synapse* 3, 30–37.

Ledray L. E. (1996) Date rape drug alert. *J. Emergency Nursing* 22, 80.

Lingenhoehl K., Brom R., Heid J., Beck P., Froestl W., Kaupmann K., Bettler B. and Mosbacher J. (1999) γ-Hydroxybutyrate is a weak agonist at recombinant $GABA_B$ receptors. *Neuropharmacology* 38, 1667–73.

Lorente P., Lacampagne A., Pouzeratte Y., Richards S., Malitschek B., Kuhn R., Bettler B. and Vassort G. (2000) γ-Aminobutyric acid type B receptors are expressed and functional in mammalian cardiomyocytes. *Proc. Natl Acad. Sci. USA* 97, 8664–69.

Madden T. E. and Johnson S. W. (1998) Gamma-hydroxybutyrate is a $GABA_B$ receptor agonist that increases a potassium conductance in rat ventral tegmental dopamine neurons. *J. Pharmacol. Exp. Ther.* 287, 261–65.

Maitre M. (1997) The gamma-hydroxybutyrate signalling system in brain: organization and functional implications. *Prog. Neurobiol.* 51, 337–61.

Maitre M., Hechler V., Vayer P., Gobaille S., Cash C. D., Schmitt M. and Bourguignon J. J. (1990) A specific γ-hydroxybutyrate receptor ligand possessing both antagonistic and anticonvulsant properties. *J. Pharmacol. Exp. Ther.* 255, 657–63.

Maldonado R., Saiardi A., Valverde O., Samad T. A., Rogues B. P. and Borrelli W. (1997) Absence of opiate rewarding effects in mice lacking dopamine D2 receptors. *Nature* 388, 586–89.

Mamelak M., Escriu J. M. and Stokan O. (1977) The effects of γ-hydroxybutyrate on sleep. *Biol. Psychiat.* 12, 273–78.

Menon K., Biegei D. and Hornykiewicz O. (1976) Pharmacological evidence for a selective antidopaminergic action of γ-hydroxybutyric acid. *J. Neural Transm.* 39, 177–86.

Moghaddam B. and Bunney B. S. (1989) Ionic composition of microdialysis perfusing solution alters the pharmacological responsiveness and basal outflow of striatal dopamine. *J. Neurochem.* 53, 652–54.

Molinoff P. B. and Axelrod J. (1971) Biochemistry of catecholamines. *Ann. Rev. Biochem.* 40, 465–500.

Morbidity and Mortality Weekly Report (1997) Gamma hydroxy butyrate use – New York and Texas, 1995–1996, Vol. 46, pp. 281–83.

Navarro J. F., Pedraza C., Martin M., Manzaneque J. M., Davila G. and Maldonado E. (1998) Tiapride-induced catalepsy is potentiated by gamma-hydroxybutyric acid administration. *Prog. Neuropsychopharmacol. Biol. Psychiatry* 22, 835–44.

Nielson J. A. and Moore K. E. (1982) Effects of chloral hydrate, gamma butyrolactone, and equithesin on the efflux of dopamine and 5-hydroxytryptamine metabolites into cerebroventricular perfusates of rats. *J. Neurochem.* 39, 235–38.

Ratomponirina C., Gobaille S., Hode Y., Kemmel V. and Maitre M. (1998) Sulpiride, but not haloperidol, up-regulates gamma-hydroxybutyrate receptors *in vivo* and in cultured cells. *Eur. J. Pharmacol.* 346, 331–37.

Roth R. H., Walters J. R. and Aghajanian G. K. (1973) Effect of impulse flow on the release and synthesis of dopamine in rat striatum. In *Frontiers in Catecholamine Research* (Usdin E. and Snyder S., eds), pp. 567–74. Pergamon, New York.

Roth R. H., Doherty J. D. and Walters J. D. (1980) Gamma-hydroxybutyrate: a role in the regulation of central dopaminergic neurons? *Brain Res.* 189, 556–60.

Schmidt-Mutter C., Muller C., Zwiller J., Gobaille S. and Maitre M. (1999) Gamma-hydroxybutyrate and cocaine administration increases mRNA expression of dopamine D1 and D2 receptors in rat brain. *Neuropsychopharmacology* 21, 662–69.

Schwartz R. H., Milteer R. and LeBeau M. A. (2000) Drug-facilitated sexual assault ('date rape'). *S. Med. Journal* 93, 558–61.

Serra M., Sanna E., Foddi C., Concas A. and Biggio G. (1991) Failure of gamma-hydroxybutyrate to alter the function of the $GABA_A$ receptor complex in the rat cerebral cortex. *Psychopharmacology* 104, 351–55.

Snead O. C. (1991) The gamma-hydroxybutyrate model of generalized absence seizures: correlation of regional brain levels of gamma-hydroxybutyric acid and gamma-butyrolactone with spike wave discharges. *Neuropharmacology* 30, 161–67.

Snead O. C. (1994) The ontogeny of [^{3}H]γ-hydroxybutyrate and [^{3}H]GABA$_B$ binding sites: relation to the development of experimental absence seizures. *Brain Res.* 659, 147–56.

Snead O. C. (1996) Relation of the [^{3}H]gamma-hydroxybutyric acid (GHB) binding site to the gamma-aminobutyric acid B (GABA$_B$) receptor in rat brain. *Biochem. Pharmacol.* 52, 1235–43.

Snead O. C. (2000) Evidence for a G protein-coupled γ-hydroxybutyric acid receptor. *J. Neurochem.* 75, 1986–96.

Snead O. C., Banerjee P. K., Burnham M. and Hampson D. (2000) Modulation of absence seizures by the GABA$_A$ receptor: a critical role for metabotropic glutamate receptor 4 (mGluR4). *J. Neurosci.* 20, 6218–24.

Solway J. and Sadove M. S. (1965) 4-Hydroxybutyrate: a clinical study. *Anesth. Analg. Curr. Res.* 44, 532–39.

Stock G., Magnusson T. and Anden N.-E. (1973) Increase in brain dopamine after axotomy or treatment with gamma hydroxybutyric acid due to elimination of the nerve impulse flow. *Naunyn-Schmied. Arch. Pharmacol.* 278, 347–61.

Timmerman W. and Westerink B. H. C. (1991) Importance of the calcium content infused during microdialysis for the effects induced by D2 agonists on the release of dopamine in the striatum of the rat. *Neurosci. Lett.* 131, 93–96.

Van Doren M. J., Matthews D. B., Janis G. C., Grobin A. C., Devaud L. L. and Morrow A. L. (2000) Neuroactive steroid 3 alpha-hydroxy-5 alpha-pregnan-20-one modulate electrophysiological and behavioral actions of ethanol. *J. Neurosci.* 20, 1982–89.

Walters J. R. and Roth R. H. (1972) Effect of γ-hydroxybutyrate on dopamine and dopamine metabolites in the rat striatum. *Biochem. Pharmacol.* 21, 2111–21.

Westerink B. H. C., Hofsteede H. M., Damsma G. and deVries J. B. (1988) The significance of extracellular calcium for the release of dopamine, acetylcholine, and amino acids in conscious rats, evaluated by brain microdialysis. *Naunyn-Schmied. Arch. Pharmacol.* 337, 373–78.

Williams S. R., Turner J. P. and Crunelli V. (1995) Gamma-hydroxybutyrate promotes oscillatory activity of rat and cat thalamocortical neurons by a tonic GABA$_B$ receptor-mediated hyperpolarization. *Neuroscience* 66, 133–41.

Xie X. and Smart T. G. (1992) γ-Hydroxybutyrate hyperpolarizes hippocampal neurones by activating GABA$_B$ receptors. *Eur. J. Pharmacol.* 212, 291–94.

7 Effects of γ-hydroxybutyrate on sleep

R. Godbout and J. Montplaisir

Introduction

Sleep is a physiological state present in all mammals and its regular manifestation is essential to survival (Rechtschaffen *et al.* 1983). Many functions have been attributed to sleep, whether it be at the physiological level (maturation of the CNS, hormonal release, protein synthesis, etc.) or at the cognitive level (learning and memory encoding, affective integration, etc.) (for a review, see Kryger *et al.* (2000)). The administration of γ-hydroxybutyrate (GHB) affects the organization of the sleep–wake cycle and, as a matter of fact, a significant proportion of clinical research involving GHB in the last forty years has been related to its effect on sleep itself. After a brief description of sleep organization, the present chapter will review the essential findings on the effects of GHB on sleep in animals and humans and will attempt to interpret the available literature in a clinically relevant, comprehensive manner.

Sleep

Although differences exist across species, the organization of sleep is basically the same in all mammals, with a predictable succession of sleep stages alternating with waking episodes of variable length. Quantification of sleep is usually achieved by means of at least two electro-physiological measures, namely the electroencephalogram (EEG) and the electromyogram (EMG). While the use of the electro-oculogram (EOG) is compulsory in humans (Rechtschaffen and Kales 1968), it is often replaced in animals by other measures of the visual system such as ponto-geniculo-occipital (PGO) waves (Farber *et al.* 1980; Sakai *et al.* 2001; Vanni-Mercier *et al.* 1994).

Sleep is divided into two distinct phases: slow-wave sleep (SWS) composed of stages 1–4 (in an ascending order of appearance) and paradoxical sleep (Borbély and Neuhaus 1978; Timo-Iaria *et al.* 1970; Rechtschaffen and Kales 1968). SWS can be further broken down into light SWS (or SWS-1: stages 1 and 2) and deep SWS (stages 3 and 4). When the EOG is recorded, SWS is rather referred to as non-rapid eye movement (non-REM) sleep, while paradoxical sleep is referred to as REM sleep.

At sleep onset and during non-REM sleep, the EOG and the EMG are decreased and stable, corresponding well with the prevailing behavioral quiescence. At the EEG level, sleep onset is characterized by a progressive synchronization and slowing of the EEG signal (5–8 Hz). The entry into stage 2 is characterized by the appearance of EEG spindle-like activity (12–14 Hz) while stages 3 and 4 will show clear delta activity (1–4 Hz). The amount of delta activity, as determined by spectral analysis, may be used to quantify the depth of SWS. Transition toward REM sleep is characterized by a complete EMG atonia and bursts of rapid eye movements. The onset of REM sleep also features a dramatic change in the EEG picture: delta waves are replaced by a low-voltage fast activity EEG resembling that of the waking state. In addition to the spectral analysis of the EEG, quantification of rapid eye movements can index the density of REM sleep. The end of a REM sleep period is often followed by a short awakening before a new non-REM–REM cycle starts again. The length of non-REM–REM cycles varies across species. For example, the non-REM–REM interval is approximately 12–16 min in rats and 90–100 min in humans (Rechtschaffen and Kales 1968; Timo-Iaria *et al.* 1970; Vivaldi *et al.* 1994).

During a typical night's sleep, light SWS accounts for approximately 50% of the total, deep SWS accounts for 20–25%, and REM sleep accounts for 25% (Vivaldi *et al.* 1994; Kryger *et al.* 2000). In rats as much as in humans, respective proportions of non-REM and REM sleep in a cycle are unevenly but predictably distributed during sleep: SWS is mostly found during the early part of the sleep period while REM sleep prevails towards the end (Vivaldi *et al.* 1994; Kryger *et al.* 2000). It appears that sleep sets a limit on metabolic expenditures to an extent necessary to balance a species' energy budget, that is, sleep occurs to enforce rest and keep energy expenditure at an affordable level (Zepelin and Rechtschaffen 1974). Total sleep time as well as SWS and REM sleep time are highly correlated with body weight and brain weight (Zepelin 2000). Table 7.1 summarizes behavioral and electrographic distinctive signs of sleep.

Current models of sleep control apply with equal relevance to rats, cats, and humans. The combined observations on the timing of dorsal raphe serotonin (5-HT) neuronal activity and locus coeruleus noradrenaline (NA) cell firing throughout the sleep–wake cycle, together with the demonstration that some cholinergic/cholinoceptive cells of the gigantocellular tegmental field fire selectively in REM sleep, have led to the formulation of the reciprocal-interaction model of Hobson and McCarley in 1975. According to this model, 5-HT and NA neurons act as inhibitory (REM-off) elements and acetylcholine (ACh) neurons act as excitatory (REM-on) elements. The mechanism by which SWS and REM sleep alternate during sleeping can be approximately sketched as follows: during the waking period, normally firing 5-HT and NA neurons progressively inhibit themselves through the release of their neurotransmitters onto their own somatodendritic autoreceptors; as the brain enters SWS, the cholinergic REM-on cells are progressively relieved from the tonic inhibition formerly exerted by 5-HT and NA inputs. As a threshold is attained, raphe and locus coeruleus REM-off neurons become totally silent and REM sleep is thus triggered. Even though REM-on cells excite themselves through release of their excitatory neurotransmitter onto their own somatodendritic autoreceptors during REM sleep, their projection fibers also release a depolarizing input onto silent REM-off cells; the latter groups of cells are thus progressively

Table 7.1 Behavioral and electrographic distinctive signs of sleep

	Waking	*SWS*	*REM sleep*
Cortical EEG	Low voltage Fast activity	High voltage Slow activity	Low voltage Fast activity
Hippocampal EEG	Presence of theta rhythm	No theta rhythm	Presence of theta rhythm
EMG	Active	Stable	Atonia and twitches
EOG	Active	Stable	Frequent REMs
PGO	Absent	Isolated	Burst firing
Pulse and respiration	Rapid and regular	Slow and regular	Irregular

turned on, and as they become more active, their inhibitory input will eventually turn off the REM-on cells. The reciprocal-interaction model has evolved dramatically over the last twenty years, with an increasing number of proposed sites and receptor subtypes, but updated versions are still based on the fundamental contribution of 5-HT and NA REM-off cells and on cholinergic REM-on cells (Jones 1998; Portas *et al.* 2000). Modeling techniques have also taken into account the uneven distributions of SWS and REM sleep throughout a major sleep episode (see above) by integrating the revised reciprocal-interaction model (Hobson *et al.* 1993; Massaquoi and McCarley 1992; McCarley and Massaquoi 1992) with more general models of the circadian rhythms of the states of vigilance (Borbély and Achermann 1992).

Effects of GHB on sleep

GHB was first synthesized in response for the clinical need to develop a GABA precursor that would cross the blood–brain barrier (Laborit 1964). The sedative effects of GHB and its low toxicity first led to its use in anesthesia. GHB had two additional advantages. First, GHB can induce a state of EEG behavioral dissociation during which the patient is responsive despite high-voltage slow activity on the EEG (Borenstein *et al.* 1969; Laborit *et al.* 1960; Hayashi 1967; Metcalf *et al.* 1966). Second, despite a rapid onset of action, the effect of GHB is short-lived. In the rat for example, behavioral modifications appear within 4–6 min following the injection (Walters and Roth 1972). Plasmatic half-life of GHB is only 60–90 min in the rat (Roth and Giarman 1966) and in humans (Ferrara *et al.* 1992). Consequently, the use of GHB as an anesthetic was first restricted to surgeries in which reflexes are needed to be preserved, for example, in deliveries or ophthalmic surgeries (Vickers 1969; Smith *et al.* 1972), but now its use is expanding to total anesthesia (Kleinschmidt *et al.* 1998, 1999). GHB was also thought to have potential use as an anxiolytic, but it was rapidly replaced by the benzodiazepines (Grove-White and Kelman 1971; Schmidt-Mutter *et al.* 1998).

Early on, the effects on sleep of GHB, and other short-chain fatty acids, were characterized in animals as well as in humans (Holmquist and Ingvar 1957; Jouvet *et al.* 1961; Laborit *et al.* 1960; Roth and Suhr 1970; Samson *et al.* 1956; White and Samson 1956), at a wide range of doses and routes of administration. Studies in animals became more sparse after the 1980s while studies in humans were almost exclusively performed in clinical populations. Whereas a majority of sleep studies performed with GHB in human subjects are based on the administration of low doses in the range of 15–50 mg/kg *per os* (next section), animal studies include results based on doses from as low as 10 mg/kg to as high as 1,200 mg/kg, administered intravenously (i.v.) or intraperitoneally (i.p.). Given the existence of low- and high-affinity GHB receptors (see Chapter 2 of this volume), results must be interpreted accordingly.

Hypnotic effects

At very high doses (250–1,200 mg/kg i.p. or i.v.), GHB induces torpor in the rat (Godschalk *et al.* 1977; Laborit *et al.* 1960; Marcus *et al.* 1967; Hayashi 1967; Metcalf *et al.* 1966), cat (Jouvet *et al.* 1961; Stock *et al.* 1978; Tabakoff and Radulovacki 1976; Winters and Kott 1979; Winters and Spooner 1965), rabbit (Godbout and Pivik 1982), monkey (Snead *et al.* 1976; Snead 1978a,b,c), and human (Laborit *et al.* 1960; Hayashi 1967; Metcalf *et al.* 1966). EEG recordings under such conditions show a progression from intermittent hypersynchronization to spike and wave activity to electric silence, suggesting an epileptogenic course. At lower doses (50–250 mg/kg i.v. or i.p.), sleep is facilitated under GHB versus saline in cats (Matsuzaki *et al.* 1964; Stock *et al.* 1978; Winters and Kott 1979) but not in rabbits

Table 7.2 Effects of GHB (10 mg/kg i.p.) on sleep at different circadian times in the rat (mean ± SEM)

Sleep variables (min)	Saline at CT2 (n = 16)	GHB at CT2 (n = 13)	p	Saline at CT8 (n = 6)	GHB at CT8 (n = 6)	p	Saline at CT14 (n = 9)	GHB at CT14 (n = 8)	p
Sleep latency	27.7 ± 7.7	22 ± 3.9	0.47	28.2 ± 4.9	22.3 ± 3	0.33	45.2 ± 6.6	37.7 ± 5.5	0.39
SWS-2 latency	23.6 ± 7.4	16.8 ± 4.6	0.42	4.4 ± 1.7	13.2 ± 6.3	0.21	7.0 ± 4.3	10.2 ± 3.7	0.58
REM latency	45.4 ± 6.4	27.6 ± 5.5	0.04	19.9 ± 2.7	29.1 ± 7.2	0.26	25.1 ± 6.3	40.0 ± 8.5	0.19

Source: Data adapted from Girodias et al. (1995a,b).

Note
The light–dark cycle was set at 12 : 12, with lights on at 06 : 00. Treatment was administered i.p. at three circadian times (CT), that is number of hours following lights on, and recordings started immediately thereafter.

(Godbout and Pivik 1982) or in rats (Girodias *et al.* 1995; Godbout *et al.* 1995). In the latter case, tests were performed in the light as well as in the dark period (see Table 7.2).

The hypnotic effect of GHB in non-clinical and clinical human participants was first demonstrated in open trials where its ability to decrease sleep onset latency and/or to increase sleep maintenance was shown (Broughton and Mamelak 1980; Delay *et al.* 1965; Hayashi 1967; Hoes *et al.* 1980; Laborit 1964; Mamelak *et al.* 1973; Metcalf *et al.* 1966; Schneider *et al.* 1963; Yamada *et al.* 1967). However, when compared to placebo in a double-blind fashion (25–50 mg/kg p.o.), polygraphic recordings did not demonstrate a shortening of sleep latency in healthy participants, whether administration of the drug was nocturnal or morning (Lapierre *et al.* 1990). On the other hand, the hypnotic effect of GHB (15–55 mg/kg p.o.) was demonstrated in clinical populations such as patients with idiopathic excessive daytime somnolence (Montplaisir and Barbezieux 1981) and in patients with insomnia or depression compared to placebo (Mamelak *et al.* 1977).

The literature thus suggests that GHB facilitates sleep onset in clinical populations with difficulties to initiate or maintain sleep but not in healthy participants, even when GHB is given at a time at which sleep latency is increased (i.e. morning administration in humans or dark period in the rat).

Sleep structure: non-REM sleep

Statistical analysis of the effects of GHB on the latency and duration of non-REM sleep stages versus saline is usually not reported in animal studies, all efforts being focused on REM sleep (see next section). Recordings performed in the rabbit (Godbout and Pivik 1982) and in the rat (Godschalk *et al.* 1977; Fig. 7.1, Tables 7.2 and 7.3) show that GHB, administered at low or high doses, does not substantially modify non-REM sleep variables.

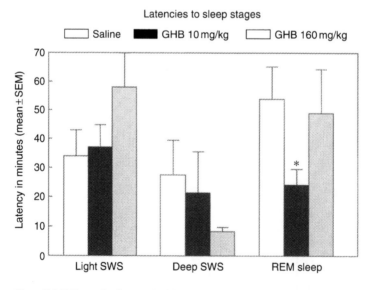

Figure 7.1 Effects of a low and a high dose of GHB on sleep stage latencies in the rat. GHB was administered at 08:00; that is 2 h after the onset of the light period. Treatment was administered i.p. and recordings started immediately after the injection. Sleep onset was always concomitant with onset of light SWS.

Source: Data adapted from Girodias *et al.* (1995, 1996a,b).

Note: * *p* < 0.05, *t*-tests for independent samples.

Table 7.3 Effects (mean ± SEM) of low and high GHB doses on proportion of sleep stages (% of total sleep time) and waking (% WASO) in the rat

	Saline (n = 13)	GHB 10 mg/kg (n = 19)	p	GHB 160 mg/kg (n = 18)	p
SWS-1 (%)	35.6 ± 3.9	38.6 ± 4.0	0.61	42.3 ± 4.5	0.28
SWS-2 (%)	45.3 ± 3.4	42.4 ± 3.8	0.61	40.5 ± 4.1	0.40
REMS (%)	19.1 ± 1.3	19.0 ± 1.1	0.91	17.2 ± 1.1	0.25
Total sleep time (min)	178.7 ± 6.6	165.8 ± 7.5	0.24	162.3 ± 9.6	0.20
WASO (min)	35.9 ± 6.8	52.7 ± 8.0	0.15	37.0 ± 6.7	0.92

Source: Data adapted from Girodias *et al.* (1995a,b).

Notes
SWS-1: light slow-wave sleep; SWS-2: deep slow-wave sleep; REMS: REM sleep; WASO: waking after sleep onset. The light–dark cycle was set at 12 : 12, with lights on at 06 : 00. Treatment was administered i.p. at 08 : 00 and recordings started immediately thereafter, lasting 4 h.

Table 7.4 Effects of GHB on sleep in five patients with idiopathic hypersomnia

	Before	After	Paired t-tests
Sleep latency (min)	14.9 ± 6.5	7.0 ± 2.3	—
Total sleep time (min)	442.6 ± 23.9	425.8 ± 27.4	—
Awakenings (no.)	22.4 ± 6.8	33.2 ± 7.4	—
Awakenings (min)	79.8 ± 14.4	56.0 ± 13.8	$p < 0.05$
Stage 1 (min)	51.4 ± 7.7	49.0 ± 9.1	—
Stage 2 (min)	245.2 ± 19.3	226.2 ± 24.0	—
SWS (min)	45.0 ± 10.6	68.4 ± 11.1	$p < 0.025$
REM sleep (min)	96.4 ± 8.6	77.6 ± 16.3	—

Data presented as mean ± SEM.

Source: Adapted from Montplaisir and Barbezieux (1981).

Latency to non-REM sleep stages is also often omitted from statistical analysis in human studies. Our work shows that GHB decreases the latency to deep SWS (stages 3 and 4) in healthy participants (Lapierre *et al.* 1990) while it does not in patients with narcolepsy (Bédard *et al.* 1989). The effects of GHB on the duration of non-REM sleep stages are more consistently reported in the human literature and point toward a facilitation in healthy participants (Lapierre *et al.* 1990; Metcalf *et al.* 1966; Van Cauter *et al.* 1997; Yamada *et al.* 1967), in insomniacs (Mamelak *et al.* 1977) including a case of fatal familial insomnia (Reder *et al.* 1995), in patients with depression (Mamelak *et al.* 1977), in patients with narcolepsy (Broughton and Mamelak 1980), and in patients with idiopathic excessive daytime somnolence (Montplaisir and Barbezieux 1981; see Table 7.4). Our original study in patients with narcolepsy (Bédard *et al.* 1989) did not report an effect of GHB on total SWS, but when the analysis is restricted to the first third of the night, it is then found that GHB also facilitates SWS in this condition (46.9 ± 5.6% versus 62.9 ± 8.6%, $p < 0.03$). It has been argued that the effect of GHB on SWS may not represent an effect on sleep mechanisms *per se* but rather on mechanisms of EEG slow wave activity (Williams *et al.* 1995). The fact that EEG slow waves are apparent following GHB in awakened subjects favors this possibility (Borenstein *et al.* 1969; Broughton and Mamelak 1980; Metcalf *et al.* 1966; Yamada *et al.* 1967), an observation confirmed by more recent reports using quantified EEG analysis (Black *et al.* 2001; Van Cauter *et al.* 1997). More details on the effects of GHB on EEG can be found in Chapter 8 of this volume.

REM sleep

The best described and most consistent finding with respect to the effect of GHB on sleep is its capacity to facilitate the triggering of REM sleep without increasing its duration. Indeed, GHB has been repeatedly reported to decrease REM sleep latency in healthy humans (Lapierre *et al.* 1990) as well as in patients with insomnia, depression (Mamelak *et al.* 1977), narcolepsy (Bédard *et al.* 1989; Broughton and Mamelak 1980; Scrima *et al.* 1990), but not in patients with idiopathic excessive daytime somnolence (Montplaisir and Barbezieux 1981).

In adult Sprague–Dawley rats, low doses of GHB (10 mg/kg i.p.) compared to saline decreased REM sleep latency when administered close to the onset but not toward the end of the light period, and have no effect during the dark period (Girodias *et al.* 1995; Godbout *et al.* 1995; Fig. 7.1 and Table 7.2). The shortening of REM sleep latency was reversed by the GHB receptor antagonist NCS-382 (Girodias *et al.* 1996a,b). On the other hand, no evidence of changes in REM sleep duration was observed after 4 h of recordings. In younger Wistar rats (160–180 g), GHB (12.5–25.0 mg/kg i.p.) also showed no effect on REM sleep duration within 4 h of recordings (Godschalk *et al.* 1977); unfortunately, REM sleep latencies were not reported in this study.

In the intact or decorticated cat, REM sleep latency is not affected with low doses of GHB but it is facilitated at higher doses ranging from 100 to 200 mg/kg i.p. (Delorme *et al.* 1966; Jouvet *et al.* 1961; Knuepfer *et al.* 1986; Matsuzaki *et al.* 1964; Matsuzaki and Takagi 1967; Vern and Hubbard 1971). GHB also facilitates REM sleep in pontine cats with intact 'REM-on' structures (Delorme *et al.* 1966). In the latter case, however, doses needed to be very high (1,000–1,500 mg/kg i.v.) and animals were already alternating directly from the waking state to REM sleep, which questions the nature of REM sleep recorded in such conditions. Finally, it has not been possible to show a facilitation of REM sleep by GHB at doses ranging from 25 to 1,200 mg/kg i.v. in the rabbit. However, this species only shows incomplete motor atonia during spontaneous REM sleep and thus different mechanisms may be involved (Pivik *et al.* 1981). This suggests that GHB, in order to induce REM sleep, must be administered in conditions in which essential control mechanisms are present (Delorme *et al.* 1966).

The mechanism by which GHB induces REM sleep is still unknown but a possible role for dopamine has been proposed (Lapierre *et al.* 1990; Stock 1982). Facilitation of cholinergic REM-on networks is probably not involved since the triggering of REM sleep by GHB is not blocked by atropine (Delorme *et al.* 1966). Moreover, GHB does not induce REM sleep in cats with a midpontine (pretrigeminal) section whereas physostigmine does, but both drugs induce REM sleep in isolated pons preparations (Matsuzaki 1968, 1969).

Chronobiologic factors

It has been repeatedly shown that the triggering of sleep using pharmacological tools is bound by chronobiologic rules, in laboratory animals as well as in humans (Gillin and Shiromani 1990). This also applies to GHB, particularly to its REM sleep-inducing properties. Our study in healthy participants shows that GHB induces REM sleep more easily when this propensity for this stage of sleep is already high such as upon morning administration or in healthy participants with shorter REM sleep latencies (Lapierre *et al.* 1990), or in patients with depression or narcolepsy (Mamelak *et al.* 1977; Bédard *et al.* 1989). It has also been observed that GHB only induced REM sleep in the pontine cat when it was administered at a time close enough to the onset of the next expected REM sleep period (Delorme *et al.* 1966).

On the other hand, our results in a nocturnal animal such as the rat point to different circadian facilitating factors since low doses of GHB facilitate REM sleep triggering only when administered close to the onset of the light period (Girodias *et al.* 1995; Godbout *et al.* 1995; Table 7.2), a period during which REM sleep propensity is not maximal (Vivaldi *et al.* 1994).

GHB in narcolepsy

At the present time, the best known clinical indication for GHB in sleep medicine is narcolepsy (AASM 1999). This condition includes daytime somnolence that culminates in sleep attacks characterized by sleep-onset REM periods, fragmentation of nocturnal sleep, and three REM sleep-related symptoms: cataplexy, a sudden loss of muscle tone related to REM sleep atonia but without loss of consciousness, hypnagogic hallucinations, and sleep paralysis (Choo and Guilleminault 1998; Guilleminault *et al.* 1998; Montplaisir *et al.* 1978).

Taken at doses between 25 and 50 mg/kg p.o., GHB is a potent anticataplectic agent; it also reduces hypnagogic hallucinations and sleep paralysis (Lammers *et al.* 1993; Mamelak *et al.* 1986). It is thought that this therapeutic effect is achieved through a consolidation of sleep, and particularly REM sleep (Broughton and Mamelak 1980; Scharf *et al.* 1985), thus reducing daytime pressure for REM sleep. In favor of this hypothesis is the fact that narcoleptic patients treated with GHB show a decrease in the number of daytime sleep-onset REM periods (Montplaisir *et al.* 2001). In a recent multicenter clinical trial of twenty-one narcoleptic patients, GHB was also found to reduce the excessive daytime sleepiness in a dose-related fashion both subjectively (decrease in the Epworth Sleepiness Scale score) and objectively (increase in the mean sleep latency on the Maintenance of Wakefulness Test) (Montplaisir *et al.* 2001). GHB was also found to restore the recuperative function of sleep in the same patients as shown by a dose-related increase in the amount of deep SWS and non-REM delta power as well as by a dose-related reduction in the number of nocturnal awakenings (Black *et al.* 2001).

GHB has been unsuccessfully tried in the canine model of narcolepsy (Foutz *et al.* 1981; Baker *et al.* 1985). High doses were used (500 mg/kg p.o.) and treatments were given at daytime. This particular case stresses the importance of controlling for doses and circadian factors when evaluating the effects of GHB on sleep and related behavior.

Summarizing, GHB is a naturally occurring molecule that affects sleep in ways that differ according to dose, circadian time, and species. Further work is needed to elucidate its mode of action on the neural networks responsible for the control of the sleep–wake cycle.

References

American Academy of Sleep Medicine (AASM) (1999) The use of gamma hydroxybutyrate (GHB) in the treatment of narcolepsy, position statement and recommendations of the AASM. http://www.aasmnet.org/PDF Files/GHB in Treating Narcolepsy.pdf.

Baker T. L. and Dement W. C. (1985) Canine narcolepsy-cataplexy syndrome: Evidence for an inherited monoaminergic-cholinergic imbalance. In *Brain Mechanisms of Sleep* (McGinty D., Drucker-Colín R., Morrison A. and Parmeggiani P. L., eds), pp. 199–234. Raven Press, New York.

Bédard M. A., Montplaisir J., Godbout R. and Lapierre O. (1989) Nocturnal γ-hydroxybutyrate. Effects on periodic leg movements and sleep organization of narcoleptic patients. *Clin. Neuropharmacol.* 12, 29–36.

Black J., Ristanovic R., Mamelak M. and Montplaisir J. (2001) Dose response effects of sodium oxybate on polysomnographic (PSG) measures in narcolepsy patients: preliminary findings. *Sleep* 24 (Suppl.), A304.

Borbély A. A. and Achermann P. (1992) Concepts and models of sleep regulation: an overview. *J. Sleep Res.* 1, 63–79.

Borbély A. A. and Neuhaus H. U. (1978) Circadian rhythm of sleep and motor activity in the rat during skeleton photoperiod, continuous darkness and continuous light. *J. Comp. Physiol* 128, 37–46.

Borenstein P., Champion C. and Cujo P. H. (1969) Etude polygraphique du 'sommeil' provoqué par le gamma hydroxybutyrate de sodium et le diazépam. *Semaine des Hôpitaux de Paris* 20, 1290–300.

Broughton R. and Mamelak M. (1980) Effects of nocturnal gamma-hydroxybutyrate on sleep/waking patterns in narcolepsy-cataplexy. *Can. J. Neurol. Sci.* 7, 23–31.

Choo K. L. and Guilleminault C. (1998) Narcolepsy and idiopathic hypersomnolence. *Clin. Chest Med.* 19, 169–81.

Delay J., Deniker P., Perier M., Ginestet D., Sempé J. C. and Verdeaux G. (1965) Effets neuro-psychiques de l'acide gamma-hydroxy-butyrique par voie orale et par voie veineuse. *Encéphale* 54, 546–54.

Delorme F., Riotte M. and Jouvet M. (1966) Conditions de déclenchement du sommeil paradoxal par les acides gras à chaîne courte chez le chat pontique chronique. *C. R. Soc. Biol.* 160, 1457–60.

Farber J., Marks G. A. and Roffwarg H. P. (1980) Rapid eye movement sleep PGO-type waves are present in the dorsal pons of the albino rat. *Science* 209, 615–17.

Ferrara S. D., Zotti S., Tedeschi L., Frison G., Castagna F., Gallimberti L., Gessa G. L. and Palatini P. (1992) Pharmacokinetics of gamma-hydroxybutyric acid in alcohol-dependant patients after single and repeated oral doses. *Br. J. Clin. Pharmacol.* 34, 231–35.

Foutz A. S., Delashaw J. B., Guilleminault C. and Dement W. C. (1981) Monoaminergic mechanisms and experimental cataplexy. *Ann. Neurol.* 10, 369–76.

Gillin J. C. and Shiromani P. (1990) Cholinergic mechanisms in sleep: basic and clinical applications. In *Sleep and Biological Rhythms. Basic Mechanisms and Applications to Psychiatry.* (Montplaisir J. and Godbout R., eds), pp. 186–208. Oxford University Press, New York.

Girodias V., Godbout R. and Webster H. H. (1995) Sleep-inducing effects of gamma-hydroxybutyrate according to dose and time of administration in the rat. *Sleep Res.* 24A, 24.

Girodias V., Godbout R., Beaulieu I., Schmitt M., Bourguignon J. J. and Webster H. H. (1996a) Triggering of paradoxical sleep with gamma-hydroxybutyrate (GHB) in the rat is blocked by the GHB receptor antagonist NCS-382. *Sleep Res.* 25, 9.

Girodias V., Beaulieu I., Webster H. H., Schmitt M., Bourguignon J. J. and Godbout R. (1996b) Reversal of gamma-hydroxybutyrate-induced facilitation of paradoxical sleep by NCS-382. *Soc. Neurosci. Abstr.* 22, 147.

Godbout R. and Pivik R. T. (1982) EEG and behavioral effects of gamma-hydroxybutyrate in the rabbit. *Life Sci.* 31, 739–48.

Godbout R., Girodias V. and Webster H. H. (1995) Effects of gamma-hydroxybutyrate on paradoxical sleep in the rat are related to dose and time of administration. *Soc. Neurosci. Abstr.* 21, 450.

Godschalk M., Dzoljic M. R. and Bonta I. L. (1997) Slow wave sleep and a state resembling absence epilepsy induced in the rat by gamma-hydroxybutyrate. *Eur. J. Pharmacol.* 44, 105–11.

Grove-White I. G. and Kelman G. R. (1971) Critical flicker frequency after small doses of methohexitone, diazepam and sodium 4-hydroxybutyrate. *Br. J. Anaesth.* 43, 110–12.

Guilleminault C., Heinzer R., Mignot E. and Black J. (1998) Investigations into the neurologic basis of narcolepsy. *Neurology* 50, S8–15.

Hayashi T. (1967) The relationship between circadian sleep and gamma-hydroxybutyrate (4HB) in brain. *Exp. Med. Surg.* 25, 148–55.

Hobson J. A., McCarley R. W. and Wyzinski P. W. (1975) Sleep cycle oscillation: reciprocal discharge by two brainstem neuronal groups. *Science* 189, 55–58.

Hobson J. A., Datta S., Calvo J. M. and Quattrochi J. (1993) Acetylcholine as a brain state modulator: triggering and long-term regulation of REM sleep. *Prog. Brain. Res.* 98, 389–403.

Hoes M. J., Vree T. and Guelen P. J. (1980) Gamma-hydroxybutyric acid as hypnotic. Clinical and pharmacokinetic evaluation of gamma-hydroxybutyric acid as hypnotic in man. *Encéphale* 6, 93–99.

Holmquist B. and Ingvar D. H. (1957) Effects of short chain fatty acid anions upon cortical blood flow and EEG in cats. *Experientia* 13, 331–33.

Jones B. E. (1998) The neural basis of consciousness across the sleep–waking cycle. *Adv. Neurol.* 77, 75–94.

Jouvet M., Cier A., Mounier D. and Valatx J. L. (1961) Effets du 4-butyrolactone et du 4-hydroxybutyrate de sodium sur l'E.E.G. et le comportement du chat. *C. R. Soc. Biol.* 155, 1313–16.

Kleinschmidt S., Grundmann U., Knocke T., Silomon M., Bach F. and Larsen R. (1998) Total intravenous anaesthesia with gamma-hydroxybutyrate (GHB) and sufentanil in patients undergoing coronary artery bypass graft surgery: a comparison in patients with unimpaired and impaired left ventricular function. *Eur. J. Anaesthesiol.* 15, 559–64.

Kleinschmidt S., Schellhase C. and Mertzlufft F. (1999) Continuous sedation during spinal anaesthesia: gamma-hydroxybutyrate vs propofol. *Eur. J. Anaesthesiol.* 16, 23–30.

Knuepfer M. M., Stumpf H. and Stock G. (1986) Baroreceptor sensitivity during desynchronized sleep. *Exp. Neurol.* 92, 323–34.

Kryger M. H., Roth T. and Dement W. C., eds (2000) *Principles and Practice of Sleep Medicine.* Saunders, Philadelphia.

Laborit H. (1964) Sodium gamma-hydroxybutyrate. *Int. J. Neuropharmacol.* 3, 433–52.

Laborit H. (1973) Gamma-hydroxybutyrate, succinic semialdehyde and sleep. *Prog. Neurobiol.* 1, 255–74.

Laborit H., Jouany J. M., Gerard J. and Fabiani F. (1960) Généralités concernant l'étude expérimentale et l'emploi clinique du γ-hydroxybutyrate de Na. *Agressologie* 1, 397–405.

Lammers G. J., Arends J., Declerck A. C., Ferrari M. D., Schouwink G. and Troost J. (1993) Gammahydroxybutyrate and narcolepsy: a double-blind placebo-controlled study. *Sleep* 16, 216–20.

Lapierre O., Montplaisir J., Lamarre M. and Bédard M. A. (1990) The effect of gamma-hydroxybutyrate on nocturnal and diurnal sleep of normal subjects: further consideration on REM sleep-triggering mechanisms. *Sleep* 13, 24–30.

McCarley R. W. and Massaquoi S. G. (1992) Neurobiological structure of the revised limit cycle reciprocal interaction model of REM cycle control. *J. Sleep Res.* 1, 132–37.

Mamelak M., Escriu J. M. and Stokan O. (1973) Sleep inducing effects of gammahydroxybutyrate. *Lancet* 2 (7824), 328–29.

Mamelak M., Escriu J. M. and Stokan O. (1977) The effects of γ-hydroxybutyrate on sleep. *Biol. Psychiatry* 12, 273–88.

Mamelak M., Scharf M. B. and Woods M. (1986) Treatment of narcolepsy with γ-hydroxybutyrate. A review of clinical and sleep laboratory findings. *Sleep* 9, 285–89.

Marcus R. J., Winters W. D., Mori K. and Spooner C. E. (1967) EEG and behavioral comparison of the effects of gamma-hydroxybutyrate, gamma-butyrolactone and short chain fatty acids in the rat. *Int. J. Neuropharmacol.* 6, 175–85.

Massaquoi S. G. and McCarley R. W. (1992) Extension of the Limit Cycle Reciprocal Interaction Model of REM cycle control. An integrated sleep control model. *J. Sleep Res.* 1, 138–43.

Matsuzaki M. (1968) Differential aspects of NA butyrate and physostigmine on brain stem activities of paradoxical sleep. *Brain Res.* 11, 251–55.

Matsuzaki M. (1969) Differential effects of sodium butyrate and physostigmine upon the activities of para-sleep in acute brain stem preparations. *Brain Res.* 13, 247–65.

Matsuzaki M. and Takagi H. (1967) Sleep induced by sodium butyrate in the cat. *Brain Res.* 4, 206–22.

Matsuzaki M., Takagi H. and Tokizane T. (1964) Paradoxical phase of sleep: its artificial induction in the cat by sodium butyrate. *Science* 146, 1328–29.

Metcalf D. R., Emde R. N. and Stripe J. T. (1966) An EEG-behavioral study of sodium hydroxybutyrate in humans. *Electroencephalogr. Clin. Neurophysiol.* 20, 506–12.

Montplaisir J. and Barbezieux M. (1981) Le gamma-hydroxybutyrate de sodium (GHB) dans le traitement de l'hypersomnie essentielle. *Can. J. Psychiatry* 26, 162–66.

Montplaisir J., Billiard M., Takahashi S., Bell I. R., Guilleminault C. and Dement W. C. (1978) Twenty-four-hour recording in REM-narcoleptics with special reference to nocturnal sleep disruption. *Biol. Psychiatry* 13, 73–89.

Montplaisir J., Ristanovic R., Mamelak M. and Black J. (2001) Effects of sodium oxybate on measures of daytime sleepiness in narcolepsy patients: preliminary evidence of dose-related improvements. *Sleep* 24 (Suppl.), A304–A305.

Pivik R. T., Sircar S. and Braun C. (1981) Nuchal muscle tonus during sleep, wakefulness and tonic immobility in the rabbit. *Physiol. Behav.* 26, 13–20.

Portas C. M., Bjorvatn B. and Ursin R. (2000) Serotonin and the sleep/wake cycle: special emphasis on microdialysis studies. *Prog. Neurobiol.* 60, 13–35.

Rechtschaffen A. and Kales A. (1968) A manual of standardized terminology, techniques and scoring systems for sleep stages of human subjects. UCLA Brain Research Institute, Los Angeles.

Rechtschaffen A., Gilliland M. A., Bergmann B. M. and Winter J. B. (1983) Physiological correlates of prolonged sleep deprivation in rats. *Science* 221, 182–84.

Reder A. T., Mednick A. S., Brown P., Spire J. P., Van Cauter E., Wollmann R. L., Cervenakova L., Goldfarb L. G., Garay A. and Ovsiew F. (1995) Clinical and genetic studies of fatal familial insomnia. *Neurology* 45, 1068–75.

Roth R. H. and Giarman N. J. (1966) Gamma-butyrolactone and gamma-hydroxybutyric acid. I – Distribution and metabolism. *Biochem. Pharmacol.* 15, 1333–48.

Roth R. H. and Suhr Y. (1970) Mechanism of the γ-hydroxybutyrate-induced increase in brain dopamine and its relationship to 'sleep'. *Biochem. Pharmacol.* 19, 3001–12.

Sakai K., Crochet S. and Onoe H. (2001) Pontine structures and mechanisms involved in the generation of paradoxical (REM) sleep. *Arch. Ital. Biol.* 139, 93–107.

Samson F. E., Dahl N. and Dahl D. R. (1956) A study on the narcotic action of the short chain fatty acids. *J. Clin. Invest.* 35, 1291–98.

Scharf M. B., Brown D., Woods M., Brown L. and Hirschowitz J. (1985) The effects and effectiveness of γ-hydroxybutyrate in patients with narcolepsy. *J. Clin. Psychiatry* 46, 222–25.

Schmidt-Mutter C., Pain L., Sandner G., Gobaille S. and Maitre M. (1998) The anxiolytic effect of gamma-hydroxybutyrate in the elevated plus maze is reversed by the benzodiazepine receptor antagonist, flumazenil. *Eur. J. Pharmacol.* 342, 21–27.

Schneider J., Thomalske G., Trautmann P., Smolarz R. and Sabbagh R. (1963) Le comportement EEG de l'homme et de l'animal soumis à l'action progressive du 4-hydroxybutyrate de sodium. *Agressologie* 4, 55–69.

Scrima L., Hartman P. G., Johnson F. H. Jr, Thomas E. E. and Hiller F. C. (1990) The effects of γ-hydroxybutyrate on the sleep of narcolepsy patients: a double-blind study. *Sleep* 13, 479–90.

Smith I., Beveridge M. E. and Wyllie A. M. (1972) 4-Hydroxybutyrate narcosis for ophthalmic surgery. *Br. J. Ophthalmol.* 56, 429–30.

Snead O. C. (1978a) Gamma hydroxybutyrate in the monkey. I. Electroencephalographic, behavioral, and pharmacokinetic studies. *Neurology* 28, 636–42.

Snead O. C. (1978b) Gamma hydroxybutyrate in the monkey. II. Effect of chronic oral anticonvulsant drugs. *Neurology* 28, 643–48.

Snead O. C. (1978c) Gamma hydroxybutyrate in the monkey. III. Effect of intravenous anticonvulsant drugs. *Neurology* 28, 1173–78.

Snead O. C., Yu R. K. and Huttenlocher P. R. (1976) Gamma hydroxybutyrate. Correlation of serum and cerebrospinal fluid levels with electroencephalographic and behavioral effects. *Neurology* 26, 51–56.

Stock G. (1982) Neurobiology of REM sleep: a possible role for dopamine. In *Sleep: Clinical and Experimental Aspects* (Ganten D. and Pfaff D., eds), pp. 1–36. Springer, New York.

Stock G., Heidt H., Buss J. and Schlör K. H. (1978) Sleep patterns in cat induced by gammahydroxybutyric acid. *Electroencephalogr. Clin. Neurophysiol.* 44, 523–27.

Tabakoff B. and Radulovacki M. (1976) γ-Hydroxybutyrate in CSF during sleep and wakefulness. *Res Commun. Chem. Pathol. Pharmacol.* 14, 587–90.

Timo-Iaria C., Negrao N., Schmidek W. R., Hoschino K., Labato de Menezes C. E. and Leme Da Rocha T. (1970) Phases and states of sleep in the rat. *Physiol. Behav.* 5, 1057–62.

Van Cauter E., Plat L., Scharf M. B., Leproult R., Cespedes S., L'Hermite-Baleriaux M. and Copinschi G. (1997) Simultaneous stimulation of slow-wave sleep and growth hormone secretion by gamma-hydroxybutyrate in normal young men. *J. Clin. Invest.* 100, 745–53.

Vanni-Mercier G., Pelisson D., Goffart L., Sakai K. and Jouvet M. (1994) Eye saccade dynamics during paradoxical sleep in the cat. *Eur. J. Neurosci.* 6, 1298–306.

Vern B. and Hubbard J. I. (1971) Reinvestigation of the effects of gamma-hydroxybutyrate on the sleep cycle of the unrestrained intact cat. *Electroencephalogr. Clin. Neurophysiol.* 31, 573–80.

Vickers M. D. (1969) Gammahydroxybutyric acid. *Intern. Anesthesiol. Clin.* 7, 75–79.

Vivaldi E. A., Wyneken U., Roncagliolo M., Ocampo A. and Zapata A. M. (1994) Measures of location and dispersion of sleep state distribution within the circular frame of a 12 : 12 light : dark schedule in the rat. *Sleep* 17, 208–19.

Walters R. J. and Roth R. H. (1972) Effects of gamma-hydroxybutyrate on dopamine and dopamine metabolites in the rat striatum. *Biochem. Pharmacol.* 21, 2111–21.

White R. P. and Samson F. E. (1956) Effects of fatty acid anions on the electroencephalogram of unanesthetized rabbits. *Am. J. Physiol.* 186, 271–74.

Williams S. R., Turner J. P. and Crunelli V. (1995) Gamma-hydroxybutyrate promotes oscillatory activity of rat and cat thalamocortical neurons by a tonic GABAB, receptor-mediated hyperpolarization. *Neuroscience* 66, 133–41.

Winters W. D. and Kott K. S. (1979) Continuum of sedation, activation and hypnosis: a comparison of low dose effects of pentobarbital, diazepam or gamma-hydroxybutyrate in the cat. *Neuropharmacology* 18, 877–84.

Winters W. D. and Spooner C. E. (1965) A neurophysiological comparison of gamma-hydroxybutyrate with pentobarbital in cats. *Electroencephalogr. Clin. Neurophysiol.* 18, 287–96.

Yamada Y., Yamamoto J., Fujiki A., Hishikawa Y. and Kaneko Z. (1967) Effect of butyrolactone and gamma-hydroxybutyrate on the EEG and sleep cycle in man. *Electroencephalogr. Clin. Neurophysiol.* 22, 558–62.

Zepelin H. (2000) Mammalian sleep. In *Principles and Practice of Sleep Medicine* (Kryger M.H., Roth T. and Dement W.C., eds) pp. 82–92. Saunders, Philadelphia.

Zepelin H. and Rechtschaffen A. (1974) Mammalian sleep, longevity, and energy metabolism. *Brain Behav. Evol.* 10, 425–70.

8 γ-Hydroxybutyrate and absence seizure activity

O. C. Snead

Introduction

γ-Hydroxybutyrate (GHB) is a metabolite of γ-aminobutyric acid (GABA) (Roth and Giarman 1969; Gold and Roth 1977; Snead *et al.* 1989) and occurs naturally in mammalian brain (Doherty *et al.* 1978; Snead and Morley 1981). When given to animals, GHB produces a predictable sequence of electrographic and behavioral events which closely resemble generalized absence seizures seen in human. Generalized absence seizures may be defined as a paroxysmal loss of consciousness of abrupt and sudden onset and offset which is associated with bursts of bilaterally synchronous 3 c/s (cycles per second) spike-and-wave discharges (SWD) recorded on the electroencephalogram (EEG). There is no aura or postictal state. This particular type of seizure usually occurs in children between the ages of 4 years and adolescence, although they may occur at either ends of that age spectrum (Berkovic *et al.* 1987; Snead 1995; Snead *et al.* 1999). Generalized absence seizures are pharmacologically unique, responding only to ethosuximide, trimethadione, valproic acid, or benzodiazepines and being resistant to, or made worse by, phenytoin, barbiturates, or carbamazepine (Snead *et al.* 1999).

The current thinking about the pathogenesis of absence seizures dates to the landmark experiments of Jasper and Droogleever-Fortuyn (1947), who demonstrated that 3 c/s stimulation of the midline and intralaminar nuclei of the thalamus in cats could produce bilaterally synchronous SWD in the cortical EEG of those animals. The relevance of this finding to human epilepsy was demonstrated by Williams (1953), who, utilizing depth electrode recordings from the thalamus of a child with absence seizures, demonstrated that bilaterally synchronous 3 c/s SWD arose from that structure. Over the next twenty years, a debate ensued in the literature as to which was preeminent in controlling the bilaterally synchronous SWD that characterized absence seizures: the cortex (Bennett 1953; Marcus and Watson 1966; Marcus *et al.* 1968), the thalamus (Pollen *et al.* 1963), or both (Gloor 1968). With the advent of a number of animal models of generalized absence seizures, this controversy has been partially resolved. Moreover, the availability of these models has advanced our understanding of the basic mechanisms of absence seizures considerably (Snead *et al.* 1999).

In order to be a valid investigative tool, an animal model of generalized absence seizures should, in addition to reflecting the clinical and pharmacological characteristics of this disorder (Snead 1995), fulfil certain other requirements (Mirski *et al.* 1986; Fariello and Golden 1987; Snead 1988; Cortez *et al.* 2001). These criteria include reproducibility and predictability as well as the ability to standardize and quantitate the model (Table 8.1). In addition, animal models of absence seizures should reflect the fact that both clinical and experimental absence seizures are exacerbated by both direct and indirect GABA agonists

Table 8.1 Comparison of features of experimental absence seizures in the GHB rat to the genetic rat model and the human condition

	GAERs	GHB	Clinical
Bilaterally synchronous SWD	+	+	+
Immobility; starring; myoclonus	+	+	+
Precise EEG/behavioral correlation	+	+	+
Reproducibility; predictability	+	+	NA
Standardized; quantitative	+	+	NA
Blocked by ETO, VPA, TMD	+	+	+
Exacerbated by $GABA_{A,B}R$ agonists	+	+	+
Blocked by $GABA_BR$ antagonists	+	+	ND
SWD frequency (Hz)	7–11	7–9	3
SWD from thalamus and cortex	+	+	+
SWD from hippocampus	–	–	ND
Appropriate ontogeny	–	+	NA
Recurrent spontaneous absences	+	–	+

GAERS: genetic absence epilepsy from Strasbourg; GHB: γ-hydroxybutyrate model; ETO: ethosuximide; VPA: valproic acid; TMD: trimethadione; SWD: spike-and-wave discharge; $GABA_BR$: $GABA_B$ receptor; NA: non-applicable; ND: no clinical data.

(van der Linden *et al.* 1981; Vergnes *et al.* 1984; Snead 1984a, 1990; Gloor and Fariello 1988; Peeters *et al.* 1989). Finally, involvement of thalamocortical circuitry and specific non-involvement of hippocampal circuitry should be demonstrated for an animal model of absence seizures to be valid (Quesney *et al.* 1977; Gloor *et al.* 1977; Gloor 1984; Gloor and Fariello 1988; Vergnes *et al.* 1987, 1990a,b; Steriade and Llinas 1988; Steriade *et al.* 1990; Banerjee *et al.* 1993; Kandel *et al.* 1996; Cortez *et al.* 2000).

The first, most useful, and best studied of the animal models of generalized absence seizures is the feline generalized penicillin epilepsy model which resulted from the observation made by Prince and Farrell (1969) that large intramuscular doses of penicillin could induce generalized SWD in the cat which were associated with behavioral manifestations that were similar to those of generalized absence seizures. The feline generalized penicillin epilepsy model fulfils all of the criteria which have been developed for animal models of absence seizures (Table 8.1). In addition, a number of rat models, both pharmacological and genetic, have been developed which also meet these criteria (Snead *et al.* 1999). These animal models encompass many species including rodents; however, when using rodent models of experimental absence seizures, one must bear in mind the work of McQueen and Woodbury (1975) which attempted to produce bilaterally synchronous SWD in the electrocorticogram of rats by a number of experimental paradigms. These included administration of pentylenetetrazole (PTZ), picrotoxin, conjugated estrogens, and bilateral intracerebral cobalt implants. No pharmacological modality produced consistent, bilaterally synchronous SWD. The authors concluded, therefore, that the rodent was not suitable for any detailed study of the pathophysiology of SWD. Since that work was published however, a number of rat models have been developed which have direct relevance to generalized absence seizures. Although rodents are usually incapable of generating 3 c/s SWD, generalized absence seizures may be defined in this species by the pharmacological, developmental, and behavioral characteristics described above. Even though rodents lack the complexity of thalamocortical circuitry apparently required to generate 3 c/s SWD, recent data in both genetic and pharmacological models of absence indicate that thalamocortical mechanisms are involved in the genesis of bilaterally synchronous SWD in this species (Vergnes *et al.* 1987, 1990a,b; Banerjee *et al.* 1993).

The γ-hydroxybutyrate model of absence seizures

The phenomenon of GHB-induced absence seizures has been well described in cats, rats, and monkeys (Snead *et al.* 1976; Godschalk *et al.* 1977; Snead 1978a,b, 1988; Bearden *et al.* 1980). Use of the pro-drug of GHB, γ-butyrolactone (GBL), has been shown to enhance the reproducibility and predictability of the GHB model of absence seizures. GBL is used because of the consistency and rapidity of onset of its effect (Bearden *et al.* 1980). It has been shown to produce exactly the same EEG and behavioral effect as that of GHB (Snead *et al.* 1980). It has been postulated that GBL itself may have intrinsic biological activity (Vayer *et al.* 1987). If true, this would negate the validity of the use of GBL as a pro-drug for GHB in the induction of SWD. We tested the hypothesis that the epileptogenic effects of GBL are due solely to its conversion to GHB (Snead 1991). The regional brain concentration of both GHB and GBL was determined in time course and dose response studies after intraperitoneal (i.p.) administration of GBL, as well as at the onset of EEG changes induced by both GHB and GBL. Also, EEG and behavior was assessed following bilateral intrathalamic microinjection of either GHB or GBL in the rat.

The i.p. administration of GBL resulted in rapid onset of bilaterally synchronous SWD in rat that correlated with an almost immediate appearance of GHB in the brain. In animals that received i.p. GHB, the EEG changes did not occur until 20 min after GHB administration when GHB levels in the brain were peaking. The threshold brain concentration of GHB for EEG changes in both GHB- and GBL-treated animals was 240 μM. GBL concentration in the brain peaked 1 min after GBL administration and fell rapidly to undetectable levels within 5 min. Bilateral microinjection of GHB into the thalamus resulted in brief bursts of SWD, while GBL administered into the thalamus had no effect. These data confirm the hypothesis that GBL is biologically inactive in the brain and support the validity of the use of GBL as a pro-drug for GHB in this model of absence seizures.

EEG and behavioral characteristics of the GHB model

Based on the above data, the GHB model has been standardized and is used routinely as an absence seizure model in the rat. A standard dose of 0.09 ml (100 mg) GBL per kilogram given i.p. reliably produces onset of bilaterally synchronous SWD within 2–5 min of GBL administration. The frequency of the SWD is 7–9 c/s. Associated with these hypersynchronous electrographic changes are behavioral arrest, facial myoclonus, and vibrissal twitching. Therefore, this model meets the first two criteria outlined in Table 8.1 in that it is predictable, reproducible, and produces electrographic and behavioral events similar to the human condition. An additional advantage of the GHB model is that it affords control of pharmacokinetic variables in any pharmacological study since the concentration of GBL and GHB can be determined in the brain and the kinetics are known (Snead 1991).

The GHB model of absence seizures is quantitated in a manner similar to other electrographic models of absence seizures (Depaulis *et al.* 1989). GHB-induced SWD may be quantitated in terms of cumulative duration (s) per 20 min epoch of time, or as a per cent of control SWD duration. In this way, the GHB model of absence may be compared with any other rodent model of generalized absence seizures using the same pharmacological paradigm (Depaulis *et al.* 1989).

Ontogeny of the GHB model

GHB in threshold doses of 100 mg/kg was observed to produce bilaterally synchronous SWD in rats initially at postnatal day (P) 18. Earlier than that time, GHB produces varying degrees of slowing and, in very young animals, a profound burst suppression seen only at doses >200 mg/kg in adult animals (Snead 1984b, 1994). This ontogeny is similar to the developmental profile of the low-dose PTZ model of typical absence seizures (Schickerova *et al.* 1984) and the AY 9944 model of atypical absence epilepsy (Cortez *et al.* 2000). It is also noteworthy that [^{3}H]GHB binding does not appear in the rat until P18, the time when GHB-induced absence occurs.

Pharmacological characteristics of SWD in the GHB model

The therapeutic profile of the GHB model is identical to that of other genetic and pharmacological models in rodents, and therefore similar to that of absence epilepsy in humans. In the GHB model, SWD are suppressed by the four main antiepileptic drugs which are effective against human absences and worsened by the two drugs which are ineffective or aggravating in humans (Godschalk *et al.* 1976; Snead 1978a,b, 1988). GHB-induced SWD are significantly decreased by the antiabsence drugs ethosuximide, trimethadione, and valproate, and enhanced by phenytoin and carbamazepine. GHB-induced absence seizures are also attenuated by benzodiazepines. Another aspect of the pharmacology of this model is that the GHB model is exacerbated by both PTZ and penicillin, drugs which produce absence-like seizures in rodents and other species (Snead 1988). Conversely, both GHB and GBL exacerbate the GAERS genetic rat model of generalized absence (Depaulis *et al.* 1988) and the PTZ model (Snead 1988).

Similarities and discrepancies between EEG and behavioral characteristics in the GHB model and absence seizures in humans

The major disadvantage of the GHB model is that it is an acute model of absence seizures rather than a chronic model of absence epilepsy such as the genetic models. However, in terms of meeting criteria which speak of relevance to human epilepsy (Table 8.1), the GHB model is almost identical to the various genetic models of absence epilepsy. Moreover, the ontogeny of the GHB model mirrors the human condition more accurately than the genetic models. GHB-induced SWD most similar to those seen in humans are produced by intravenous administration of GHB to prepubescent monkeys. In this animal an intravenous dose of 200 mg/kg GHB results in 2.5 c/s SWD associated with behavioral immobility, head drops, starring, pupillary dilation, eyelid fluttering, rhythmic eye movements, and stereotyped automatisms (Snead 1978a).

Thalamocortical circuitry and absence seizures in the GHB model

In both rodent and feline models of absence seizures, the evidence suggests that the mechanisms that underlie the SWD bursts which characterize this seizure type may be related to the thalamocortical mechanism that mediates spindles and recruiting responses (Gloor *et al.* 1990). In the cat, the oscillatory neuronal firing pattern associated with each SWD burst seems to be initiated by the cortex and entrain the thalamus (Avoli *et al.* 1983). However, in the rat both the thalamus and cortex seem to be required for absence seizure activity since neither structure alone can sustain bilaterally synchronous SWD in any model studied in this species (Vergnes *et al.* 1987, 1990a,b). Since the human and animal data both strongly suggest that generalized absence seizures arise from aberrant thalamocortical rhythms, it may prove helpful to consider functional aspects of thalamocortical circuitry before reviewing experimental data apropos the GHB model of absence seizures.

Functional aspects of thalamocortical circuitry

The EEG is state dependent because the electrical activity of mammalian forebrain as recorded on the EEG varies with the state of consciousness. When the animal is alert, the EEG is characterized by desynchronization or a replacement of synchronized rhythms by lower amplitude and faster wave forms. Alternatively, certain altered states of consciousness (i.e. slow wave sleep) are associated with synchronous EEG activity, that is high-amplitude oscillations with relatively slow frequencies (Avoli *et al.* 1983; Steriade *et al.* 1990). These state related alterations in EEG are a reflection of fundamental and dynamic underlying changes in the activity of forebrain neurons in response to interplay between the intrinsic activity of thalamocortical circuitry with ascending neurotransmitter systems which project upon thalamocortical structures (Steriade *et al.* 1993).

Thalamic neurons have the unique ability to shift between the oscillatory and tonic firing mode. In this way, the traffic of stimuli to the cortex from the exterior world via the thalamus is regulated. The alert behavioral state is characterized by a desynchronized EEG caused by tonic firing of thalamocortical neurons which allows for faithful signal transmission from the external environment of the organism to the cortex via the thalamus. Conversely, when the firing pattern of these neurons shifts to an oscillatory, rhythmic, burst firing mode, the threshold for EPSP occurrence in the thalamus is raised, signal transmission to the cortex is dampened, and consciousness is altered, that is depressed (Steriade and Llinas 1988; Steriade *et al.* 1990, 1993; McCormick 1992; McCormick and Bal 1997; Sherman and Guillery 1996).

Oscillatory neuronal behavior within thalamocortical circuits relies upon the intrinsic ability of a group of thalamic neurons within the nucleus reticularis thalami (NRT) to impose their oscillatory behavior upon thalamocortical circuitry. The NRT forms a shell which surrounds much of the dorsal and lateral extent of the dorsal thalamus (Jones 1985) and is composed of GABAergic neurons which project heavily to one another and to almost all thalamic relay nuclei. NRT neurons receive excitatory, glutamatergic inputs from axon collaterals of thalamocortical fibers that traverse the NRT on their way from thalamic relay nuclei to the cortex and of corticothalamic fibers that project back from lamina VI of the cerebral cortex to thalamic relay nuclei (Steriade *et al.* 1984; Yen *et al.* 1985; Cornwall *et al.* 1990; Spreafico *et al.* 1991; Bal and McCormick 1993; McCormick and Bal 1997). That the NRT is uniquely situated to influence the flow of information between the thalamus and cerebral cortex is demonstrated by the fact that NRT cells show rhythmic burst firing during periods of synchronized EEG activity and tonic, single spike firing during wakefulness. The transition from slow wave sleep to the waking state is associated with an abolition of rhythmic burst firing and the appearance of tonic, single spike activity in NRT neurons (Steriade *et al.* 1986, 1993; Steriade and Llinas 1988; Bal and McCormick 1993; Contreras *et al.* 1993; Cox *et al.* 1997; Kim and McCormick 1998; Seidenbecher *et al.* 1998; Destexhe *et al.* 1999).

The cellular event which underlies the ability of NRT neurons to shift between an oscillatory and firing mode and therefore dictate EEG synchronization/desynchronization, and ultimately consciousness, appears to be the low-threshold Ca^{2+} spike. This event is triggered via $GABA_B$-mediated late IPSPs which, in turn, give rise to a rebound burst of high-frequency action potentials (von Krosigk *et al.* 1993; Huguenard and Prince 1994a). These low-threshold Ca^{2+} potentials represent a key membrane property involved in burst firing excitation and are associated with the oscillatory activity observed in thalamocortical cells during synchronized sleep (Steriade and Llinas 1988; Crunelli and Leresche 1991). A major role of low-threshold Ca^{2+} potentials in the pathogenesis of generalized absence seizures is suggested by the observation that the antiepileptic drugs ethosuximide and trimethadione may owe their antiabsence activity to their ability to decrease low-threshold calcium currents (Huguenard and Prince 1994a; Coulter *et al.* 1989a,b, 1990). It should be noted that, recently, the hypothesis that rhythmic sequences of $GABA_B$R IPSPs alternating with low-threshold Ca^{2+} potentials play a definitive role in the pathogenesis of absence seizures has been challenged (Pinault *et al.* 1998; Leresche *et al.* 1998).

Thalamocortical circuitry involvement in the GHB model of absence seizures

The regional distribution of GHB-induced SWD has been determined by the use of EEG mapping and lesional studies. In order to carry out the EEG mapping studies, bipolar depth electrodes were implanted in discrete regions of the thalamus, cortex, and hippocampus in rat. With the advent of GHB-induced absence seizures, the ventroposterolateral (VPL), ventroposteromedial (VPM), medial, and reticular nuclei (RT) of the thalamus discharged synchronously with layers I–IV of the cerebral cortex. No SWD were recorded from deeper layers (V–VI) of the cerebral cortex. Hippocampal structures were completely silent during the GHB-induced SWD (Banerjee *et al.* 1993).

The effect of bilateral electrolytic lesions of various thalamic nuclei on the GHB-induced absence seizures has also been determined. Bilateral lesions in mediodorsal (MD) and intralaminar thalamic nuclei abolished GHB-induced SWD from both cortex and thalamus. Bilateral lesions of the VPL and RT suppressed but did not eliminate GHB-induced SWD. The emanation of SWD from superficial layers of cortex during GHB-induced absence suggests that the projections from MD thalamic nuclei to those superficial cortical laminae rich in [^{3}H]GHB binding sites form an integral part of thalamocortical circuitry involved in GHB-induced absence seizures (Banerjee *et al.* 1993; Banerjee and Snead 1994). In agreement with this hypothesis are the thalamic *in vivo* microdialysis data reviewed below.

Glutamatergic and GABAergic mechanisms of GHB-induced absence seizures

Glutamate   Glutamate receptor-mediated, recurrent excitation between thalamocortical and corticothalamic pathways, both of which project excitatory axon collaterals to the NRT, are integral to the modulation and regulation of thalamocortical rhythmicity (McCormick 1992; McCormick and von Krosigk 1992; McCormick *et al.* 1993; Block 1994; Fox *et al.* 1996). Intraperitoneal or intracerebroventricular injections of both *N*-methyl-D-aspartate (NMDA) and NMDA antagonists dose dependently suppressed SWD in the GHB model. Bilateral infusions of the same drugs in the lateral relay nuclei of the thalamus had the same type of suppressive effects (Banerjee and Snead 1992, 1995a). Also, systemic administration of GBL has been associated with a decrease in K^+-evoked glutamate release in the thalamus, the onset and duration of which correlates with that of GHB-induced absence seizures. However, basal release of glutamate was unaltered in these experiments (Banerjee and Snead 1995b). Using the same experimental design, Hu *et al.* (2000a) saw no alteration in basal or in K^+-evoked glutamate release in superficial laminae of cerebral cortex, that region of cortex from which SWD emanate in the GHB model of absence seizures.

We have explored the involvement of the glutamate receptor subunit B (GluR2) in the mechanism of absence seizures induced by GHB. The expression of GluR2 protein and the distribution of GluR2 in rat brain were examined following the onset of GHB-induced absence seizures. The data indicate that GluR2 protein expression significantly decreases following the onset of GHB-induced absence seizures. The suppression of GluR2 expression was quite prolonged and outlasted the duration of the absence seizure activity by many hours. Absence seizure-induced alteration of GluR2 protein levels was accompanied by a redistribution of GluR2 expression from lamina V to lamina IV in the cerebral cortex. To address whether the observed absence seizure-induced alteration of GluR2 protein levels contributed to subsequent absence seizure severity, we analyzed the duration and latency of absence seizures induced by GHB 72 h following an initial GHB-induced absence seizure, a time when suppression of GluR2 was maximal. The second GHB-induced absence seizure was significantly more prolonged than the first. These data suggest that GluR2 expression is altered in response to absence seizures in the GHB model. Absence seizure-induced down-regulation of GluR2 may contribute to the potentiation of subsequent absence seizures in animals following an initial GHB-induced absence seizure. As well, the data raises the possibility that absence seizure-induced down-regulation of GluR2 is involved in the mechanisms of the maintenance of recurrent absence seizure activity once it is initiated and, therefore, may contribute to the chronicity of seizures in absence epilepsy (Hu and Snead 1998).

In a related study, we examined the susceptibility of GluR2 mutant null mice (GluR2$^{-/-}$) to absence seizures induced by GHB in comparison to wild-type controls (GluR2$^{+/=}$). GBL was given systemically to induce the absence seizures. The latency from GBL administration to onset of seizure was significantly prolonged in GluR2$^{-/-}$ mice when compared to GluR2$^{+/+}$ mice. The duration of SWD was also significantly decreased in the GluR2$^{-/-}$ mice compared to wild-type controls. Ninety minutes following GBL administration, wild-type animals continued to exhibit intermittent SWD bursts while GluR2$^{-/-}$ mice had returned to baseline. These data suggest that the GluR2 subunit may be required for the initiation and maintenance of absence seizures induced by GHB (Hu *et al.* 2000b). We have also examined the GHB model of absence seizures in mGluR4 knockout mice, but have observed no differences between mutant and wild-type mice in either latency to seizure onset or SWD duration (Snead *et al.* 2000).

GABA   During SWD recorded from the thalamus and cortex in experimental absence seizures, there is preservation of GABAergic function as well as sparing of classical IPSPs and the absence of typical depolarizing shifts seen in focal epileptogenesis (Gloor 1984; Gloor and Fariello 1988; Kostopoulos 1986; Avoli *et al.* 1990). Preservation of GABAergic inhibition contributes to synchronization of SWD because the pacing of SWD and bursts of action potentials are dependent upon rhythmically recurring inhibition in both the cortex and thalamus (Gloor and Fariello 1988) which is driven by NRT GABAergic neurons (Steriade *et al.* 1993; Bal and McCormick 1993; von Krosigk *et al.* 1993; McCormick and Bal 1997). Enhancement of GABAergic inhibition in the brain potentiates clinical (van der Linden *et al.* 1981; Parker *et al.* 1998) and all experimental forms of generalized absence and may even be sufficient to produce it under certain conditions (Vergnes *et al.* 1984; Snead 1984a, 1990, 1998; Gloor and Fariello 1988; Myslobodsky *et al.* 1988; Peeters *et al.* 1989; Budziszewska *et al.* 1999; Hosford and Wang 1998).

$GABA_A$ receptors ($GABA_AR$) GHB-induced absence seizures have been shown to be potentiated by $GABA_AR$ agonists, but they are not blocked by $GABA_AR$ antagonists such as bicuculline (Snead 1984a). Low doses of $GABA_AR$ antagonists, including bicuculline, in rodents induce absence-like seizures which are similar to those induced by GHB (Depaulis *et al.* 1989; Matejovska *et al.* 1998). However, autoradiographic binding studies have indicated no changes in binding to the $GABA_AR$ before, or during, GHB-induced absence, indicating that the $GABA_AR$ is not implicated in the genesis of GHB-induced absence (Snead and Liu 1993).

Benzodiazepines are known to augment $GABA_AR$-mediated neurotransmission in the brain (Vicini 1999; Mehta and Ticku 1999). In the light of the data discussed above, benzodiazepines would be predicted to exacerbate absence seizures, yet they block GHB-induced absence seizures and are useful therapeutically in absence epilepsy. This apparent paradox may be explained by the fact that benzodiazepines have been shown to enhance $GABA_AR$-mediated inhibition within the NRT *in vivo*. This action would be predicted to suppress $GABA_BR$-mediated inhibition in thalamic relay neurons (Huguenard and Prince 1994a; Gibbs *et al.* 1996; Ulrich and Huguenard 1997) with a resultant decrease of SWD duration in generalized absence seizures. Conversely, low doses of $GABA_AR$ antagonists induce absence seizures by blocking intra-NRT $GABA_AR$-mediated inhibition (Huguenard and Prince 1994b; Ulrich and Huguenard 1997; Snead *et al.* 2000) rather than acting upon the corticothalamic pathways where GHB appears to exert its effect to induce absence seizures (Banerjee and Snead 1995b; Hu *et al.* 2000a).

Although there are no data to support the hypothesis that $GABA_AR$-mediated mechanisms are involved in the pathogenesis of GHB-induced absence seizures, there are data which suggest that GHB-induced absence seizures may result in a significant alteration of the $\alpha4$ subunit of the $GABA_AR$. During the course of GHB-induced absence seizures, the neurosteroids alphaxalone and tetrahydrodeoxycorticosterone failed to increase [^{3}H]flunitrazepam and *t*-butylbicyclophosphoro[^{35}S]thionate binding in the thalamic relay nuclei. However, as the GHB seizures ceased, the effect of neurosteroids on binding in the thalamus was restored. Further, there was a marked decrease in $GABA_AR$ $\alpha4$ mRNA level in thalamic relay nuclei while the GHB seizures were terminating and the neurosteroid modulation of binding was returning to the baseline level. It appears from these studies that GHB absence seizure-induced decrease in the levels of relatively neurosteroid-insensitive $GABA_AR$ $\alpha4$ mRNA in the thalamus may be related to cessation of absence seizures in the GHB model and the restoration of neurosteroid modulation of $GABA_AR$ binding (Banerjee *et al.* 1998a,b).

$GABA_B$ receptors ($GABA_BR$) The discovery that the late, long-lasting, K^+-dependent IPSP was mediated by the $GABA_BR$ and that $GABA_BR$-mediated IPSPs in the thalamus activated low-threshold calcium potentials which led to burst firing and oscillatory behavior in thalamic neurons gave rise to the hypothesis that $GABA_BR$-mediated mechanisms might be operative in the pathogenesis of generalized absence seizures (Crunelli and Leresche 1991; Snead 1995). This hypothesis was given further credence by three separate groups of experiments. A specific $GABA_BR$ antagonist, CGP 35348, has been shown to dose-dependently attenuate or block SWD in (1) the GAERS rat model (Liu *et al.* 1992; Marescaux *et al.* 1992), (2) the GHB and PTZ rat model (Snead 1992), and (3) the *lh/lh* mouse, a proposed model of absence epilepsy (Hosford *et al.* 1992). In addition, other $GABA_BR$ antagonists have

been shown to be effective against PTZ- and GHB-induced absence seizures as well as in the *lh/lh* mouse model (Hosford *et al.* 1995; Snead 1996a). Pretreatment with the GABA$_B$R agonist, baclofen, exacerbated the SWD in the GHB and PTZ models of absence much more potently than the GABA$_A$R agonist, muscimol (Snead 1992, 1996a).

These experiments, taken in conjunction with the *in vitro* data cited above which demonstrates the role of GABA$_B$R-mediated IPSPs in regulating thalamocortical oscillatory behavior via low-threshold calcium currents (Steriade *et al.* 1993; Bal and McCormick 1993; von Krosigk *et al.* 1993), strongly suggest that GABA$_B$R-mediated mechanisms might be involved in the pathogenesis of absence seizures. Indeed, the animal model data suggest that the GABA$_B$ receptor may be the final common pathway in absence seizures. However, GHB-induced absence seizures are also blocked by the GHB antagonist NCS-382 as well as by GABA$_B$R antagonists (Snead 1996a), giving rise to the question of whether the GHB receptor (GHBR) or the GABA$_B$R is primarily involved in the pathogenesis of GHB-induced absence seizures.

Are GHB-induced absence seizures mediated by the GABA$_B$R, the GHBR, or both?

A high-affinity binding site for GHB has been identified in the brain (Benavides *et al.* 1982; Snead and Liu 1984), and it has been proposed that this binding site represents a specific receptor belonging to the G-protein superfamily (Ratomponirina *et al.* 1995; Kemmel *et al.* 1998). There appear to be a high- and a low-affinity component to the GHB binding site with a K_d of 30–580 nM and 2.3–16 μM, respectively (Maitre 1997), and GHB binding is sensitive to pertussis toxin (Ratomponirina *et al.* 1995; Snead 1996b). There is no GHB binding in cerebellum.

The question as to whether the GHB binding site is a GABA$_B$R or a specific GHBR remains unresolved to date. GHB has been hypothesized to be a weak agonist at the GABA$_B$R based on binding (Bernasconi *et al.* 1999; Mathivet *et al.* 1997) and electrophysiological (Erhardt *et al.* 1998; Madden and Johnson 1998) studies. However, electrophysiological experiments which indicate that GHB induces long-lasting IPSPs and rebound Ca^{2+} spikes in a manner similar to the GABA$_B$R agonist (−) baclofen (Williams *et al.* 1995; Xie and Smart 1992) indicate that the concentration of GHB required to mimic the postsynaptic effects of baclofen is in the millimolar range. This concentration of GHB is considerably higher than the concentration of GHB in the brain which is the threshold for absence seizures (240 μM), but is commensurate with the concentration of GHB achieved in the brain with systemic administration of high doses (>500 mg/kg) of GHB to rats which are associated with burst suppression on the EEG, loss of righting reflex, and 'anesthesia' (Snead 1991). Also, when interpreting the data concerning the 'GABA$_B$R agonist' properties of GHB in terms of the concentrations of GHB used, one should bear in mind that the normal brain concentrations of GHB in rat brain are of the order of 2 μM (Snead and Morley 1981). Moreover, the electrophysiologic effects of GHB on the postsynaptic GABA$_B$R, while antagonized by a GABA$_B$R antagonist, are not blocked by a specific GHB antagonist (Williams *et al.* 1995; Emri *et al.* 1996). Additional lines of evidence that mitigate against the hypothesis that GHB is a physiologically significant GABA$_B$R agonist, or that GABA$_B$R-mediated mechanisms are responsible for GHB-induced absence seizures, are the inability of baclofen to displace [^{3}H]GHB binding (Snead 1996b) and the differences between [^{3}H]GHB and [^{3}H]GABA$_B$ binding in rat brain in their ontogeny and regional distribution (Snead 1994), as well as the finding that there is an exact correlation between the developmental appearance of [^{3}H]GHB binding and the emergence of GHB-induced absence seizures (Snead 1994).

There are few data that address the molecular aspects of the GABA$_B$R in relation to GHB. GABA$_B$R1a and GABA$_B$R1b were cloned in 1997 (Kaupmann *et al.* 1997). Recently, it has been shown by a number of laboratories that GABA$_B$R2 is co-expressed with GABA$_B$R1 in many brain regions and that inwardly rectifying potassium channels are activated by GABA$_B$R agonists only upon co-expression of GABA$_B$R1 with GABA$_B$R2 (Kaupmann *et al.* 1998; White *et al.* 1998; Jones *et al.* 1998; Kuner *et al.* 1999). Lingenhoehl *et al.* (1999) have shown that GHB can activate GABA$_B$R1/R2 co-expressed with Kir3 channels in Xenopus oocytes with an EC$_{50}$ of approximately 5 mM. This affinity is too low to explain a high-affinity GHB binding site on GABA$_B$R. Further, an EC$_{50}$ of 5 mM in these experiments makes it unlikely that GABA$_B$R is the receptor that mediates the depolarizing effects of low doses of GHB (Diana *et al.* 1991; Godbout *et al.* 1995), the inhibition of Ca^{2+} conductances in neurohybridoma cells in culture (Kemmel *et al.* 1998), or the induction of absence seizures in whole animals (Snead 1991) since the concentration of GHB utilized to induce these effects is too low to activate recombinant GABA$_B$R1/R2.

Much of the investigation of GABA$_B$R-mediated mechanisms in absence seizures has focused upon postsynaptic mechanisms. However, the postsynaptic GABA$_B$R in the thalamus and cortex differs functionally from the presynaptic GABA$_B$R within this circuitry. The postsynaptic GABA$_B$R mediates the late IPSP in the thalamus which activates burst firing and oscillatory behavior in thalamic neurons (Crunelli and Leresche 1991). The presynaptic GABA$_B$R regulates excitability within thalamocortical circuitry by the modulation of glutamate and GABA release (Mott and Lewis 1994). Therefore, we sought to test the alternative hypothesis that GHB acts via a GHB-specific, G-protein-coupled presynaptic receptor which is different from the GABA$_B$R.

The effect of GHB on regional and subcellular brain adenylyl cyclase in adult and developing rats was determined and compared to that of the GABA$_B$R agonist (−) baclofen. Also, using [^{35}S]GTPγS binding and low K_m GTPase activity as markers, we determined the effects of GHB and (−) baclofen on G-protein activity in the brain. Neither GHB nor baclofen had an effect on basal cAMP. GHB significantly decreased forskolin-stimulated cAMP by 40–50% in the cortex and hippocampus, but not in the thalamus or cerebellum, while (−) baclofen had an effect throughout the brain. The effect of GHB on adenylyl cyclase was observed in presynaptic and not postsynaptic subcellular tissue preparations, but the effect of baclofen was observed in both subcellular preparations. The GHB-induced alteration in forskolin-induced cAMP formation was blocked by a specific GHB antagonist, but not by a specific GABA$_B$R antagonist. The (−) baclofen-induced alteration in forskolin-induced cAMP formation was blocked by a specific GABA$_B$R antagonist, but not by a specific GHB antagonist. The negative coupling of GHB to adenylyl cyclase appeared at P21, a developmental time point that is concordant with the developmental appearance of [^{3}H]GHB binding in cerebral cortex, but the effects of (−) baclofen were present by P14. GHB and baclofen both stimulated [^{35}S]GTPγS binding and low K_m GTPase activity by 40–50%. The GHB-induced effect was blocked by GHB antagonists but not by GABA$_B$R antagonists and was seen only in the cortex and hippocampus. The (−) baclofen-induced effect was blocked by GABA$_B$R antagonists but not by GHB antagonists and was observed throughout the brain. These data support the hypothesis that GHB induces a G-protein-mediated decrease in adenylyl cyclase via a GHB-specific G-protein-coupled presynaptic receptor which is different from the GABA$_B$R (Snead 2000).

The hypothesis of presynaptic mechanism of GHB-induced absence is given additional credence by experiments which have demonstrated a robust decrease in the basal release of GABA in the thalamus and cortex in the GHB model; however, there was no concomitant alteration in basal glutamate release. Moreover, the GHB-induced presynaptic inhibition of GABA release in the thalamus and cortex occurred at concentrations of GHB that are threshold for the occurrence of absence seizures, and showed a time course similar to that of the absence seizures induced by GHB (Banerjee and Snead 1995b; Hu *et al.* 2000a).

Summary and conclusions

In summary, GHB has the ability to induce absence seizures in a number of species. GHB-induced absence seizures involve thalamocortical circuitry and result in significant molecular perturbations of both GluR2 and $GABA_AR$. The mechanism by which GHB induces absence appears to be a presynaptic decrease in GABA release within thalamocortical circuitry without a corresponding alteration of glutamate release. The resultant increase in excitation within thalamocortical circuitry triggers hypersynchronous activity within this neural network and absence seizures. It is unclear whether GHB acts at a specific presynaptic GHBR or a presynaptic GHB-sensitive $GABA_BR$ to produce this effect. The data suggest that the low- and high-affinity GHB binding sites are different from the $GABA_BR$ and that some of the pharmacological effects of GHB are mediated by a presynaptic GHBR which is different from the $GABA_BR$. However, the dose response data suggest that as the concentration of GHB in the brain increases, the $GABA_BR$ is recruited. These dose response data are directly relevant to the mechanisms not only of GHB-induced absence seizures, but also to the mechanism of the addictive properties of GHB.

The normal brain concentrations of GHB in rat brain are of the order of 2 μM, while the K_d of GHB for the low-affinity binding site is 1.5 μM. In a number of species, including monkey, there is a progression of EEG and behavioral changes with increasing doses of GHB. Low doses of GHB (10–25 mg/kg) produce memory problems with few EEG changes. As the dosage increases from 50 to 100 mg/kg, absence-like seizures emerge. Both these abnormalities can be reversed with the GHB antagonist NCS-382 and $GABA_BR$ antagonists (Snead 1994). The brain concentration of GHB associated with absence seizures is 100–200 μM (Snead 1991). At a dosage of GHB of 150–300 mg/kg in rat and a brain GHB concentration ≥400 μM, EEG slowing appears which is associated with a stuporous state. These GHB-induced changes are only partially reversed by NCS-382, but completely reversed by $GABA_BR$ antagonists (Snead 1994). With GHB or GBL doses in excess of 300 mg/kg in the rat, cat, and monkey, burst suppression appears on the EEG that is associated with deep coma. As the dosage of GHB increases past 400 mg/kg, the isoelectric period in the burst suppression pattern in the EEG gets more and more prolonged. These doses result in millimolar concentrations of GHB in the brain (Snead *et al.* 1976; Snead 1991). It takes millimolar concentrations of GHB to activate $GABA_BR$ (Lingenhoehl *et al.* 1999; Williams *et al.* 1995; Emri *et al.* 1996). As the dose of GHB exceeds 1,000 mg/kg in rats, mortality appears. The LD_{50} of GHB in this species is 1,500 mg/kg.

Taken together the dose response, pharmacokinetic, binding, molecular, and electrophysiology data suggest that GHB-induced absence seizures involve GHBR-mediated mechanisms, but that those effects of GHB that require higher doses (i.e. 'anesthesia', EEG burst suppression, psychotropic effects, and abuse potential) are mediated by $GABA_BR$.

References

Avoli M., Gloor P., Kostopoulos G. and Gotman J. (1983) An analysis of penicillin-induced generalized spike and wave discharges using simultaneous recordings of cortical and thalamic single neurons. *J. Neurophysiol.* 50, 819–37.

Avoli M., Gloor P. and Kostopoulos G. (1990) Focal and generalized epileptiform activity in the cortex: in search of differences in synaptic mechanisms, ionic movements, and long lasting changes in neuronal excitability. In *Generalized Epilepsy: Neurobiological Approaches* (Avoli M., Gloor P., Kostopoulos G. and Naquet R., eds), pp. 213–31. Birkhauser, Boston.

Bal T. and McCormick D. A. (1993) Mechanisms of oscillatory activity in guinea pig nucleus reticularis thalami *in vitro*: a mammalian pacemaker. *J. Physiol.* 468, 669–91.

Banerjee P. K. and Snead O. C. (1992) Excitatory amino acid-mediated mechanisms in the γ-hydroxybutyrate model of absence. *Neuropharmacology* 31, 1009–19.

Banerjee P. K. and Snead O. C. (1994) Thalamic mediodorsal and intralaminar nuclear lesions disrupt the generation of experimentally induced generalized absence-like seizures in rats. *Epilepsy Res.* 17, 193–205.

Banerjee P. K. and Snead O. C. (1995a) Thalamic NMDA receptors in the γ-hydroxybutyrate model of absence seizures: a cerebral microinjection study in rats. *Neuropharmacology* 34, 43–53.

Banerjee P. K. and Snead O. C. (1995b) Presynaptic γ-hydroxybutyric acid (GHB) and γ-aminobutyric acid$_B$ receptor-mediated release of GABA and glutamate (GLU) in rat thalamic ventrobasal nucleus (VB): a possible mechanism for the generation of absence-like seizures induced by GHB. *J. Pharmacol. Exp. Ther.* 273, 1534–43.

Banerjee P. K., Hirsch E. and Snead O. C. (1993) γ-Hydroxybutyric acid induced spike and wave discharges in rats: relation to high affinity [^{3}H]-γ-hydroxybutyric acid binding sites in the thalamus and cortex. *Neuroscience* 56, 11–21.

Banerjee P. K., Olsen R. W., Tillakaratne N. J. K., Brailowsky S., Tobin A. J. and Snead O. C. (1998a) Absence seizures decrease steroid modulation of *t*-[^{35}S]butylbicyclo-phosphorothionate (TBPS) binding in thalamic relay nuclei. *J. Pharmacol. Exp. Pharmacol.* 287, 766–72.

Banerjee P. K., Tillakaratne N. J. K., Brailowsky S., Tobin A. J., Olsen R. W. and Snead O. C. (1998b) Alterations in GABA$_A$ receptor α1 and α4 subunit mRNA levels in thalamic relay nuclei following absence-like seizures in rats. *Exp. Neurol.* 154, 213–23.

Bearden L. J., Snead O. C., Healy C. T. and Pegram G. V. (1980) Antagonism of gamma hydroxybutyrate-induced frequency shifts in the cortical EEG of rats by dipropylacetate. *Electroencephalogr. Clin. Neurophysiol.* 49, 181–83.

Benavides J., Rumigny J. F., Bourguignon J. J., Wermuth C. G., Mandel P. and Maitre M. (1982) High affinity binding sites for γ-hydroxybutyric acid in rat brain. *Life Sci.* 30, 953–61.

Bennett F. E. (1953) Intracarotid and intravertebral metrazol in petit mal epilepsy. *Neurology* 3, 668–73.

Berkovic S. F., Andermann F., Andermann E. and Gloor P. (1987) Concepts of absence epilepsies: discrete syndromes or biological continuum? *Neurology* 37, 993–1000.

Bernasconi R., Mathivet P., Bischoff S. and Marescaux C. (1999) Gamma-hydroxybutyric acid: an endogenous neuromodulator with abuse potential. *Trends Pharmacol. Sci.* 20, 135–41.

Block F. (1994) Stimulation of *N*-methyl-D-aspartate receptors in the rat nucleus reticularis thalami suppresses somatosensory evoked potentials. *Brain Res.* 636, 143–46.

Budziszewska B., Van Luijtelaar G., Coenen A. M., Leskiewicz M. and Lason W. (1999) Effects of neurosteroids on spike-wave discharges in the genetic epileptic WAG/Rij rat. *Epilepsy Res.* 33, 23–29.

Contreras D., Dossi R. C. and Steriade M. (1993) Electrophysiological properties of cat reticular thalamic neurones *in vivo*. *J. Physiol.* 470, 273–94.

Cornwall J., Cooper J. D. and Phillipson O. T. (1990) Projections to the rostral reticular thalamic nucleus in the rat. *Exp. Brain Res.* 80, 157–71.

Cortez M. A., McKerlie C. and Snead O. C. (2001) A model of atypical absence seizures: EEG, pharmacology, and developmental characterization. *Neurology* 56, 341–49.

Coulter D. A., Huguenard J. R. and Prince D. A. (1989a) Specific petit mal anticonvulsants reduce calcium currents in thalamic neurons. *Neurosci. Lett.* 98, 74–78.

Coulter D. A., Huguenard J. R. and Prince D. A. (1989b) Characterization of ethosuximide reduction of low-threshold calcium current in thalamic neurons. *Ann. Neurol.* 25, 582–93.

Coulter D. A., Huguenard J. R. and Prince D. A. (1990) Differential effects of petit mal anticonvulsants and convulsants on thalamic neurones: calcium current reduction. *Br. J. Pharmacol.* 100, 800–05.

Cox C.L., Huguenard J. R. and Prince D. A. (1997) Nucleus reticularis neurons mediate diverse inhibitory effects in thalamus. *Proc. Natl Acad. Sci. USA* 94, 8854–59.

Crunelli V. and Leresche N. (1991) A role for GABA$_B$ receptors in excitation and inhibition of thalamocortical cells. *Trends Neurosci.* 14, 16–21.

Depaulis A., Bourguignon J., Marescaux C., Vergnes M., Schmitt M., Micheletti G. and Warter J. M. (1988) Effect of γ-hydroxybutyrate and γ-butyrolactone derivatives on spontaneous generalized non-convulsive seizures in the rat. *Neuropharmacology* 27, 6863–69.

Depaulis A., Snead O. C., Marescaux C. and Vergnes M. (1989) Suppressive effects of intranigral injection of muscimol in three models of generalized non-convulsive epilepsy induced by chemical agents. *Brain Res.* 498, 64–72.

Destexhe A., McCormick D. A. and Sejnowski T. J. (1999) Thalamic and thalamocortical mechanisms underlying 3 Hz spike-and-wave discharges. *Prog. Brain Res.* 121, 289–307.

Diana M., Meuer G., Mura A., Fadda F., Passino N. and Gessa G. (1991) Low doses of gamma-hydroxybutyric acid stimulate the firing rate of dopaminergic neurons in unanesthetized rats. *Brain Res.* 566, 208–11.

Doherty J. D., Hattox S. E., Snead O. C. and Roth R. H. (1978) Positive identification of endogenous gamma-hydroxybutyrate in human and bovine brain and its regional distribution in human, guinea pig, and rhesus monkey brain. *J. Pharmacol. Exp. Ther.* 207, 130–39.

Emri Z., Antal K. and Crunelli V. (1996) Gamma-hydroxybutyric acid decreases thalamic sensory excitatory postsynaptic potentials by an action on presynaptic GABA$_B$ receptors. *Neurosci. Lett.* 216, 121–24.

Erhardt S., Andersson B., Nissbrandt H. and Engberg G. (1998) Inhibition of firing rate and changes in the firing pattern of nigral dopamine neurons by γ-hydroxybutyric acid (GHBA) are specifically induced by activation of GABA$_B$ receptors. *Naunyn-Schmied. Arch. Pharmacol.* 357, 611–19.

Fariello R. G. and Golden G. T. (1987) The THIP-induced model of bilateral synchronous spike and wave in rodents. *Neuropharmacology* 26, 161–65.

Fox K., Schilaggar B. I., Glazewski S. and O'Leary D. D. M. (1996) Glutamate receptor blockade at cortical synapses disrupts development of thalamocortical and columnar organization in somatosensory cortex. *Proc. Natl Acad. Sci. USA* 93, 5584–89.

Gibbs J. W., Berkow Schroder G. and Coulter D. A. (1996) GABA$_A$ receptor function in developing rat thalamic reticular neurons: whole cell recordings of GABA-mediated currents and modulation by clonazepam. *J. Neurophysiol.* 76, 2568–79.

Gloor P. (1968) Generalized cortico-reticular epilepsies. Some considerations on the pathophysiology of generalized bilaterally synchronous spike and wave discharge. *Epilepsia* 9, 249–63.

Gloor P. (1984) Electrophysiology of generalized epilepsy. In *Electrophysiology of Epilepsy* (Schwartzkroin P. A. and Wheal H., eds), pp. 107–36. Academic Press, New York.

Gloor P. and Fariello R. G. (1988) Generalized epilepsy: some of its cellular mechanisms differ from those of focal epilepsy. *Trends Neurosci.* 11, 63–68.

Gloor P., Quessney L. F. and Zumsteim H. (1977) Pathophysiology of generalized penicillin epilepsy in the cat: the role of cortical and subcortical structures. II. Topical application of penicillin to the cerebral cortex and to subcortical structures. *Electroencephalogr. Clin. Neurophysiol.* 43, 79–94.

Gloor P., Avoli M. and Kostopoulos G. (1990) Thalamo-cortical relationships in generalized epilepsy with bilaterally synchronous spike and wave discharge. In *Generalized Epilepsy: Neurobiological Approaches* (Avoli M., Gloor P., Kostopoulos G. and Naquet R., eds), pp. 190–212. Birkhauser, Boston.

Godbout R., Jelenic P., Labrie C., Schmitt M. and Bourguignon J. J. (1995) Effect of gamma-hydroxybutyrate and its antagonist NCS-382 on spontaneous cell firing in the prefrontal cortex of rat. *Brain Res.* 673, 157–60.

Godschalk M., Dzoljic M. R. and Bonta I. L. (1976) Antagonism of gamma-hydroxybutyrate-induced hypersynchronization in the ECoG of rat by anti-petit mal drugs. *Neurosci. Lett.* 3, 1173–78.

Godschalk M., Dzoljic M. R. and Bonta I. L. (1977) Slow wave sleep and a state resembling absence epilepsy induced in the rat by γ-hydroxybutyrate. *Eur. J. Pharmacol.* 144, 105–11.

Gold B. I. and Roth R. H. (1977) Kinetics of *in vivo* conversion of γ-[^{3}H]aminobutyric acid to γ-[^{3}H]hydroxybutyric acid by rat brain. *J. Neurochem.* 28, 1069–73.

Hosford D. A. and Wang Y. (1998) Utility of the lethargic (lh/lh) mouse model of absence seizures in predicting the effects of lamotrigine, vigabatrin, tiagabine, gabapentin, and topiramate against human absence seizures. *Epilepsia* 38, 408–14.

Hosford D. A., Clark S., Cao Z., Wilson W. A., Lin F. H.., Morrisett R. A. and Huin A. (1992) The role of GABA$_B$ receptor activation in absence seizures of lethargic (lh/lh) mice. *Science* 257, 398–401.

Hosford D. A., Liu C. C., Wang Y. and Snead O. C. (1995) Characterization of the anti-absence efficacy of SCH50911, a GABA$_B$ receptor antagonist, in the lethargic mouse, γ-hydroxybutyric acid, and pentylenetetrazole models of absence seizures. *J. Pharmacol. Exp. Ther.* 274, 1399–403.

Hu R. Q. and Snead O. C. (1998) Up-regulation of GluR2 subunit of the glutamate receptor in rat cortex and thalamus in γ-hydroxybutyric acid (GHB) model of absence seizures. *Soc. Neurosci. Abs.* 24, 1936.

Hu R. Q., Banerjee P. K. and Snead O. C. (2000a) Regulation of γ-aminobutyric acid (GABA) and glutamate release mediated by the GABA$_B$/γ-hydroxybutyric acid (GHB) receptor complex in rat cerebral cortex in the GHB model of absence seizures in rat. *Neuropharmacology* 39, 427–39.

Hu R. Q., Cortez M. A., Roder J. and Snead O. C. (2000b) GluR2 knockout mice have a decreased sensitivity to absence seizures. *Soc. Neurosci. Abs.* 26, 1778.

Huguenard J. R. and Prince D. A. (1994a) Intrathalamic rhythmicity studied *in vitro*: nominal T-current modulation causes robust antioscillatory effects. *J. Neurosci.* 14, 5485–502.

Huguenard J. R. and Prince D. A. (1994b) Clonazepam suppresses GABA$_B$-mediated inhibition in thalamic relay neurons through effects in nucleus reticularis. *J. Neurophysiol.* 71, 2576–81.

Jasper H. H. and Droogleever-Fortuyn J. (1947) Experimental studies on the functional anatomy of petit mal epilepsy. *Res. Publ. Assoc. Nerv. Ment. Dis.* 26, 272–98.

Jones E. G. (1985) *The Thalamus.* Plenum, New York.

Jones K. A., Borowsky B., Tamm J. A., Craig D. A., Durkin M. M., Dai M., Yao W.-J., Johnson M., Gunwaldsen C., Huang L.-Y., Tang C., Shen Q., Salon J. A., Morse K., Laz T., Smith K. E., Nagarathnam D., Noble S. A., Branchek T. A. and Gerald C. (1998) GABA$_B$ receptors function as a heteromeric assembly of the subunits GABA$_B$R1 and GAB$_B$R2. *Nature* 396, 674–78.

Kandel A., Bragin A., Carpi D. and Buzsaki G. (1996) Lack of hippocampal involvement in a rat model of petit mal epilepsy. *Epilepsy Res.* 23, 123–27.

Kaupmann K., Huggel K., Hjeid J., Flor P. J., Bishoff S., Mickel S. J., McMaster G., Angst C., Bittiger H., Froestl W. and Bettler B. (1997) Expression cloning of GABA$_B$ receptors uncovers similarity to metabotropic glutamate receptors. *Nature* 386, 239–46.

Kaupmann K., Malitschek B., Schuler V., Heid J., Froestl W., Beck P., Mosbacher J., Bishoff S., Kulik A., Shigemoto R., Karschin A. and Bettler B. (1998) GABA$_B$ receptor subtypes assemble into functional heteromeric complexes. *Nature* 396, 683–86.

Kemmel V., Taleb O., Perard A., Andriamampandry C., Siffert J. C., Mark J. and Maitre M. (1998) Neurochemical and electrophysiological evidence for the existence of a functional γ-hydroxybutyrate system in NCB-20 neurons. *Neuroscience* 86, 989–1000.

Kim K. and McCormick D. A. (1998) The functional influence of burst and tonic firing mode on synaptic interactions in the thalamus. *J. Neurosci.* 18, 1–17.

Kostopoulos G. (1986) Neuronal sensitivity to GABA and glutamate in generalized epilepsy with spike and wave discharges. *Exp. Neurol.* 92, 20–36.

Kuner R., Köhr G., Grünewald S., Eisenhardt G., Bach A. and Kornau H. C. (1999) Role of heteromer formation in GABA$_B$ receptor function. *Science* 283, 74–77.

Leresche N., Parri H. R., Erdemli G., Guyon A., Turner J. P., Williams S. R., Asprodine E. and Crunelli V. (1998) On the action of the anti-absence drug ethosuximide in the rat and cat thalamus. *J. Neurosci.* 18, 4842–53.

Lingenhoehl K., Brom R., Heid J., Beck P., Froes W., Kaupmann K., Bettler B. and Mosbacher J. (1999) γ-Hydroxybutyrate is a weak agonist at recombinant GABA$_B$ receptors. *Neuropharmacology* 38, 1667–73.

Liu Z., Vergnes M., Depaulis A. and Marescaux C. (1992) Involvement of intrathalamic GABA$_B$ neurotransmission in the control of absence seizures in the rat. *Neuroscience* 48, 87–93.

McCormick D. A. (1992) Neurotransmitter actions in the thalamus and cerebral cortex and their role in neuromodulation of thalamocortical activity. *Prog. Neurobiol.* 39, 337–88.

McCormick D. A. and Bal T. (1997) Sleep and arousal: thalamocortical mechanisms. *Ann. Rev. Neurosci.* 20, 185–215.

McCormick D. A. and von Krosigk M. (1992) Corticothalamic activation modulates thalamic firing through glutamate 'metabotropic' receptors. *Proc. Natl Acad. Sci. USA* 89, 2774–78.

McCormick D. A., Wang Z. and Huguenard J. (1993) Neurotransmitter control of neocortical neuronal activity and excitability. *Cerebral Cortex* 3, 387–98.

McQueen J. K. and Woodbury D. M. (1975) Attempts to produce spike wave complexes in the electro-corticogram of the rat. *Epilepsia* 16, 295–99.

Madden T. E. and Johnson S. W. (1998) Gamma-hydroxybutyrate is a GABA$_B$ receptor agonist that increases potassium conductance in rat ventral tegmental dopamine neurons. *J. Pharmacol. Exp. Ther.* 287, 261–65.

Maitre M. (1997) The γ-hydroxybutyrate signaling system in brain: organization and functional implications. *Prog. Neurobiol.* 51, 337–61.

Marcus E. M. and Watson C. W. (1966) Bilateral synchronous spike wave electrographic patterns in the cat. *Arch. Neurol.* 14, 406–08.

Marcus E. M., Watson C. W. and Simon S. A. (1968) An experimental model of some varieties of petit mal epilepsy. Electrical-behavioral correlation of acute bilateral epileptogenic foci in cerebral cortex. *Epilepsia* 9, 233–48.

Marescaux C., Vergnes M., Depaulis A. and Bernasconi R. (1992) GABA$_B$ receptor involvement in the control of genetic absence in rats. *Epilepsy Res.* (Suppl.) 9, 1121–38.

Matejovska I., Veliskova J. and Velisek L. (1998) Bicuculline-induced rhythmic EEG episodes as a model of absence seizures: gender differences and the effects of ethosuximide and baclofen treatment. *Epilepsia* 39, 1243–52.

Mathivet P., Bernasconi R., De Barry J., Marescaux C. and Bittiger H. (1997) Binding characteristics of γ-hydroxybutyric acid as a weak but selective GABA$_B$ receptor agonist. *Eur. J. Pharmacol.* 321, 67–75.

Mehta A. K. and Ticku M. K. (1999) An update on GABA$_A$ receptors. *Brain Res. Rev.* 29, 196–217.

Mirsky A. F., Duncan C. C. and Myslobodsky M. S. (1986) Petit mal epilepsy: a review and integration of recent information. *J. Clin. Neurophysiol.* 3, 179–208.

Mott D. D. and Lewis D. V. (1994) The pharmacology and function of central GABA$_B$ receptors. *Int. J. Neurobiol.* 36, 97–223.

Myslobodsky M. S., Ackermann R. F. and Engel J. (1988) Effects of γ-acetylenic GABA and γ-vinyl GABA on metrazol activated and kindled seizures. *Pharmacol. Biochem. Behav.* 11, 265–71.

Parker A. P., Agathonikou A., Robinson R. O. and Panayiotopoulos C. P. (1998) Inappropriate use of carbamazepine and viagabatrin in typical absence seizures. *Devel. Med. Child Neurol.* 40, 517–19.

Peeters B. W. M. M., van Rijn C. M., Vossen J. M. H. and Coenen A. M. L. (1989) Effects of GABAergic agents on spontaneous non-convulsive epilepsy, EEG, and behaviour, in the WAG/Rij inbred strain of rats. *Life Sci.* 45, 1171–76.

Pinault D., Leresche N., Charpier S., Deniau J.-M., Marescaux C., Vergnes M. and Crunelli V. (1998) Intracellular recordings in thalamic neurones during spontaneous spike and wave discharges in rats with absence epilepsy. *J. Physiol.* 509, 449–56.

Pollen D. A., Perot P. and Reid K. H. (1963) Experimental bilateral wave and spike from thalamic stimulation in relation to level of arousal. *Electroencephalogr. Clin. Neurophysiol.* 15, 1017–28.

Prince D. and Farrell D. (1969) 'Centrencephalic' spike wave discharges following parenteral penicillin injection in the cat. *Neurology* 19, 309–10.

Quesney L. F., Gloor P., Kratzenberg E. and Zumstein J. (1977) Pathophysiology of generalized penicillin epilepsy in the cat: the role of cortical and subcortical structures. I. Systemic application of penicillin. *Electroencephalogr. Clin. Neurophysiol.* 42, 640–55.

Ratomponirina C., Hode Y., Hechler V. and Maitre M. (1995) γ-Hydroxybutyrate receptor binding in rat brain is inhibited by guanyl nucleotides and pertussis toxin. *Neurosci. Lett.* 189, 51–53.

Roth R. H. and Giarman N. J. (1969) Conversion *in vivo* of γ-aminobutyric acid to γ-hydroxybutyrate in the rat. *Biochem. Pharmacol.* 18, 247–50.

Schickerova R., Mares P. and Trojan S. (1984) Correlation between electro-corticographic and motor phenomena induced by pentamethylenetetrazol during ontogenesis in rats. *Exp. Neurol.* 84, 153–64.

Seidenbecher T., Staak R. and Pape H.-C. (1998) Relations between cortical and thalamic cellular activities during absence seizures in rats. *Eur. J. Neurosci.* 10, 1102–12.

Sherman S. M. and Guillery R. W. (1996) Functional organization of thalamocortical relays. *J. Neurophysiol.* 76, 1367–95.

Snead O. C. (1978a) Gamma-hydroxybutyrate in the monkey. I. Electroencephalographic, behavioral, and pharmacokinetic studies. *Neurology* 28, 636–42.

Snead O. C. (1978b) Gamma-hydroxybutyrate in the monkey. II. Effect of chronic oral anticonvulsant drugs. *Neurology* 28, 643–48.

Snead O. C. (1984a) γ-Hydroxybutyric acid, γ-aminobutyric acid, and petit mal epilepsy. In *Neurotransmitters, Seizures, and Epilepsy II* (Fariello R. G., Morselli P. L., Lloyd K. G., Quesney L. F. and Engel J., eds), pp. 37–47. Raven Press, New York.

Snead O. C. (1984b) Ontogeny of γ-hydroxybutyric acid. II. Electroencephalographic effects. *Dev. Brain Res.* 15, 89–96.

Snead O. C. (1988) The γ-hydroxybutyrate model of generalized absence seizures: further characterization and comparison to other absence models. *Epilepsia* 29, 361–68.

Snead O. C. (1990) The ontogeny of GABAergic enhancement of the γ-hydroxybutyrate model of generalized absence seizures. *Epilepsia* 31, 253–58.

Snead O. C. (1991) The γ-hydroxybutyrate model of absence seizures: correlation of regional brain levels of γ-hydroxybutyric acid and γ-butyrolactone with spike wave discharges. *Neuropharmacology* 30, 161–67.

Snead O. C. (1992) Evidence for $GABA_B$-mediated mechanisms in experimental absence seizures. *Eur. J. Pharmacol.* 213, 343–49.

Snead O. C. (1994) The ontogeny of [^{3}H]γ-hydroxybutyrate and [^{3}H]$GABA_B$ binding sites: relation to the development of experimental absence seizures. *Brain Res.* 659, 147–56.

Snead O. C. (1995) Basic mechanisms of generalized absence seizures. *Ann. Neurol.* 37, 146–47.

Snead O. C. (1996a) Antiabsence activity of specific $GABA_B$ and γ-hydroxybutyric acid antagonists. *Pharmacol. Biochem. Behav.* 53, 73–80.

Snead O. C. (1996b) Relation of the [^{3}H]γ-hydroxybutyric acid (GHB) binding site to the γ-aminobutyric acid$_B$ ($GABA_B$) receptor in rat brain. *Biochem. Pharmacol.* 52, 1235–43.

Snead O. C. (1998) Ganaxalone exacerbates experimental absence seizures in rat. *Ann. Neurol.* 44, 688–90.

Snead O. C. (2000) Evidence for a G protein coupled γ-hydroxybutyric acid receptor. *J. Neurochem.* 75, 1986–96.

Snead O. C. and Liu C. C. (1984) γ-Hydroxybutyric acid binding sites in rat and human brain synaptosomal membranes. *Biochem. Pharmacol.* 33, 2587–90.

Snead O. C. and Liu C. C. (1993) $GABA_A$ receptor function in the γ-hydroxybutyrate model of generalized absence seizures. *Neuropharmacology* 32, 401–09.

Snead O. C. and Morley B. J. (1981) Ontogeny of gamma-hydroxybutyric acid. I. Regional concentration in developing rat, monkey, and human brain. *Brain Res.* 227, 579–89.

Snead O. C., Yu R. K. and Huttenlocher P. R. (1976) Gamma-hydroxybutyrate: correlation of serum and cerebrospinal fluid levels with electroencephalographic and behavioral effects. *Neurology* 26, 51–56.

Snead O. C., Bearden L. J., Healy C. T. and Pegram V. (1980) Effect of acute and chronic anticonvulsant administration on endogenous γ-hydroxybutyrate in rat brain. *Neuropharmacology* 19, 47–52.

Snead O. C., Furner R. and Liu C. C. (1989) *In vivo* conversion of γ-aminobutyric acid and 1,4 butanediol to γ-hydroxybutyric acid in rat brain: studies using stable isotopes. *Biochem. Pharmacol.* 38, 4375–80.

Snead O. C., Depaulis A., Vergnes M. and Marescaux C. (1999) Absence epilepsy: advances in experimental animal models. In *Jasper's Basic Mechanisms of the Epilepsies*, 3rd edn (Delgado-Escueta A. V., Wilson W., Olsen R. W. and Porter R. J., eds), pp. 253–78. Raven Press, New York.

Snead O. C, Banerjee P. K., Burnham M. and Hampson D. R. (2000) Modulation of absence seizures by the GABA$_A$ receptor: a critical role for mGluR4. *J. Neurosci.* 20, 6218–24.

Spreafico R., Battaglia G. and Frassoni C. (1991) The reticular thalamic nucleus (RTN) of the rat: cytoarchitectural, Golgi, immunocytochemical, and horseradish peroxidase study. *J. Comp. Neurol.* 304, 478–90.

Steriade M. and Llinas R. R. (1988) The functional states of the thalamus and the associated neuronal interplay. *Physiol. Rev.* 68, 649–742.

Steriade M., Parent A. and Hada J. (1984) Thalamic projections of nucleus reticularis thalami of cat: a study using retrograde transport of horseradish peroxidase and fluorescent tracers. *J. Comp. Neurol.* 229, 531–47.

Steriade M., Domich L. and Oakson G. (1986) Reticularis thalami neurons revisited: activity changes during shifts in states of vigilance. *J. Neurosci.* 6, 68–81.

Steriade M., Gloor P., Llinas R. R., Lopes da Silva F. H. and Mesulam M. M. (1990) Basic mechanisms of cerebral rhythmic activities. *Electroencephalogr. Clin. Neurophysiol.* 76, 481–508.

Steriade M., McCormick D. A. and Sejnowski T. J. (1993) Thalamocortical oscillations in the sleeping and aroused brain. *Science* 262, 679–85.

Ulrich D. and Huguenard J. R. (1997) GABA$_A$ receptor-mediated rebound burst firing and burst shunting in thalamus. *J. Neurophysiol.* 78, 1748–51.

van der Linden G. J., Meinardi H., Meijer S. W. A., Bossi L. and Gomeni C. A. (1981) A double blind crossover trial with progabide (SL76002) against placebo in patients with secondary generalized epilepsy. In *Advances in Epileptology: XIIth International Epilepsy Symposium* (Dam M., Gram L. and Penry J. K., eds), pp. 141–44. Raven Press, New York.

Vayer P., Mandel P. and Maitre M. (1987) γ-Hydroxybutyrate, a possible neurotransmitter. *Life Sci.* 41, 1547–57.

Vergnes M., Marescaux C., Micheletti G., Depaulis A., Rumbach L. and Warter J. M. (1984) Enhancement of spike and wave discharges by GABA-mimetic drugs in rat with spontaneous petit mal-like epilepsy. *Neurosci. Lett.* 44, 91–94.

Vergnes M., Marescaux C., Depaulis A., Micheletti G. and Warter J. M. (1987) Spontaneous spike and wave discharges in thalamus and cortex in a rat model of genetic petit mal-like seizures. *Exp. Neurol.* 96, 127–36.

Vergnes M., Marescaux C., Depaulis A., Micheletti G. and Warter J. M. (1990a) The spontaneous spike and wave discharges in Wistar rats: a model of genetic generalized non convulsive epilepsy. In *Generalized Epilepsy: Neurobiological Approaches* (Avoli M., Gloor P., Kostopoulos G. and Naquet R., eds), pp. 238–53. Birkhauser, Boston.

Vergnes M., Marescaux C. and Depaulis A. (1990b) Mapping of spontaneous spike and wave discharges in Wistar rats with genetic generalized non-convulsive epilepsy. *Brain Res.* 523, 87–91.

Vicini S. (1999) New perspectives in the functional role of GABA$_A$ channel heterogeneity. *Mol. Neurobiol.* 19, 97–110.

von Krosigk M., Bal T. and McCormick D. A. (1993) Cellular mechanisms of a synchronized oscillation in the thalamus. *Science* 261, 361–64.

White J. H., Wise A., Main M. J., Green A., Fraser N. J., Disney G. H., Barnes A. A., Emson P., Ford S. M. and Marshall F. H. (1998) Heterodimerization is required for the formation of a functional GABA$_B$ receptor. *Nature* 396, 674–78.

Williams D. (1953) A study of thalamic and cortical rhythms in petit mal. *Brain* 76, 50–69.

Williams S. R., Turner J. P. and Crunelli V. (1995) Gamma-hydroxybutyrate promotes oscillatory activity of rat and cat thalamocortical neurons by a tonic GABA$_B$ receptor-mediated hyperpolarization. *Neuroscience* 66, 133–41.

Xie X. and Smart T. G. (1992) Gamma-hydroxybutyrate hyperpolarizes hippocampal neurones by activating GABA$_B$ receptors. *Eur. J. Pharmacol.* 212, 291–94.

Yen C. T., Conley M., Hendry S. H. C. and Jones E. G. (1985) The morphology of physiologically identified GABAergic neurons in the somatic sensory part of the thalamic reticular nucleus in the cat. *J. Neurosci.* 5, 2254–68.

9 Behavioral pharmacology of γ-hydroxybutyrate

G. Colombo, C. Lobina and G. L. Gessa

Introduction

The exogenous administration of γ-hydroxybutyric acid (GHB), a putative neurotransmitter or neuromodulator of the central nervous system of mammals, has been reported to produce a number of pharmacological effects (for review see Vayer *et al.* 1987; Tunnicliff 1992; Cash 1994; Maitre 1997; Bernasconi *et al.* 1999). However, the majority of studies conducted to date have examined its effects at a cellular level, while there is a scarcity of systematic information concerning GHB behavioral effects.

In spite of the infancy of the behavioral pharmacology of GHB, the drug is known to produce a number of behavioral effects (e.g. anxiolysis, reinforcement) that are interesting because of (a) the potential therapeutic use of GHB (e.g. in the treatment of alcohol depen- dence – see Addolorato *et al.* 2000; Agabio and Gessa, Chapter 10 of this volume; Gallimberti *et al.* 2000) and (b) the possible role and contribution of the GHB receptor system in the mediation of that particular phenomenon.

The present review focuses exclusively on those studies investigating GHB effects on behavior in laboratory animals, where behavior is intended as any response of the animal to the drug under experimental conditions that involve the voluntariness of the animal. In other words, the animal has a free choice between two or more options, such as pressing a right or a left lever, running an arm or its opposite in a maze, staying in a particular compartment or in a different one. Autonomic responses (such as catalepsy and hypnosis), that are often desig- nated as behavioral effects of a given drug, will not be discussed in this chapter.

The anxiolytic effect of GHB

The anxiolytic effect of GHB, as well as of its precursor γ-butyrolactone (GBL), has been consistently observed throughout the few studies that have addressed this issue to date. In the first study (Kršiak *et al.* 1974), mice were initially kept in isolation for 23 days and subse- quently exposed to contact with non-aggressive, previously group-housed mice. The defen- sive responses displayed by the formerly isolated mice were recorded as measures of anxiety. The acute, intraperitoneal administration of a non-sedative dose of GHB (50 mg/kg) resulted in (a) reduction of the frequency of occurrence of passive defensive behaviors, such as moving away from the previously grouped mouse, alertness, tail shaking and defensive pos- ture, and (b) increase in the number of interaction behaviors (e.g. approaching and sniffing the partner).

The pharmacological profile of the anxiolytic effect of GHB has recently been investigated by Schmidt-Mutter *et al.* (1998). This study employed the elevated plus maze, a validated and widely accepted model of experimental anxiety, based on the natural avoidance and aversion of rodents for open, anxiogenic spaces and their innate preference for closed, more comforting environments (see Pellow *et al.* 1985; Rodgers and Cole 1994; Hogg 1996). The maze consists of four arms, positioned to form a plus sign: two facing arms have no walls (the so-called 'open arms'), while the other two opposite arms present high walls ('closed arms'). The entire maze is elevated above the floor (usually 50 cm high in the apparatus used for rats). The proportion of spontaneous explorations in the open and closed spaces of the maze provides a behavioral measure of anxiety in rats and mice. In the study by Schmidt-Mutter *et al.* (1998), GHB was administered intraperitoneally to rats at the non-sedative doses of 50, 150 and 250 mg/kg. GHB exerted a robust anxiolytic effect: indeed, both time spent in and number of entries into the open arms of the maze were dose-dependently increased by GHB administration; at the dose of 250 mg/kg, the magnitude of GHB effect was similar to that elicited by 2 mg/kg of the benzodiazepine, diazepam, included in the study as reference compound. Interestingly, further experiments were conducted to investigate the receptor system(s) involved in the mediation of the anxiolytic effect of GHB. To this aim, the possible antagonism exerted by the selective GHB receptor antagonist, NCS-382, the non-selective opioid receptor antagonist, naloxone, and the benzodiazepine receptor antagonist, flumazenil, was tested. Previous studies demonstrated that NCS-382 was capable of antagonizing the discriminative (Colombo *et al.* 1995a), sedative and cataleptic (Schmidt *et al.* 1991) effects of GHB, while naloxone has been shown to reverse different pharmacological effects of GHB, including catalepsy (Snead and Bearden 1980) and reduction of immobility at the despair test (Zerbib *et al.* 1992). However, in the study by Schmidt-Mutter *et al.* (1998), neither NCS-382 (tested at the single dose of 300 mg/kg i.p.) nor naloxone (10 mg/kg i.p.) altered the anxiolytic effect exerted by 150 mg/kg GHB. In contrast, pretreatment with a *per se* non-effective dose of flumazenil (10 mg/kg i.p.) significantly reduced both time in and number of entries into the open arms of the maze in GHB-treated rats (Fig. 9.1). These results suggest that (a) the $GABA_A$–benzodiazepine receptor complex is involved in the mediation of the anxiolytic effect of GHB, and (b) no unique mechanism apparently mediates the different pharmacological effects of GHB. The neurochemical basis of the $GABA_A$–benzodiazepine receptor complex mediation of the anti-anxiety effect of GHB does not appear to be a direct interaction of GHB with either $GABA_A$- or benzodiazepine-receptors, because GHB failed to bind to both recognition sites (Olsen *et al.* 1981; Bernasconi *et al.* 1992) as well as to alter the chloride uptake through the associated channel (Serra *et al.* 1991). It is more likely that $GABA_A$-mediated effects result from (a) the conversion of GHB into GABA, which in turn binds to the $GABA_A$ receptor (Hechler *et al.* 1997), and (b) GHB-induced stimulation of GABA release in specific brain regions (Gobaille *et al.* 1999). An alternative explanation may come from the results of a recent experiment demonstrating that the acute administration of 150 and 300 mg/kg GHB (i.p.) produced a two- to threefold increase in the levels of the endogenous neurosteroids 3α-hydroxy-5α-pregnan-20-one (allopregnanolone (AP)) and $3\alpha,21$-dihydroxy-5α-pregnan-20-one (allotetrahydro-deoxycorticosterone (THDOC)) in different areas of the rat brain (Barbaccia *et al.* 2002). AP and THDOC are potent positive modulators of the $GABA_A$ receptor complex, and their administration has been reported to exert anxiolytic, rewarding and sedative effects in laboratory animals (see Lambert *et al.* 1995).

A subsequent study (Agabio *et al.* 1998) investigated the anxiolytic effect of GHB in Sardinian alcohol-preferring (sP) rats, selectively bred for high alcohol preference and consumption. This study was of some interest because (a) GHB, as noted below, significantly reduced voluntary alcohol intake in sP rats, and (b) sP rats have been featured as a possible

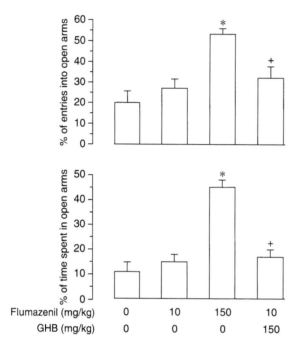

Figure 9.1 Anxiolytic effect of GHB and reversal by the benzodiazepine receptor antagonist, flumazenil, in Long Evans rats tested at the elevated plus maze. Flumazenil (0 and 10 mg/kg) and GHB (0 and 150 mg/kg) were injected i.p. 30 min before testing at the elevated plus maze. Exposure to the elevated plus maze lasted 5 min. Percentage of (i) entries into (top panel) and (ii) time spent in (bottom panel) the open arms of the maze were used as measures of anxiety. Each bar is the ±SEM of $n = 8$. *: $P < 0.01$ in comparison to 0 mg/kg flumazenil + 0 mg/kg GHB; +: $P < 0.01$ in comparison to 0 mg/kg flumazenil + 150 mg/kg GHB.

Source: Adapted from Schmidt-Mutter *et al.* (1998) with permission.

model for the association between alcohol drinking and anxiety. Indeed, when tested at the elevated plus maze, sP rats displayed an approximately tenfold higher degree of anxiety-like behaviors than their alcohol-avoiding counterpart (Sardinian alcohol-non-preferring (sNP) rats), and voluntary alcohol intake partially reversed their innate anxiety profile (Colombo *et al.* 1995b); in other words, anxiety appeared to be a genetic trait likely predisposing sP rats to alcohol drinking, while alcohol was voluntarily consumed for self-medicating anxiety. In the investigation testing the anxiolytic effect of GHB, alcohol-experienced sP rats were exposed to the elevated plus maze 20 min after GHB administration. GHB was injected intraperitoneally at the same dose (300 mg/kg) that effectively reduced voluntary alcohol intake in sP rats without exerting any sedative effect. GHB-treated rats spent approximately sevenfold more time in and made approximately fourfold more entries into the open arms of the maze than saline-treated rats. In contrast, no significant difference was monitored between the two rat groups in the number of entries into the closed arms (a measure of general locomotor activity). These results demonstrate that GHB, administered at the dose of 300 mg/kg, exerted a robust anxiolytic effect in sP rats. It is likely that the anxiolytic effect of GHB may replace the tension-reducing effect of alcohol usually sought by sP rats, and this substitution would contribute to the reducing effect of GHB on voluntary alcohol intake (see below).

An investigation by McIntire and Liddell (1984) demonstrated that GBL also induced anxiolysis in rats. In this study, animals were trained to lever press for food reinforcement; however, food presentation coincided with exposure to a brief electric shock. Administration of 75–150 mg/kg GBL (i.p.) resulted in a significant increase, in comparison to saline-treated rats, in the rate of punished lever pressing. The anti-conflict effect exerted by GBL was comparable to that induced by doses of pentobarbital in the 3.5–14 mg/kg range.

The above results, showing the capability of GHB and GBL to exert anti-anxiety effects in experimental models of anxiety, are in agreement with clinical observations reporting that GHB exerted anxiolytic effects in healthy (Ferrara *et al.* 1999) and alcoholic (Addolorato *et al.* 1999) subjects as well as with anecdotal reports of calmness and relaxation after the recreational intake of GHB (Galloway *et al.* 1997, 2000).

The positive reinforcing effects of GHB

The positive reinforcing properties and abuse potential of a drug are usually assessed in laboratory animals by determining its capability to (a) be self-administered, (b) induce conditioned place preference and (c) elicit discriminative stimulus effects similar to those produced by abused drugs. These experimental procedures have been shown to possess high predictive validity for abuse and dependence potential of drugs in humans (see Brady 1991; Altman *et al.* 1996). This section will address the studies demonstrating the ability of GHB to be self-administered and produce conditioned place preference, while description of the studies investigating the discriminative stimulus effects of GHB is deferred to the next section.

Self-administration of GHB

Recent investigations demonstrated that GHB is easily self-administered both orally and intravenously by rats and mice. However, more equivocal data come from studies investigating the capability of GHB to serve as a reinforcer in monkeys.

The first study addressing the abuse potential of GHB in laboratory animals investigated whether Wistar rats would voluntarily consume orally GHB (Colombo *et al.* 1995c). Rats were initially exposed to a GHB solution (1% w/v in water) as the sole fluid available for 14 consecutive days, to allow the rats to become accustomed to the salty taste of the solution and experience the GHB central effects. Subsequently, GHB was presented under the two-bottle free choice regimen with water for additional 20 weeks. During the latter phase, all rats tested showed preference for the GHB solution and consumed daily doses of GHB (averaging approximately 700 mg/kg) expected to produce salient psychopharmacological effects. Periods of robust preference for, and intake of, GHB were alternated with periods of voluntary, virtually complete abstinence from GHB. On the days in which GHB was preferred over water, GHB intake occurred in two to three discrete episodes, regularly distributed during the nocturnal phase of the light/dark cycle (Fig. 9.2); in each drinking bout, GHB intake peaked up to 300 mg/kg. Taking into account the short half-life of GHB (approximately 30 min in the rat plasma after parenteral administration (Lettieri and Fung 1979)), this fractioning of GHB daily ingestion is suggestive of the rat capability to adjust both timing of administration and dose of GHB to be consumed presumably in order to experience specific pharmacological effects.

These data were later confirmed by a study testing preference for, and consumption of, the 1% GHB solution in drug-naive, alcohol-preferring sP and non-preferring sNP rats

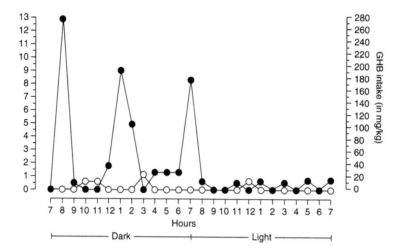

Figure 9.2 Illustrative pattern of oral self-administration of GHB in a single Wistar rat. GHB (1% w/v in tap water) was offered under the two-bottle free choice regimen with tap water for 24 h/day. The hourly intake of GHB (●) is expressed on the left ordinate as milliliters of solution consumed by the rat and on the right ordinate axis as mg/kg of the drug. Water (○) intake is expressed on the left ordinate axis as milliliters consumed by the rat.

Source: Adapted from Colombo *et al.* (1995c).

(Colombo *et al.* 1998a). Presentation of the GHB solution for 14 consecutive days as the sole fluid available resulted in robust preference and intake of GHB without any significant line difference. It was hypothesized that the 14 day period of forced GHB drinking might have acted as such a powerful procedure for induction of GHB preference and intake that it led to the unmasking of the reinforcing properties of GHB and acquisition of GHB drinking behavior also in the supposedly less susceptible sNP rats. In a subsequent experiment, the period of forced GHB drinking was restricted to 3 days. GHB preference and consumption during the following free choice regimen were initially low in both rat lines. However, after approximately 10 days of free access, daily GHB intake increased to pharmacologically relevant amounts in sP rats (daily GHB intake settling at 500–700 mg/kg), presumably because episodic binges of GHB intake occurring during the first 10 days of the free-choice phase resulted in disclosure and experience of the reinforcing properties of GHB. In contrast, voluntary daily consumption of GHB remained at negligible levels (steadily lower than 250 mg/kg) in sNP rats. Thus sP rats, selectively bred for high alcohol preference and consumption, appear to possess a genetically based higher sensitivity, when compared to sNP rats, to the reinforcing properties not only of alcohol but also of GHB. Consequently, it can be speculated that GHB exerts reinforcing properties similar to those produced by alcohol, adding further support to the hypothesis that GHB and alcohol may have a similar pharmacological profile (we will return to this in the following).

A subsequent study demonstrated that GHB maintained intravenous self-administration in mice (Martellotta *et al.* 1998a). Animals self-injected GHB acutely by nose-poking in single 30 min sessions; under this specific procedure, every nose-poke resulted in the injection, through the tail vein, of doses of GHB in the 0.01–0.5 mg/kg range. As commonly observed in studies testing the intravenous self-administration of drugs of abuse (see Katz 1989), self-administration of GHB was an inverted U-shape function of GHB concentration.

At the concentration of 0.1 mg/kg/injection, mice made an average of approximately forty nose-pokings during the 30 min session. Interestingly, in the study by Martellotta *et al.* (1998a) self-administration of GHB was completely blocked by pretreatment with the specific GHB receptor antagonist, NCS-382, suggesting that the reinforcing properties of GHB are secondary to activation of the GHB receptor.

The few studies that tested self-administration of GHB in primates generated more ambiguous results. Beardsley *et al.* (1996) used four rhesus monkeys previously trained to self-administer phencyclidine. Once phencyclidine was replaced by GHB, only in one monkey did the number of intravenous infusions of GHB exceed control levels (i.e. mean infusions of saline) and this increase occurred solely at the dose of 3 mg/kg/injection GHB. A subsequent study by Woolverton *et al.* (1999) tested a wide range of GHB concentrations (0.01–10 mg/kg/injection) in three rhesus monkeys previously trained to self-administer the barbiturate, methohexital. The number of GHB injections/session was significantly higher than that of saline injections only in one monkey at the GHB concentration of 3.2 mg/kg/injection and in a second monkey at the GHB concentration of 10 mg/kg/injection, the latter resulting at the end of the 130 min session in the infusion of 250 mg/kg GHB and occurrence of sedative effect. Based upon these two studies in monkeys (Beardsley *et al.* 1996; Woolverton *et al.* 1999), it might be concluded that GHB is a weak positive reinforcer and has little potential for recreational use and abuse.

The reason for these discrepancies between rodent and monkey studies in the ability of GHB to maintain self-administration at present escapes our understanding. From the monkey data reviewed above, it appears that the reinforcing effects and the sedative/hypnotic effect of GHB were modestly separated; therefore, it is possible that the sedative/hypnotic effect of self-administered GHB may have masked the reinforcing properties of the drug. Further studies testing wider concentration ranges of GHB are warranted to possibly clarify this issue.

GHB-induced conditioned place preference

The conditioned place preference procedure is a behavioral technique widely employed to investigate the rewarding properties of psychoactive drugs (see Tzcshentke 1998). In the conditioned place preference paradigm, laboratory rodents are trained to associate the interoceptive cues produced by a drug with the external stimuli (olfactory, visual, tactile) of a specific environment and the absence of those effects with the stimuli of a different environment. Two-compartment apparatus, easily distinguishable for smell (e.g. vinegar or peppermint), wall color and design, wire-made or smooth floor, are usually employed. After a proper number of conditioning sessions, animals are given a choice between the two environments: if the animal increases the time spent in the drug-paired context, it is conceivable that the drug possesses rewarding properties. All drugs of abuse, with very few exceptions, have been reported to induce conditioned place preference (see Hoffman 1989).

The rodent self-administration data (see above) were predictive of the induction of place preference after conditioning with GHB. Martellotta *et al.* (1997) addressed this issue in an investigation that employed Sprague–Dawley rats, alternately exposed to GHB- and to water-associated compartments for an overall twenty daily sessions of 30 min each. On the test day, water-treated rats did not exhibit any preference for either compartment; in contrast, rats treated with doses of GHB comprised between 87.5 and 350 mg/kg (i.g.) significantly increased, with respect to the preconditioning session, the time spent in the GHB-paired box (Fig. 9.3), demonstrating the ability of GHB to induce conditioned place preference and

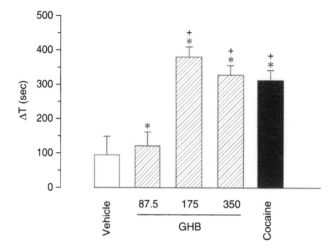

Figure 9.3 Conditioned place preference induced by GHB in Sprague–Dawley rats. Rats were alternatively exposed to GHB- and water-associated compartments for total twenty daily sessions of 30 min each. GHB was administered i.g. at the doses of 0, 87.5, 175 and 350 mg/kg 30 min before exposure to the apparatus. Cocaine, used as reference compound, was injected i.p. at the dose of 10 mg/kg immediately before exposure to apparatus. On the test day, rats had free access to both compartments for 15 min. Δ values express the difference in time spent in the GHB-associated (or cocaine-associated) compartment between test session and a 15 min session conducted before the conditioning phase. Each bar is the ±SEM of $n = 8$. ∗: $P < 0.01$ in comparison to same group on the preconditioning session; +: $P < 0.01$ in comparison to vehicle-treated rats.

Source: Adapted from Martellotta *et al.* (1997) with permission.

suggesting that it possesses rewarding properties. Interestingly enough, the magnitude of conditioned place preference produced by 175 and 350 mg/kg GHB was comparable to that produced by 10 mg/kg cocaine (Fig. 9.3).

Conclusions

The results of the above studies showing that, at least in rats and mice, GHB is self-administered in pharmacologically relevant doses (Colombo *et al.* 1995c, 1998a; Martellotta *et al.* 1998a) and can induce conditioned place preference (Martellotta *et al.* 1997), suggest that GHB possesses positive reinforcing properties. These data are in close agreement with a number of clinical observations (although most of them anecdotal) and reports from US agencies showing that GHB is used as a recreational drug, because of the salient feelings of euphoria, disinhibition, anxiolysis and relaxation produced by low to moderate doses (see Galloway *et al.* 2000).

The discriminative stimulus effects of GHB

The drug discrimination procedure is a sensitive and reliable behavioral technique which allows investigation on receptor systems mediating the discriminative stimulus effects, that is the animal correlate of human subjective effects, of a particular psychoactive drug (see NIDA 1991; Samele *et al.* 1991; Goudie and Leathley 1993; Kamien *et al.* 1993; Stolerman *et al.* 1995). In drug discrimination procedures, laboratory animals are trained to recognize the interoceptive cues of a given dose of a certain drug (the so-called training drug) and associate these with a specific behavior (usually motivated by food reinforcement in a condition of food-restriction). Namely, animals are trained to carry out a particular task (e.g. pressing a lever in a two-lever operant procedure or running an arm of a T-maze) each time they detect the effects of the training drug (drug condition) and behave differentially (e.g. pressing the second lever or running the opposite arm) when those effects are absent (control or non-drug condition). Thus, animals learn to associate the internal state induced by the training drug with an observable and measurable behavior; in other words, the response behavior is the means by which the animal 'self-reports' whether it feels 'drugged' or not. Once animals have been successfully trained to discriminate the training drug, substitution and blockade tests can be conducted. In substitution tests, administration of a second drug (named testing drug) in place of the training drug, which still produces the training drug-associated behavior, is indicative of the similarity of the discriminative stimulus effects of the testing drug to those of the training drug. In blockade tests, combination of both testing and training drugs, resulting in the selection of the non-drug-associated behavior, accounts for a reduced perception of the discriminative stimulus effects of the training drug.

A large number of studies have shown that (a) drugs acting in a similar manner at a specific class of receptors possess similar discriminative stimulus effects, and (b) antagonists at a specific receptor block the discriminative stimulus effects of receptor agonists (see Overton 1982; Colpaert 1986; Holtzman 1990; Goudie and Leathley 1993). Thus, when the mechanism of action of the testing drug is known, the results of the substitution and blockade tests may provide relevant information on the receptor system(s) involved in the mediation of the discriminative stimulus effects of the training drug. Furthermore, subjective effects of a drug play a major role in its abuse potential (see Overton 1982; Preston and Bigelow 1991): in this light, identification of possible similarities between the discriminative stimulus effects of a given drug and those elicited by proven addictive drugs may be predictive of its abuse liability.

Based on these premises, the drug discrimination technique appears to be a particularly useful tool for investigating the to-date elusive neural substrates underlying the pharmacological effects of GHB as well as its abuse potential. Although the ability of GHB in controlling discriminative responding in laboratory animals was demonstrated for the first time almost twenty years ago (Winter 1981), the pharmacological profile of the discriminative stimulus effects of GHB has been poorly investigated to date.

GABA-mediated component of the GHB cue

Two studies (Winter 1981; Colombo *et al.* 1998b) suggested that (a) the discriminative stimulus effects of GHB are composed of different cues, each one being the effect of GHB on a specific neurotransmitter system, and (b) $GABA_A$- and $GABA_B$-mediated cues are major ingredients of the mixed stimulus of GHB. Furthermore, in addition to varying quantitatively (i.e. with respect to intensity), the discriminative stimulus effects of GHB may vary qualitatively (i.e. involving different proportions of the component cues) as the training doses of GHB are increased.

The $GABA_A$ receptor agonist, muscimol, and the benzodiazepines, chlordiazepoxide (Winter 1981) and diazepam (Colombo *et al.* 1998b), were found to substitute partially for GHB in rats trained to discriminate low to moderate doses of GHB. Indeed, the intraperitoneal administration of non-sedative doses of muscimol, chlordiazepoxide and diazepam resulted in 50–80% mean selection of the GHB-associated responding in rats trained to discriminate 200 mg/kg i.p. (Winter 1981) or 300 mg/kg i.g. (Colombo *et al.* 1998b) GHB from water. Furthermore, the GABA$_A$ receptor antagonist, bicuculline, partially antagonized the GHB cue, as indicated by the reduction, to approximately 50%, of mean responding on GHB-associated condition (Winter 1981). When tested in rats trained to discriminate a higher dose of GHB (namely, 700 mg/kg i.g.) from water, diazepam elicited minimal responding in the GHB-associated condition (Colombo *et al.* 1998b).

The GABA$_B$ receptor agonist, baclofen, dose-dependently substituted for both low (200 and 300 mg/kg i.g.) and high (700 mg/kg i.g.) doses of GHB (Winter 1981; Colombo *et al.* 1998b). Although complete substitution (conventionally defined as an average of 80% or more of training drug-appropriate responding) was observed in both rat groups trained to discriminate 300 and 700 mg/kg GHB, the potency of baclofen in substituting for GHB was somewhat higher in the high than in low training dose group (Colombo *et al.* 1998b). The GABA$_B$ receptor antagonist, CGP 35348, administered before GHB, attenuated the interoceptive cues elicited by both 300 and 700 mg/kg GHB (Colombo *et al.* 1998b). Consistent with the results of the baclofen substitution study, CGP 35348 was more potent and effective in blocking the discriminative stimulus effects of 700 mg/kg GHB than the effects of 300 mg/kg GHB.

These results suggest that (a) positive modulation of the GABA$_A$ receptor is a relevant part of the interoceptive stimuli produced by 300 mg/kg GHB while it is minimally involved in the mediation of the discriminative stimulus effects of 700 mg/kg GHB, and (b) the GABA$_B$-mediated cue is a prominent component of the discriminative stimulus effects of GHB, however being more salient at the dose of 700 than 300 mg/kg GHB. These results are consistent with data from drug mixture studies indicating that, as the relative amount of one of the components is increased, its contribution to the discriminative stimulus effects of the mixture increases, while that of the other component decreases (Stolerman *et al.* 1987; Garcha and Stolerman 1989; Stolerman and Mariathasan 1990; Mariathasan *et al.* 1991; Mariathasan and Stolerman 1993). Furthermore, it has been shown that complete blockade of the discriminative stimulus effects of a drug mixture occurs only when the stimuli produced by each component are blocked; indeed, blockade of only one component at most only partially attenuates the discriminative stimulus effects of the mixture, and such attenuation is proportional to the prominence of the component (White and Stolerman 1994). Thus, the complete blockade by CGP 35348 of the discriminative stimulus effects of 700 mg/kg GHB is consistent with the hypothesized prominence of the GABA$_B$ component of the GHB cue at this dose, which may overshadow the perception of the other components. That is, the behavior of rats trained to discriminate 700 mg/kg GHB from water appeared to be guided predominantly by GABA$_B$-mediated cues; once these cues were blocked, overshadowed components could become apparent, but since the rats were unable to recognize the cues to which they were trained, they selected the water-associated arm of the maze. The partial blockade by bicuculline (Winter 1981) and CGP 35348 (Colombo *et al.* 1998b) of the discriminative stimulus effects of 200 and 300 mg/kg GHB, respectively, together with the partial substitution elicited by muscimol, chlordiazepoxide, diazepam and baclofen, suggest that GABA$_A$- and GABA$_B$-mediated cues are both recognizable in the discriminative stimulus effects of low GHB doses.

The benzodiazepine- and baclofen-like discriminative stimulus effects of GHB are likely due to (a) its conversion into GABA, binding to both GABA$_A$ and GABA$_B$ receptors (Hechler *et al.* 1997) or (b) stimulation of GABA release in specific brain regions (Gobaille *et al.* 1999). The prominence of the GABA$_B$ component in the discriminative stimulus effects of GHB observed in the study by Colombo *et al.* (1998b) is consistent with the higher sensitivity of GABA$_B$ receptors to GABA, with respect to GABA$_A$ receptors (Hechler *et al.* 1997). Furthermore, GHB has been reported to be a weak agonist at the GABA$_B$ binding site (Mathivet *et al.* 1997; Lingenhoehl *et al.* 1999), and displacement of GABA$_B$ receptor agonists has been observed with concentrations of GHB in the range of those that are obtained after the *in vivo* administration of moderate to high doses of GHB. Finally, it has been proposed that the GABA$_B$-like effects of GHB may also be secondary to activation of a GHB recognition site related to, although separate from, a GABA$_B$ receptor, forming a presynaptic GABA$_B$/GHB receptor complex that regulates neurotransmitter release (Snead 1996). Furthermore, the ability of GHB to increase the levels of AP and THDOC, endogenous neurosteroids which positively modulate the GABA$_A$ receptor complex (Barbaccia *et al.* 2002), may contribute to the benzodiazepine-like discriminative stimulus effects of GHB.

GHB receptor-mediated component of the GHB cue

A recent investigation (Colombo *et al.* 1995a) demonstrated that combination of NCS-382 (25 and 50 mg/kg i.p.) and GHB resulted in the complete blockade of the discriminative stimulus effects of both 300 and 700 mg/kg GHB (i.g.) in rats. These results suggest that stimulation of GHB receptors constitutes a salient component of the GHB cue. However, re-evaluation of these data may be necessary, since a subsequent study (Colombo *et al.* 1999) found that NCS-382 dramatically reduced alcohol absorption from the gastrointestinal system and, as predictable, discrimination of intragastrically administered alcohol. Thus, further studies are needed to verify whether the reported attenuation of the discrimination of orally administered GHB by NCS-382 (Colombo *et al.* 1995a) might have been due to a reducing effect of NCS-382 on GHB absorption from the gastrointestinal tract.

Similarities of the GHB cue with those of abusive drugs

In his pioneering study, Winter (1981) reported that morphine and lysergic acid diethylamide elicited intermediate responding in rats trained to discriminate 200 mg/kg GHB i.p. from saline; furthermore, in the same rats, neither *d*-amphetamine nor phencyclidine substituted for GHB. More recently, this laboratory found that also cocaine (Fig. 9.4) and the cannabinoid receptor agonist, WIN 55,212-2 (Colombo *et al.* 1998b), failed to substitute for GHB in rats trained to discriminate GHB from water.

When GHB was used as testing drug, and its ability of substituting for specific drugs of abuse was tested, it failed to consistently substitute for heroin and phencyclidine (Beardsley *et al.* 1996), for cocaine (this laboratory: Fig. 9.5, left panel) and Δ^9-tetrahydrocannabinol (Browne and Weissman 1981) in rats, and for pentobarbital, *d*-amphetamine and triazolam in rhesus monkeys (Woolverton *et al.* 1999). Furthermore, when tested in blockade experiments, GHB failed to attenuate the discriminative stimulus effects produced by cocaine in rats trained to discriminate cocaine from saline (Beardsley *et al.* 1996; this laboratory: Fig. 9.5, right panel).

Several lines of evidence, demonstrating that GHB and alcohol share different pharmacological similarities, suggest that GHB may exert its effects on alcohol dependence by mimicking alcohol actions in the central nervous system (see Gessa *et al.* 2000). In other words, GHB efficacy in controlling alcohol craving, consumption and withdrawal syndrome would

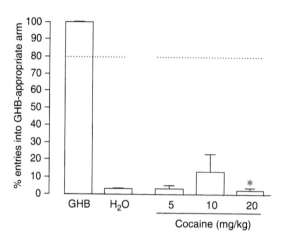

Figure 9.4 Failure of cocaine to substitute for the discriminative stimulus effects of GHB in rats trained to discriminate GHB from water in a T-maze food-reinforced drug discrimination paradigm (Colombo *et al.* 1996). The apparatus was a T-shaped maze, made of black plexiglas and comprising a central stem (start point) and two opposite runways (right and left arm). A sliding door divides the start point from the arms. A recessed food cup is placed at the far end (goal area) of each arm and sunflower seeds are used as reinforcements to motivate responding of food-deprived rats. Each daily session consists of ten consecutive trials (runs). Rats are trained to run the GHB-appropriate arm (the left arm for half the rats and the right arm for the other half) after the i.g. administration of 300 mg/kg GHB (30 min pretreatment time) and the water-appropriate arm (the opposite arm) after the i.g. administration of tap water. Only the selection of the condition-appropriate arm was food-reinforced. Sessions under GHB- and water-condition were daily alternated. Five consecutive correct training sessions define the criterion for acquisition of the discrimination. Once rats are trained to criterion, substitution tests with cocaine were performed. In test sessions, the choice of each arm was food-reinforced. Cocaine was dissolved in 1 ml/kg saline and administered i.p. at the doses of 5, 10 and 20 mg/kg 15 min before the start of the session. Each bar indicates the average percentage of entries into the GHB-appropriate arm following the administration of a particular dose of cocaine. GHB and H_2O points represent the average per cent of entries into the GHB-appropriate arm during training sessions with GHB and water. The dashed line indicates the limit ($\geq$80%) of GHB-appropriate responding for complete substitution for the discriminative stimulus effects of GHB. Bars are the mean $\pm$ SEM of two determinations in each of the four rats employed. The asterisk indicates occurrence of motor disrupting effects.

be exerted through a substitution mechanism, similar to that underlying methadone use in heroin addiction. The results of two recent drug discrimination studies from this laboratory (Colombo *et al.* 1995d; Agabio *et al.* 1995) strengthened this substitution hypothesis. In these studies, rats were trained to discriminate either alcohol (1 or 2 g/kg i.g.) or GHB (300 or 700 mg/kg i.g.) from water in a T-maze, food-reinforced drug discrimination procedure. The results of these investigations, summarized in Fig. 9.6, indicated that GHB and alcohol cross-substituted. However, the symmetrical generalization between the two drugs occurred only within narrow dose ranges: indeed, 300 mg/kg was the only dose of GHB capable of substituting for alcohol and this substitution occurred only in rats trained to discriminate 1 g/kg alcohol (Fig. 9.6, top panel); conversely, 1 g/kg was the only dose of alcohol capable of

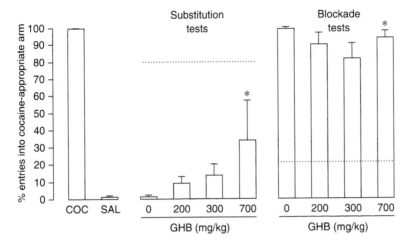

Figure 9.5 Failure of GHB to substitute for and block the discriminative stimulus effects of cocaine in rats trained to discriminate cocaine from saline in a T-maze food-reinforced drug discrimination paradigm. See legend of Fig. 9.4 for description of the apparatus and training procedure. In training sessions, pretreatment time was 15 min. Cocaine was dissolved in 1 ml/kg saline and administered i.p. at the dose of 10 mg/kg. GHB was dissolved in 5–10 ml/kg distilled water and administered i.p. at the doses of 0, 200, 300 and 700 mg/kg. In substitution tests, GHB was administered 30 min before the start of the session; in blockade tests, GHB was administered 30 min before cocaine injection. Each bar in the 'substitution' panel indicates the average percentage of entries into the cocaine-appropriate arm following the administration of a particular dose of GHB; the dashed line indicates the limit ($\geq 80\%$) of cocaine-appropriate responding for complete substitution for the discriminative stimulus effects of cocaine. Each bar in the 'blockade' panel indicates the average percentage of entries into the cocaine-appropriate arm following the combination of a particular dose of GHB and the training dose of cocaine; the dashed line indicates the limit ($\leq 20\%$) of cocaine-appropriate responding for complete blockade of the discriminative stimulus effects of cocaine. COC and SAL points represent the average per cent of entries into the cocaine-appropriate arm during training sessions with cocaine and saline. Bars are the mean ± SEM of two determinations in each of the eight rats employed. The asterisk indicates occurrence of motor disrupting effects.

substituting for GHB and this substitution occurred only in rats trained to discriminate 300 mg/kg GHB (Fig. 9.6, bottom panel). These results suggest that 1 g/kg alcohol and 300 mg/kg GHB elicit discriminative stimulus effects that are perceived as similar by rats. Interestingly, the maximal efficacy of GHB in reducing voluntary alcohol intake in selectively bred alcohol-preferring P and sP rats is observed at doses ranging between 200 and 300 mg/kg (see below). Previous studies aimed at determining the daily drinking pattern of both alcohol-preferring P (Waller *et al.* 1982; Murphy *et al.* 1986) and sP (Agabio *et al.* 1996) rats showed that most of their alcohol drinking occurred in distinct binges, each one averaging approximately 1 g/kg alcohol. Thus, the suppressing effects of GHB on voluntary alcohol intake and the amount of alcohol usually ingested by high alcohol-consuming P and sP rats in each drinking bout occur at doses similar to those that cross-substituted in the drug discrimination experiments. These data suggest that the reducing effects of GHB on voluntary alcohol intake are due, at least in part, to the substitution of GHB for the reinforcing effects of alcohol.

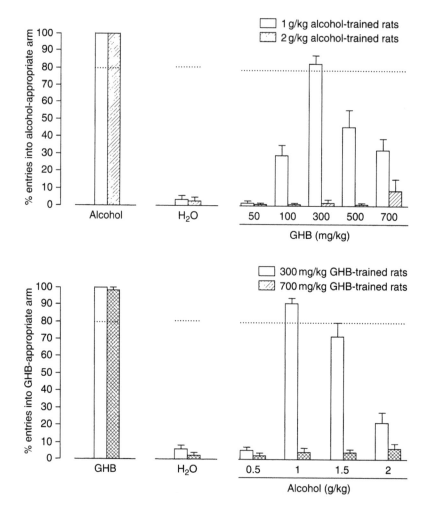

Figure 9.6 Cross-substitution between the discriminative stimulus effects of alcohol and GHB. Each bar in the top panel indicates the average percentage of entries into the alcohol-appropriate arm following the administration of various doses of GHB (0–700 mg/kg i.g.) in rats trained to discriminate either 1 ($n = 6$) or 2 g/kg ($n = 5$) alcohol (i.g.) from water in a T-maze, food-reinforced drug discrimination procedure. See legend of Fig. 9.4 for description of the apparatus and training procedure. Alcohol and H_2O points represent the average per cent of entries into the alcohol-appropriate arm during training sessions with alcohol and water. Each bar in the bottom panel indicates the average percentage of entries into the GHB-appropriate arm following the administration of various doses of alcohol (0–2 g/kg i.g.) in rats trained to discriminate either 300 ($n = 5$) or 700 mg/kg ($n = 6$) GHB (i.g.) from water. GHB and H_2O points represent the average per cent of entries into the GHB-appropriate arm during training sessions with GHB and water. The dashed line indicates the limit ($\geq 80\%$) of alcohol-appropriate responding (top panel) and GHB-appropriate responding (bottom panel) for complete substitution of the discriminative stimulus effects of alcohol and GHB, respectively. Bars are the mean ± SEM of two determinations in each rat.

Sources: Adapted from Colombo *et al.* (1995d) and Agabio *et al.* (1995).

Collectively, the above data predict that GHB has few, if any, subjective effects similar to those of abusive drugs such as morphine, lysergic acid diethylamide, *d*-amphetamine, phencyclidine and cocaine. These results might lead to the conclusion that GHB has no or minimal abuse potential. However, GHB possesses alcohol-like and, to a minor extent, benzodiazepine-like discriminative stimulus effects (at least in rat studies). This similarity, besides constituting the possible mechanism of GHB action on alcohol dependence (see Gessa *et al.* 2000), is consistent with anecdotal reports of GHB producing anxiolysis, relaxation and alcohol-like subjective effects (Galloway *et al.* 1997) and may contribute to its abuse potential.

Effect of GHB on self-administration of alcohol and cocaine

GHB reducing effect on alcohol intake

Recent clinical studies (for review, see Addolorato *et al.* 2000; Colombo and Gessa 2000; Gallimberti *et al.* 2000) have featured GHB as an effective agent in the pharmacotherapy of alcohol craving and consumption. These studies were instigated by preclinical evidence demonstrating the ability of GHB to markedly reduce voluntary alcohol intake in alcohol-preferring sP rats, selectively bred for high alcohol preference and consumption. Interestingly, sP rats were found to possess a high predictive validity in the search for drugs potentially effective in controlling intake of alcoholic beverages, as suggested by the evidence that agents reported to attenuate alcohol consumption in human alcoholics (e.g. disulfiram and naltrexone) also reduced voluntary alcohol intake in these animals (this laboratory, unpublished results).

In the studies with GHB, sP rats were offered alcohol (10% v/v in tap water) and tap water under the standard, two-bottle free choice regimen with unlimited access for 24 h/day. Under this paradigm, sP rats usually (a) consume approximately 6 g/kg alcohol per day, (b) avoid water almost completely (the daily alcohol solution versus water preference ratio is constantly higher than 80%), (c) tritiate daily alcohol intake in separate binges and (d) consume pharmacologically relevant amounts of alcohol at each binge, giving rise to blood alcohol levels up to 100 mg% (see Colombo 1997).

The acute administration of non-sedative doses of GHB (namely, 200 and 300 mg/kg i.p.), 15–20 min before the start of the dark phase of the light/dark cycle, resulted in a significant and dose-dependent reduction of voluntary alcohol intake, up to 60% compared to saline-treated rats (Fig. 9.7). However, the reducing effect of GHB on alcohol intake was evident solely during the first 30 min, returning to control group levels immediately later (60 min observation time in Fig. 9.7). In contrast, water intake was not affected by GHB administration.

The reducing effect of GHB on voluntary alcohol intake in sP rats has been closely replicated in alcohol-preferring rats from the P line. Indeed, the i.p. injection of 300 mg/kg GHB induced a robust reduction of alcohol intake exclusively during the initial 15 min of the daily 2 h drinking session (June *et al.* 1995).

The brief duration of the GHB reducing effect on alcohol intake in sP (Fig. 9.7) and P (June *et al.* 1995) rats is consistent with the brief duration of GHB action observed in human alcoholics as well as the need, often encountered in clinical practice, to administer the medication six times a day to achieve the best treatment outcome (see Addolorato *et al.* 2000). Furthermore, the short onset of the reducing effect of GHB on alcohol consumption is in close agreement with the replacement hypothesis on the mechanism of GHB action on alcohol dependence (see Gessa *et al.* 2000). Accordingly, GHB exerts central pharmacological effects

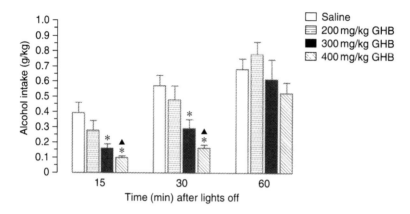

Figure 9.7 Reducing effect of GHB on voluntary alcohol intake in selectively bred sP rats. Alcohol (10% v/v in tap water) and tap water were offered under the standard, two-bottle free choice regimen with unlimited access for 24 h/day. GHB was administered i.p. at the doses of 0, 200, 300 and 400 mg/kg 15–20 min before the start of the dark phase of the light/dark cycle. Alcohol intake was monitored 15, 30 and 60 min after the start of the dark phase. Each bar is the $\pm$SEM of $n = 9$. *: $P < 0.05$ in comparison to vehicle-treated rats; $\blacktriangle$: occurrence of sedation.

Source: Adapted from Agabio *et al.* (1998).

similar to those of alcohol that maintain alcohol drinking behavior in alcohol-preferring rats; thus, administration of GHB and perception of its alcohol-like effects render further alcohol ingestion superfluous, resulting in an immediate reduction of alcohol intake. The complete and symmetrical generalization between the discriminative stimulus effects of 1 g/kg alcohol (i.e. the amount of alcohol usually consumed by alcohol-preferring rats in each drinking bout) and 300 mg/kg GHB (the non-sedative, most effective dose of GHB in reducing alcohol intake in alcohol-preferring rats), reported in Fig. 9.6, adds further support to the substitution hypothesis.

GHB reducing effect on cocaine self-administration

A recent study (Martellotta *et al.* 1998b) investigated the effect of GHB on self-administration of cocaine. Two different sets of rats were trained to intravenously self-administer cocaine (0.25–0.5 mg/kg/injection) by nose-poking and lever-pressing, respectively, in daily sessions of 2–3 h. The intragastric administration of 350 and 700 mg/kg GHB in the nose-poking rats and the intraperitoneal administration of 200 and 400 mg/kg GHB in the lever-pressing rats induced a dose-dependent reduction of the number of operant responding for GHB; under both paradigms, GHB reducing effect on cocaine self-administration was the result of a dose-dependent increase in the inter-reinforcement intervals (i.e. the time interposed between two self-infusions of cocaine). This effect was particularly evident during the first half of the self-administration session, when operant responding for cocaine after 700 mg/kg GHB i.g. and 400 mg/kg GHB i.g. was suppressed almost completely. No episode of sedation was reported.

It has been repeatedly demonstrated that laboratory rodents are capable of precisely regulating both timing and dose of self-administration of a specific drug (see Caine *et al.* 1993); for instance, lowering the dose per injection results in an increased number of infusions and a reduced inter-injection time, while increasing the drug dose induces a proportional decrease in the number of injections and increase in the time interval between injections. In both cases, at the end of the self-administration session, drug intake remains rather constant. In the study by Martellotta *et al.* (1998b), analysis of the cocaine self-administration patterns after GHB administration is suggestive of an effect of GHB similar to an increase in the dose of cocaine dispensed at each reinforcement; in other words, GHB appears to substitute for the reinforcing properties of cocaine, rendering cocaine infusion less urgent. This conclusion contrasts somewhat with the results of the drug discrimination study indicating the inability of GHB to substitute for the discriminative stimulus effects of cocaine (Fig. 9.4). The results by Martellotta *et al.* (1998b) are suggestive of a possible therapeutic use of GHB for cocaine addiction; it would be of interest if clinical studies were conducted to verify this hypothesis.

Acknowledgment

The authors are grateful to Mrs Anne Farmer for language editing of the manuscript.

References

Addolorato G., Balducci G., Capristo E., Attilia M. L., Taggi G., Gasbarrini G. and Ceccanti M. (1999) Gamma-hydroxybutyric acid (GHB) in the treatment of alcohol withdrawal syndrome: a randomized comparative study versus benzodiazepine. *Alcoholism: Clin. Exp. Res.* 23, 1596–604.

Addolorato G., Caputo F., Capristo E., Stefanini G. F. and Gasbarrini G. (2000) Gamma-hydroxybutyric acid: efficacy potential abuse and dependence in the treatment of alcohol addiction. *Alcohol* 20, 217–22.

Agabio R., Balaklievskaia N., Colombo G., Fadda F., Gessa G. L., Lobina C. and Reali R. (1995) Cross-substitution between the discriminative stimulus effects of gamma-hydroxybutyric acid and ethanol. *Alcologia* 7, 211–14.

Agabio R., Cortis G., Fadda F., Gessa G. L., Lobina C., Reali R. and Colombo G. (1996) Circadian drinking pattern of Sardinian alcohol-preferring rats. *Alcohol Alcohol.* 31, 385–88.

Agabio R., Colombo G., Loche A., Lobina C., Pani M. L., Reali R. and Gessa G. L. (1998) Gamma-hydroxybutyric acid (GHB) reducing effect on ethanol intake: evidence in favour of a substitution mechanism. *Alcohol Alcohol.* 33, 465–74.

Altman J., Everitt B. J., Glautier S., Markou A., Nutt D., Oretti R., Phillips G. D. and Robbins T. W. (1996) The biological, social and clinical bases of drug addiction: commentary and debate. *Psychopharmacology* 125, 285–345.

Barbaccia M. L., Colombo G., Affricano D., Carai M. A. M., Vacca G., Melis S., Purdy R. H. and Gessa G. L. (2002) $GABA_B$ receptor-mediated increase of neurosteroids by γ-hydroxybutyric acid. *Neuropharmacology*, in press.

Beardsley P. M., Balster R. L. and Harris L. S. (1996) Evaluation of the discriminative stimulus and reinforcing effects of gammahydroxybutyrate (GHB). *Psychopharmacology* 127, 315–22.

Bernasconi R., Lauber J., Marescaux C., Vergnes M., Martin P., Rubio V., Leonhardt T., Reymann N. and Bittiger H. (1992) Experimental absence seizures: potential role of γ-hydroxybutyric acid and $GABA_B$ receptors. *J. Neural Transm.* 35 (Suppl.), 155–77.

Bernasconi R., Mathivet P., Bischoff S. and Marescaux C. (1999) Gamma-hydroxybutyric acid: an endogenous neuromodulator with abuse potential? *Trends Pharmacol. Sci.* 20, 135–41.

Brady J. V. (1991) Animal models for assessing drugs of abuse. *Neurosci. Biobehav. Rev.* 15, 35–43.

Browne G. B. and Weissman A. (1981) Discriminative stimulus properties of Δ^9-tetrahydrocannabinol: mechanistic studies. *J. Clin. Pharmacol.* 21, 227S–234S.

Caine S. B., Lintz R. and Koob G. F. (1993) Intravenous drug self-administration techniques in animals. In *Behavioural Neuroscience, A Practical Approach*, Vol. 2 (Sahgal A., ed.), pp. 117–44. IRL Press, Oxford.

Cash C. D. (1994) Gamma-hydroxybutyrate: an overview of the pros and cons for it being a neurotransmitter and/or a useful therapeutic agent. *Neurosci. Biobehav. Rev.* 18, 291–304.

Colombo G. (1997) Ethanol drinking behaviour in Sardinian alcohol-preferring rats. *Alcohol Alcohol.* 32, 443–53.

Colombo G. and Gessa G. L. (2000) Gamma-hydroxybutyric acid in alcohol preference, dependence and withdrawal. *Addict. Biol.* 5, 389–403.

Colombo G., Agabio R., Bourguignon J.-J., Fadda F., Lobina C., Maitre M., Reali R., Schmitt M. and Gessa G. L. (1995a) Blockade of the discriminative stimulus effects of gamma-hydroxybutyric acid (GHB) by the GHB receptor antagonist NCS-382. *Physiol. Behav.* 58, 587–90.

Colombo G., Agabio R., Lobina C., Reali R., Zocchi A., Fadda F. and Gessa G. L. (1995b) Sardinian alcohol-preferring rats: a genetic animal model of anxiety. *Physiol. Behav.* 57, 1181–85.

Colombo G., Agabio R., Balaklievskaia N., Diaz G., Lobina C., Reali R. and Gessa G. L. (1995c) Oral self-administration of γ-hydroxybutyric acid in the rat. *Eur. J. Pharmacol.* 285, 103–07.

Colombo G., Agabio R., Lobina C., Reali R., Fadda F. and Gessa G. L. (1995d) Symmetrical generalization between the discriminative stimulus effects of gamma-hydroxybutyric acid and ethanol: occurrence within narrow dose ranges. *Physiol. Behav.* 57, 105–11.

Colombo G., Agabio R., Balaklievskaia N., Lobina C., Reali R., Fadda F. and Gessa G. L. (1996) T-maze and food-reinforcement: an inexpensive drug discrimination procedure. *J. Neurosci. Methods* 67, 83–87.

Colombo G., Agabio R., Diaz G., Fà M., Lobina C., Reali R. and Gessa G. L. (1998a) γ-Hydroxybutyric acid (GHB) intake in ethanol-preferring sP and non-preferring sNP rats. *Physiol. Behav.* 64, 197–202.

Colombo G., Agabio R., Lobina C., Reali R. and Gessa G. L. (1998b) Involvement of $GABA_A$ and $GABA_B$ receptors in the mediation of discriminative stimulus effects of gamma-hydroxybutyric acid. *Physiol. Behav.* 64, 293–302.

Colombo G., Agabio R., Bourguignon J.-J., Lobina C., Loche A., Maitre M., Reali R. and Gessa G. L. (1999) Reduction of blood ethanol levels by the gamma-hydroxybutyric acid receptor antagonist, NCS-382. *Alcohol* 17, 93–95.

Colpaert F. C. (1986) Drug discrimination: behavioral, pharmacological, and molecular mechanisms of discriminative stimulus effects. In *Behavioral Analysis of Drug Dependence* (Goldberg S. R. and Stolerman I. P., eds), pp. 161–93. Academic Press, London.

Ferrara S. D., Giorgetti R., Zancaner S., Orlando R., Tagliabracci A., Cavarzeran F. and Palatini P. (1999) Effects of single dose of gamma-hydroxybutyric acid and lorazepam on psychomotor performance and subjective feelings in healthy volunteers. *Eur. J. Clin. Pharmacol.* 54, 821–27.

Gallimberti L., Spella M. R., Soncini C. A. and Gessa G. L. (2000) Gamma-hydroxybutyric acid (GHB) in the treatment of alcohol and heroin dependence. *Alcohol* 20, 257–62.

Galloway G. P., Frederick S. L., Staggers F. E. Jr., Gonzales M., Stalcup S. A. and Smith D. E. (1997) Gamma-hydroxybutyrate: an emerging drug of abuse that causes physical dependence. *Addiction* 92, 89–96.

Galloway G. P., Frederick S. L., Seymour R., Contini S. E. and Smith D. E. (2000) Abuse and therapeutic potential of gamma-hydroxybutyrate. *Alcohol* 20, 263–69.

Garcha H. S. and Stolerman I. P. (1989) Discrimination of a drug mixture in rats: role of training dose, and specificity. *Behav. Pharmacol.* 1, 25–31.

Gessa G. L., Agabio R., Carai M. A. M., Lobina C., Pani M., Reali R. and Colombo G. (2000) Mechanism of the anti-alcohol effect of gamma-hydroxybutyric acid (GHB). *Alcohol* 20, 271–76.

Gobaille S., Hechler V., Andriamampandry C., Kemmel V. and Maitre M. (1999) γ-Hydroxybutyrate modulates synthesis and extracellular concentration of gamma-aminobutyric acid in discrete rat brain regions *in vivo*. *J. Pharmacol. Exp. Ther.* 290, 303–09.

Goudie A. J. and Leathley M. J. (1993) Drug-discrimination assays. In *Behavioural Neuroscience: A Practical Approach*, Vol. 2 (Sahgal A., ed.), pp. 145–67. IRL Press, Oxford.

Hechler V., Ratomponirina C. and Maitre M. (1997) γ-Hydroxybutyrate conversion into GABA induces displacement of $GABA_B$ binding that is blocked by valproate and ethosuximide. *J. Pharmacol. Exp. Ther.* 281, 753–60.

Hoffman D. H. (1989) The use of place conditioning in studying the neuropharmacology of drug reinforcement. *Brain Res. Bul.* 23, 373–87.

Hogg S. (1996) A review of the validity and variability of the Elevated Plus-Maze as an animal model of anxiety. *Pharmacol. Biochem. Behav.* 54, 21–39.

Holtzman S. G. (1990) Discriminative stimulus effects of drugs: relationship to potential for abuse. In *Modern Methods in Pharmacology, Vol. 6, Testing and Evaluation of Drugs of Abuse* (Adler M. W. and Cowan A., eds), pp. 193–210. Wiley-Liss, New York.

June H. L., Williams J. A., Cason C. R., Devaraju S., Lin M., Murphy J. M., Lewis M. J., Lumeng L. and Li T.-K. (1995) Low doses of gamma-hydroxybutyric acid (GHB) attenuate ethanol intake in alcohol-preferring (P) rats. *Alcoholism: Clin. Exp. Res.* 19 (Suppl. 2), 14A.

Kamien J. B., Bickel W. K., Hughes J. R., Higgins S. T. and Smith B. J. (1993) Drug discrimination by humans compared to nonhumans: current status and future directions. *Psychopharmacology* 111, 259–70.

Katz J. L. (1989) Drugs as reinforcers: pharmacological and behavioural factors. In *The Neuropharmacological Basis of Reward* (Liebman J. M. and Cooper S. J., eds), pp. 164–213. Oxford University Press, Oxford.

Kršiak M., Novakova D., Paclt I. and Ostrovskaya R. U. (1974) Effect of sodium hydroxybutyrate on behavior of mice after prolonged isolation. *Bull. Exp. Biol. Med.* 77, 288–91.

Lambert J. J., Belelli D., Hill-Venning C. and Peters J. A. (1995) Neurosteroids and GABA$_A$ receptor function. *Trends Pharmacol. Sci.* 16, 295–303.

Lettieri J. T. and Fung H.-L. (1979) Dose-dependent pharmacokinetics and hypnotic effects of sodium γ-hydroxybutyrate in the rat. *J. Pharmacol. Exp. Ther.* 208, 7–11.

Lingenhoehl K., Brom R., Heid J., Beck P., Froestl W., Kaupmann K., Bettler B. and Mosbacher J. (1999) γ-Hydroxybutyrate is a weak agonist at recombinant GABA$_B$ receptors. *Neuropharmacology* 38, 1667–73.

McIntire K. D. and Liddell B. J. (1984) Gamma-butyrolactone increases the rate of punished lever pressing by rats. *Pharmacol. Biochem. Behav.* 20, 307–10.

Maitre M. (1997) The gamma-hydroxybutyrate signalling system in brain: organization and functional implications. *Prog. Neurobiol.* 51, 337–61.

Mariathasan E. A. and Stolerman I. P. (1993) Overshadowing of nicotine drug discrimination in rats: a model for behavioural mechanisms of drug interactions? *Behav. Pharmacol.* 4, 209–15.

Mariathasan E. A., Garcha H. S. and Stolerman I. P. (1991) Discriminative stimulus effects of amphetamine and pentobarbitone separately and as mixture in rats. *Behav. Pharmacol.* 2, 405–15.

Martellotta M. C., Fattore L., Cossu G. and Fratta W. (1997) Rewarding properties of gamma-hydroxybutyric acid: an evaluation through place preference paradigm. *Psychopharmacology* 132, 1–5.

Martellotta M. C., Cossu G., Fattore L., Gessa G. L. and Fratta W. (1998a) Intravenous self-administration of gamma-hydroxybutyric acid in drug-naive mice. *Eur. Neuropsychopharmacol.* 8, 293–96.

Martellotta M. C., Balducci C., Fattore L., Cossu G., Gessa G. L., Pulvirenti L. and Fratta W. (1998b) Gamma-hydroxybutyric acid decreases intravenous cocaine self-administration in rats. *Pharmacol. Biochem. Behav.* 59, 697–702.

Mathivet P., Bernasconi R., De Barry J., Marescaux C. and Bittiger H. (1997) Binding characteristics of the γ-hydroxybutyric acid as a weak but selective GABA$_B$ receptor antagonist. *Eur. J. Pharmacol.* 321, 67–75.

Murphy J. M., Gatto G. J., Waller M. B., McBride W. J., Lumeng L. and Li T.-K. (1986) Effects of scheduled access on ethanol intake by the alcohol-preferring (P) line of rats. *Alcohol* 3, 331–36.

NIDA Research Monograph 116 (1991) *Drug Discrimination: Applications to Drug Abuse Research* (Glennon R. A., Järbe T. U. C. and Frankenheim J., eds). U.S. DHHS, Rockville.

Olsen R. W., Bergman R. O., Van Ness P. C., Lummis S. C., Watkins A. E., Napias C. and Greenlee D. V. (1981) γ-Aminobutyric acid receptor binding in mammalian brain. Heterogeneity of binding sites. *Mol. Pharmacol.* 19, 217–27.

Overton D. A. (1982) Application and limitations of the drug discrimination method for the study of drug abuse. In *Drug Discrimination: Applications in CNS Pharmacology* (Colpaert F. C. and Slanger J. L., eds). pp. 291–340. Elsevier, Amsterdam.

Pellow S., Chopin P., File S. E. and Briley M. (1985) Validation of open : closed entries in an elevated plus-maze as a measure of anxiety in the rat. *J. Neurosci. Methods* 1, 149–67.

Preston K. L. and Bigelow G. E. (1991) Subjective and discriminative stimulus effects of drugs. *Behav. Pharmacol.* 2, 293–313.

Rodgers R. J. and Cole J. C. (1994) The elevated plus-maze: pharmacology, methodology and ethology. In *Ethology and Psychopharmacology* (Cooper S.J. and Hendrie C.A., eds), pp. 9–44. John Wiley, Chichester.

Samele C., Shine P. J. and Stolerman I. P. (1991) A bibliography of drug discrimination research, 1989–1991. *Behav. Pharmacol.* 3, 171–92.

Schmidt C., Gobaille S., Hechler V., Schmitt M., Bourguignon J.-J. and Maitre M. (1991) Anti-sedative and anti-cataleptic properties of NCS-382, a γ-hydroxybutyrate receptor antagonist. *Eur. J. Pharmacol.* 203, 393–97.

Schmidt-Mutter C., Pain L., Sandner G., Gobaille S. and Maitre M. (1998) The anxiolytic effect of γ-hydroxybutyrate in the elevated plus maze is reversed by the benzodiazepine receptor antagonist, flumazenil. *Eur. J. Pharmacol.* 342, 21–27.

Serra M., Sanna E., Foddi C., Concas A. and Biggio G. (1991) Failure of gamma-hydroxybutyrate to alter the function of the $GABA_A$ receptor complex in the rat cerebral cortex. *Psychopharmacology* 104, 351–55.

Snead O. C. (1996) Antiabsence seizure activity of specific $GABA_B$ and gamma-hydroxybutyric acid receptor antagonists. *Pharmacol. Biochem. Behav.* 53, 73–79.

Snead O. C. and Bearden L. J. (1980) Naloxone overcomes the dopaminergic, EEG, and behavioral effects of γ-hydroxybutyrate. *Neurology* 30, 832–38.

Stolerman I. P. and Mariathasan E. A. (1990) Discrimination of an amphetamine–pentobarbitone mixture by rats in an AND-OR paradigm. *Psychopharmacology* 102, 557–60.

Stolerman I. P., Rauch R. and Norris E. A. (1987) Discriminative stimulus effects of a nicotine–midazolam mixture in rats. *Psychopharmacology* 93, 250–56.

Stolerman I. P., Samele C., Kamien J. B., Mariathasan E. A. and Hague D. S. (1995) A bibliography of drug discrimination research, 1992–1994. *Behav. Pharmacol.* 6, 643–68.

Tunnicliff G. (1992) Significance of γ-hydroxybutyric acid in the brain. *Gen. Pharmacol.* 23, 1027–34.

Tzcshentke T. M. (1998) Measuring reward with the conditioned place preference paradigm: a comprehensive review of drug effects, recent progress and new issues. *Prog. Neurobiol.* 56, 613–72.

Vayer P., Mandel P. and Maitre M. (1987) Gamma-hydroxybutyrate, a possible neurotransmitter. *Life Sci.* 52, 1382–87.

Waller M. B., McBride W. J., Lumeng L. and Li T.-K. (1982) Induction of dependence on ethanol by free-choice drinking in alcohol-preferring rats. *Pharmacol. Biochem. Behav.* 16, 501–07.

White J.-A. W. and Stolerman I. P. (1994) Antagonism of a nicotine plus midazolam discriminative cue in rats. *Behav. Pharmacol.* 5, 351–55.

Winter J. C. (1981) The stimulus properties of gamma-hydroxybutyrate. *Psychopharmacology* 73, 371–75.

Woolverton W. L., Rowlett J. K., Winger G., Woods J. H., Gerak L. R. and France C. P. (1999) Evaluation of the reinforcing and discriminative stimulus effects of gamma-hydroxybutyrate in rhesus monkeys. *Drug Alcohol Depend.* 54, 137–43.

Zerbib R., Pierrefiche G., Ferran C. and Laborit H. (1992) Potential antidepressant activity of gamma-hydroxybutyrate in the mouse 'behavioral despair' test: correlation with the central dopaminergic system. *Res. Comm. Psychol. Psychiat. Behav.* 17, 109–22.

Note added to proofs

A recent report by Metcalf and colleagues (*Pharmacol. Biochem. Behav.* 70, 31–41, 2001) described the lack of symmetrical generalization between the discriminative stimulus effects of 300 mg/kg GHB and 1 g/kg alcohol in rats. These data are inconsistent with those collected in the authors' laboratory and extensively described in the present chapter. Differences in the methodologies used in the two studies may be responsible for these discrepancies.

10 Therapeutic uses of γ-hydroxybutyrate

R. Agabio and G. L. Gessa

Introduction

Since the early sixties and in the wake of the pioneering studies conducted by Laborit (1964), γ-hydroxybutyric acid (GHB) has been tested for different uses and in the treatment of a number of pathologies, almost exclusively in Europe. This chapter is aimed at reviewing the outcomes of these studies, some of them perhaps possessing a historical significance rather than a robust indication of therapeutic usefulness for GHB, while others feature GHB as a drug of some efficacy in the therapy of alcoholism and narcolepsy and as an anaesthetic adjuvant. Accordingly, at present GHB (a) is marketed in some European countries, including Italy, for the treatment of alcohol withdrawal and craving, and (b) has been approved by the American Food and Drug Administration (FDA) as a Treatment Investigational New Drug for further evaluation of its efficacy in the treatment of narcolepsy. Interestingly enough, the studies available to date depict for GHB a profile of safety and good tolerability, with few episodes of side effects and abuse potential, in apparent contrast with the increasing number of recent reports on GHB abuse in the United States (see Galloway *et al.* 2000).

GHB efficacy in the treatment of alcohol dependence

Medical interventions in the field of alcoholism are primarily aimed at (a) relieving the consequences of alcohol withdrawal syndrome, (b) controlling craving for alcohol (usually defined as the preoccupation with, thought about, and urge for alcohol), (c) arresting or at least reducing intake of alcohol and (d) maintaining sobriety for as long as possible. Pharmacotherapy is conceived to provide a substantial contribution to these goals, facilitating the psychological support and social rehabilitation of alcoholic patients.

Despite the significant advances achieved over the last twenty years in the neurobiology of alcohol (Eckardt *et al.* 1998; Woodward 1999), the elusive mechanisms of alcohol action in brain are still far from being fully elucidated. The incomplete knowledge of how alcohol acts in the brain and why drinking is pleasurable and compelling in some individuals has so far hampered the rational design and development of selective drugs capable of correcting the possible neurochemical alterations underlying alcohol dependence. In contrast, most pharmacotherapeutic agents currently available to physicians do not appear to possess relevant therapeutic efficacy or give problems of tolerance, dependence or even abuse liability themselves (see Litten and Allen 1998; Garbutt *et al.* 1999; Swift 1999).

Different lines of experimental evidence (Gessa *et al.* 2000) have demonstrated the effectiveness of GHB in animal models of alcoholism. Briefly, (a) the acute administration of GHB to rats rendered physically dependent on alcohol suppressed the intensity of alcohol withdrawal signs, and (b) non-sedative doses of GHB, administered acutely, produced a dose-dependent reduction (up to 70% compared to placebo-treated controls) of voluntary alcohol intake in selectively bred alcohol-preferring rats. Furthermore, (1) cross-tolerance to the motor impairing effects of GHB and alcohol has been observed, suggesting the presence of common adaptive changes in neural substrates to chronic alcohol and GHB, and (2) alcohol and GHB possess similar discriminative stimulus effects (i.e. the animal correlate of human subjective feelings elicited by a psychoactive drug) in rats. On the basis of these results, it has been suggested that GHB exerts its reducing effects on alcohol withdrawal syndrome and intake by mimicking alcohol actions in the central nervous system (CNS) (Gessa *et al.* 2000).

The results led to the studies, reviewed below, that evaluated the efficacy of GHB in reducing alcohol withdrawal symptoms as well as alcohol craving and consumption in alcoholic subjects.

Alcohol withdrawal syndrome

Prolonged drinking of high quantities of alcohol induces compensatory changes in CNS functioning. Upon abrupt cessation of alcohol drinking, these maladaptive changes lead to development of alcohol withdrawal syndrome. This syndrome is a distressing and even life-threatening condition affecting several alcohol-dependent patients. The most frequent signs and symptoms, due to some levels of dysfunction of the autonomic nervous system, are the following: increase of heart rate, increase in respiratory rate, elevation in body temperature, anorexia, nausea, vomiting, sadness, psychosomatic symptoms, sweating and tremors (Edwards 1990). Approximately 5% and 1% of alcoholics experience acute seizures (Pieninkeroinen *et al.* 1992) and *delirium tremens* (Morton *et al.* 1994), respectively. The latter condition may last several days and is characterized by severe hyperactivity of the autonomic nervous system (i.e. marked elevation in blood pressure, pulse rate, respiratory rate and body temperature), with increase of hand tremor, severe confusion and relative common hallucinations (Morton *et al.* 1994). *Delirium tremens* may lead to death.

Pharmacological medications during alcohol withdrawal syndrome are aimed at decreasing (a) intensity of overall signs and symptoms and (b) risk of occurrence of convulsions and *delirium tremens*. Benzodiazepines (e.g. diazepam, chlordiazepoxide, oxazepam, lorazepam) are the drugs of choice; alternative pharmacotherapies include the α-adrenergic agonists, clonidine and lofexidine, and the β-blockers, propranolol and atenolol (Garbutt *et al.* 1999; Swift 1999). However, (a) the existence of cross-tolerance and dependence with alcohol (particularly for benzodiazepines), (b) their pharmacokinetic characteristics (long-acting drugs inducing drug buildup and subsequent lethargy, drowsiness and ataxia; short acting drugs resulting in an excessively short protection), and (c) the limited efficacy on specific signs and symptoms may limit their clinical use.

The ability of GHB in reducing the severity of alcohol withdrawal symptomatology was initially evaluated in a randomized double-blind study that recruited twenty-three alcoholic patients meeting the DSM III-R criteria for alcohol withdrawal syndrome (Gallimberti *et al.* 1989). Exclusion criteria were: presence of convulsions, *delirium tremens*, concurrent severe illness or abuse of other drugs. Patients were divided into two groups and treated with 50 mg/kg/day GHB ($n = 11$) and placebo ($n = 12$). GHB and placebo were given orally as a syrup. Six withdrawal signs and symptoms (namely: tremors, sweating, nausea, anxiety, depression and restlessness) were evaluated in each patient 30 min before and 1, 2, 3, 5 and 7 h after GHB administration. Each symptom was scored on a 4-point scale (0–3, paralleling the increased severity); therefore, the individual score ranged from 0 to 18 points. As shown

Table 10.1 Reducing effect of acutely administered GHB (50 mg/kg *per os*) on the severity of alcohol withdrawal syndrome in alcoholics

Treatment group (no. of patients)	Before treatment (30 min)	After treatment (h)				
		1	*2*	*3*	*5*	*7*
GHB (11)	12.6 ± 3.9	7.2 ± 3.9*	4.2 ± 3.1**	2.1 ± 1.6**	1.5 ± 1.7**	2.6 ± 1.3**
Control (12)	11.8 ± 5.7	11.8 ± 4.7+	11.3 ± 3.5+	12.6 ± 9.2+	13.6 ± 6.5+	14.7 ± 4.3+

Source: Adapted from Gallimberti *et al.* (1989).

Values are mean ± SD. *$P < 0.05$; **$P < 0.01$ (Pratt's test for comparison of scores before and after treatment); +$P < 0.05$ (Mann–Whitney test for comparison of control and GHB groups).

in Table 10.1, the mean score of alcohol withdrawal syndrome in the placebo group increased significantly over the observation period, averaging 11.8 immediately before placebo administration and 14.7 after seven hours. In contrast, treatment with GHB resulted in a dramatic decrease in alcohol withdrawal severity over the 7 h observation period; indeed, withdrawal score initially averaged 12.6 (immediately before drug administration), was already significantly reduced to 7.2 one hour after GHB administration and progressively decreased up to 1.5, as recorded at the 5 h observation time. The only side effects due to GHB were slight and transient dizziness, lasting about 15 min. Dizziness appeared 30 min after GHB administration in 60% of patients on day 1 of treatment and in 30% of patients on day 2. No patient in the control group reported dizziness.

The second study (Addolorato *et al.* 1999a) used a randomized, single-blind design and compared the effect of the benzodiazepine, diazepam and GHB on alcohol withdrawal syndrome. Both drugs were administered orally, for 6–10 consecutive days, to thirty alcoholic patients, according to the DSM IV criteria, suffering from alcohol withdrawal syndrome. The total dose of 0.5–0.75 mg/kg/day of diazepam and of 50 mg/kg/day of GHB were divided into six and three daily administrations, respectively. Evaluation of withdrawal symptoms was performed by the Clinical Institute Withdrawal Assessment for Alcohol (revised) scale (CIWA-Ar). Treatments with diazepam and GHB resulted in a significant and comparable reduction in CIWA-Ar score, particularly on days 2 and 3. A shorter onset of the anxiolytic effect was monitored in GHB- than diazepam-treated patients. The major disadvantage in the clinical use of GHB is generally considered to be the short duration of its effects; however, in the study by Addolorato *et al.* (1999a), it is noteworthy that protection from alcohol withdrawal syndrome elicited by three daily administrations of GHB was similar to that exerted by a double fractioning of the daily dose of the reference medication, diazepam. Transient, slight vertigo and drowsiness were the only side effects appeared in both groups.

A further double-blind placebo-controlled investigation (Nimmerrichter *et al.* 2002) with ninety-eight patients found that GHB (50 and 100 mg/kg/day, administered orally four times per day for six consecutive days) was as effective and rapid as clomethiazole (1,000 mg/day), chosen as reference medication, in reducing the intensity of alcohol withdrawal symptoms and signs. There was no difference, among the three treatments tested, also in terms of craving-reducing effect. Drug efficacy was not significantly different between the two groups of GHB, while the higher dose produced more side effects.

Finally, an open-label study (Moncini *et al.* 2000) assessed the efficacy of GHB in twenty-two patients suffering from alcohol withdrawal syndrome. A dose of GHB ranging from 50 to 150 mg/kg/day in three daily administrations was given orally for six consecutive days.

Treatment with GHB resulted in a virtually total suppression of alcohol withdrawal score in all patients; a significant reduction was already apparent after the first administration of GHB. Two patients reported slight and transient diarrhea as well as gastric upset shortly after the first drug administration.

None of the above studies (Gallimberti *et al.* 1989; Addolorato *et al.* 1999a; Nimmerrichter *et al.* 2002; Moncini *et al.* 2000) reported somnolence following GHB administration; furthermore, neither symptoms of withdrawal from GHB nor craving for GHB were reported after discontinuation of treatment.

The results of these studies depict for GHB a profile of effectiveness, tolerability and safety in the treatment of alcohol withdrawal syndrome. However, we believe that the studies reviewed above require confirmation in more controlled trials possibly examining larger samples of alcoholic patients.

Alcohol craving and consumption

The core features of alcohol dependence appear to be the compulsive desire, or craving, for alcohol and loss of control over drinking (i.e. a higher consumption of alcohol than the patient intended despite awareness of negative consequences) (see Morse and Flavin 1992). Historically, however, pharmacotherapies were mainly focused on relieving the symptomatology of alcohol withdrawal syndrome; only recently, medications to help achieve and maintain abstinence and control alcohol craving have been developed. To this regard, the opioid receptor antagonist, naltrexone (O'Brien *et al.* 1996; O'Malley *et al.* 1996; Weinrieb and O'Brien 1997), and the putative $GABA_A$ and NMDA receptor modulator, acamprosate (Sass *et al.* 1996; Pelc *et al.* 1997), have been shown to possess some efficacy in reducing relapse rate and controlling the craving for alcohol; naltrexone and acamprosate are presently marketed in several countries for the treatment of alcoholism. More recently, a preliminary study reported promising results on the reducing effect of the $GABA_B$ agonist, baclofen, on alcohol craving and consumption in alcoholics (Addolorato *et al.* 2000a).

To date, three studies have assessed the efficacy of GHB on craving for alcohol as well as consumption of alcoholic beverages. The first survey (Gallimberti *et al.* 1992) included eighty-two patients with a history of alcoholism, according to DSM III-R criteria, of 5 years or more. Subjects were divided into two groups and received, according to a randomized double-blind experimental design, 50 mg/kg/day GHB and placebo for three consecutive months. GHB and placebo were administered orally (in form of syrup) three times per day. Intensity of craving for alcohol was evaluated using a yes-or-no questionnaire; alcohol consumption was monitored on the basis of patients' self-reports and reports by relatives.

Seventy-one patients (thirty-six and thirty-five in the GHB and placebo group, respectively) completed the study. As shown in Table 10.2, over the 3-month treatment period, the number of daily drinks and days of abstinence was not significantly altered in the placebo group; in contrast, GHB-treated patients reduced the number of daily drinks by approximately 60% (from 12.1 to 4.7 average daily drinks before and 3 months after the start of GHB treatment, respectively) and increased the overall number of days of abstinence approximately fourfold (from 5.6 to 25.9 average number of abstinence days before and 3 months after the start of GHB treatment, respectively). GHB treatment also resulted in a significant reduction of craving for alcohol. This effect was already apparent at the first monthly monitoring (when mean alcohol craving score was reduced from 8.5 (pretreatment value) to 2.1) and persisted throughout the entire treatment period (averaging 3.3 and 3.1 at the 2 and 3 month observation times, respectively). Placebo treatment produced only a modest

Table 10.2 Reducing effect of GHB administration (50 mg/kg/day) on alcohol consumption and craving

Response	*During the 3 months before the treatment*			*During the 3 months after the treatment*		
	Placebo	*GHB*	*P*	*Placebo*	*GHB*	*P*
Daily drinks	11.4 ± 0.6	12.1 ± 0.5	NS	9.3 ± 0.7	4.7 ± 0.4	*
% of abstinent days	4.9 ± 0.4	5.6 ± 0.5	NS	8.4 ± 1.6	25.9 ± 3.1	**
Craving score	8.5 ± 0.3	8.9 ± 0.5	NS	7.6 ± 0.3	3.1 ± 0.6	***

Source: Adapted from Gallimberti *et al.* (1992).

Values are mean ± SEM from thirty-five placebo and thirty-six GHB treated subjects. *$P < 0.01$ with respect to placebo value by Student's test; **$P < 0.001$ with respect to placebo value by Student's test; ***$P < 0.001$ with respect to basal value by Student's test.

Table 10.3 Number of patients reporting abstinence, controlled or excessive drinking at the end of the third month of treatment with GHB (50 mg/kg/day)

	Placebo (n = 35)	*GHB (n = 36)*
Abstinence	2	11
Controlled drinking	6	15
Excessive drinking	27	10

Source: Adapted from Gallimberti *et al.* (1992).

reduction in craving score during the first month and then returned to pretreatment level. As shown in Table 10.3, at the end of the treatment, abstinent patients were 30% and 6% in GHB- and placebo-groups, respectively, while controlled drinking (defined as a daily alcohol consumption lower than 40 g) occurred in 42% and 17% in GHB- and placebo-groups, respectively. Adverse side effects in GHB-treated patients were limited to few episodes of transient dizziness, vertigo and headache that occurred within the first 3 days of treatment.

These results were later confirmed by an open study conducted by Addolorato *et al.* (1996) based on 179 alcohol-dependent outpatients. GHB was administered orally, at the dose of 50 mg/kg/day, fractioned in three times a day, for 24 consecutive weeks. Seventy subjects dropped out, while 109 subjects completed the 6-month study; 78% of the latter group (i.e. eighty-four subjects) remained abstinent throughout the entire treatment. Six and twelve months after suspension of GHB treatment, the number of abstinent subjects was 43 (51%) and 30 (36%), respectively. Similar to the observations made by Gallimberti *et al.* (1992), tolerability was good: vertigo, sleepiness and tiredness were the only side effects reported; these occurred in a few patients and resolved within the first 2–3 weeks of GHB intake. Finally, the study by Addolorato *et al.* (1996) reported, for the first time, cases of abuse of GHB among patients taking the drug for therapeutic reasons and under medical control; indeed, eleven patients (10%) showed craving for GHB and voluntarily increased the daily dosage up to six to seven times the recommended amount. Later, the same research group reported an episode of GHB withdrawal syndrome (characterized by anxiety, tremors, sweating, nausea and tachycardia) in an alcoholic patient who self-increased the prescribed dose of GHB and took approximately 18 g/day for 4 months (Addolorato *et al.* 1999b). Complete remission of the withdrawal symptomatology occurred within 2 h of administration of 20 mg diazepam.

Further confirmation of the results featuring GHB as an efficacious medication for the treatment of alcoholism is provided by a small double-blind survey recently completed by Moncini *et al.* (2000). At the end of the 6-month treatment with 50 mg/kg/day GHB ($n = 7$) or placebo ($n = 6$), alcohol craving score was significantly lower in GHB- than placebo-treated alcoholic individuals.

Finally, in keeping with the rapid metabolism of GHB in human alcoholics (50 mg/kg GHB is completely eliminated within 4–6 h (Ferrara *et al.* 1992; Palatini *et al.* 1993)) and its short-lasting effect on alcohol intake in animal studies (June *et al.* 1995; Agabio *et al.* 1998), recent investigations demonstrated an increase in GHB efficacy after a greater dosage fractioning than that usually prescribed, three administrations per day (Addolorato *et al.* 1998a,b). Indeed, these studies found that thirty-seven non-responders to the standard dosage regimen (3 administrations/day) of the daily dose (50 mg/kg) of GHB benefited from a further fractioning to six administrations per day: this tritiation of the daily dose of GHB resulted in a significant reduction in alcohol craving score, and over the 8 weeks of treatment twenty-six subjects reached and maintained complete abstinence.

Viewed as a whole, the results of these studies are indicative of the efficacy, safety and tolerability of GHB when used to control alcohol craving and consumption and promote abstinence. Abuse of GHB among alcoholics taking GHB for therapeutic purposes appears to be a limited phenomenon, involving a percentage of patients varying between 10% (Addolorato *et al.* 1996) and 15% (Gallimberti *et al.* 2000) in the two studies addressing this issue. Proper fractioning of the daily dose and assignment of the medication to a responsible care giver of the patient has been suggested to further reduce this figure (Addolorato *et al.* 2000b; Beghè and Carpanini 2000). However, further double-blind studies, examining larger samples of alcoholic patients, are needed to definitively assess the effectiveness of GHB in the treatment of alcohol dependence. These studies should also possibly evaluate whether the efficacy of GHB may vary across the spectrum of the different subtypes of alcoholism (see Cloninger 1987; Babor *et al.* 1992; Lesch and Walter 1996).

GHB efficacy in the treatment of heroin dependence

Through use of GHB in the management of alcohol withdrawal syndrome in a number of alcoholic patients who concomitantly abused heroin, investigators from Gallimberti and Gessa's research groups hypothesized that GHB could reduce the severity of withdrawal syndrome also from heroin. Subsequently, they performed a double-blind, placebo-controlled trial to verify whether GHB was effective in suppressing the withdrawal syndrome in heroin- and methadone-dependent subjects.

In this study (Gallimberti *et al.* 1993), patients were twenty-two heroin addicts with a 3–6 year history of heroin use, and nineteen subjects under maintenance treatment with methadone (30–60 mg/day for at least 6 months before the start of the survey). All patients had expressed their interest in discontinuing opiate consumption. They were hospitalized for 8 days. Withdrawal syndrome was assessed using a rating scale comprised of twenty-one items (i.e. signs and symptoms associated to opiate withdrawal syndrome such as craving, anxiety/restlessness, tremors, yawning, goose flesh) were rated as present (score 1) or absent (score 0). GHB (25 mg/kg) and placebo were administered orally every 2–6 h for eight consecutive days.

All patients showed an increase in withdrawal score during the 3 h immediately before treatment. The first administration of GHB resulted in a marked reduction of withdrawal score in both heroin- and methadone-dependent subjects. GHB effect had a rapid onset (<15 min) and short duration (at the 3 h observation interval, withdrawal score tended to increase). When the withdrawal items were evaluated singly, GHB was found to be effective in reducing all signs and symptoms with exception of diarrhea and insomnia. All patients reported relief from subjective distress. On the following days, withdrawal score in GHB-treated patients remained constantly reduced in comparison to placebo-treated subjects. On the eighth day, GHB administration was interrupted and patients were observed for a period of 5–6 h. Subsequently, patients received an intravenous injection of naloxone (0.4 mg); no withdrawal sign and symptom was observed.

The above results were later confirmed in a subsequent small study (Gallimberti *et al.* 1994), where 180–300 mg/kg/day GHB *per os* in six daily administrations alleviated opiate withdrawal symptoms in two opiate addicts who abruptly interrupted long-term methadone treatment. The results of the above studies by Gallimberti *et al.* (1993, 1994) suggest the efficacy of GHB in suppressing opiate withdrawal in humans.

In contrast, minimal effects of GHB on naloxone-precipitated opiate withdrawal were described by Rosen *et al.* (1996). In this study, opiate-dependent inpatients maintained on the opioid levorphanol received an acute, oral dose of GHB (15 or 30 mg/kg) or placebo followed 1 h later by an intravenous injection of naloxone (0.4 mg). GHB administration resulted in no significant protection in any withdrawal measure. The different GHB dose and timing of drug administration (postwithdrawal in the study by Gallimberti *et al.* (1993); before naloxone in the study by Rosen *et al.* (1996)) might account for these discrepancies.

Finally, combination of GHB (1,750 mg administered orally three times per day, approximately 75 mg/kg/day) plus naltrexone (50 mg/day) induced a significant reduction in heroin craving score and dropouts, in comparison to placebo plus naltrexone, in heroin addicts (Gerra *et al.* 1994). Intake of other addictive drugs (namely, benzodiazepines, cannabis, cocaine and alcohol) was also reduced.

Although the results of the above studies were vaguely promising, further confirmation is required before affirmation of the efficacy of GHB in the treatment of opiate withdrawal syndrome.

GHB efficacy in the treatment of narcolepsy

Introductory notes on narcolepsy

At present, one of the most important therapeutic uses of GHB is represented by the treatment of narcolepsy. This sleep disorder is a relative rare condition (occurring in 0.02–0.09% of the population (Scharf and Fletcher 1989)) characterized by (a) excessive daytime sleepiness with sudden and irresistible sleep attacks, and (b) presence of the following three frightening auxiliary symptoms: paralysis upon lying down or at waking (sleep paralysis), visual or auditory dream-like hallucinations at sleep onset (hypnagogic hallucinations) and temporary loss of muscle tone while awake (cataplexy) or asleep (see Association of Sleep Disorder Centers and the Association for the Psychophysiological Study of Sleep 1979).

Electroencephalographic registrations of sleep are usually classified as: stage 0 or wakefulness state (characterized by the electroencephalographic pattern of alpha activity); stage 1 or drowsy, light sleep (low voltage without spindle); stage 2 or spindle (spindling against a low voltage background); stages 3 and 4 (otherwise named delta stage), that is, deep sleep (slow, high voltage activity); REM sleep (characterized by low voltage electroencephalographic activity, rapid eye movements and disappearance of tonic muscle activity). The first REM sleep period appears approximately 90 min after sleep onset (REM latency) (Sinton and McCarley 2000). Electroencephalographic registrations during sleep in narcoleptic patients include (a) instability of both REM as well as non-REM (NREM) sleep, (b) sleep onset with REM periods, (c) increased number of awakening episodes, (d) increased total wake time after sleep onset, (e) increased time spent in stage 1 and (f) shortened REM sleep latency (Passouant *et al.* 1968; Montplaisir *et al.* 1978; Guilleminault 1986; Zorick *et al.* 1986). It has been shown that auxiliary symptoms (sleep paralysis, hypnagogic hallucinations and cataplexy) are based upon abnormal REM sleep mechanisms (Broughton 1971; Zarcone 1973).

The most common strategy in the pharmacotherapy of narcolepsy includes (a) a stimulant medication (such as amphetamine or methylphenidate) to reduce the frequency and duration of diurnal sleepiness and sleep attacks (Parkers 1976; Mitler *et al.* 1986) and (b) an anti-REM sleep agent (usually, a tricyclic antidepressant such as imipramine or protriptyline) in an attempt to avoid the occurrence of auxiliary symptoms (Scharf *et al.* 1988; Scrima *et al.* 1989). However, the outcomes are still unsatisfactory: occurrence of several adverse effects, drug tolerance and rebound cataplexy are common problems encountered with both types of drugs and result in low compliance.

GHB as a hypnotic in healthy subjects

In the early 1960s, Henri Laborit suggested the potential use of GHB as a hypnotic agent (Laborit 1964). Indeed, in his pioneering study, the intravenous administration of 50 mg/kg GHB induced sleep in healthy subjects. Sleep onset ranged between 5 and 10 min.

Following the initial observation by Laborit (1964), two different studies tested the efficacy of GHB in producing a sleep-like state (Metcalf *et al.* 1966; Yamada *et al.* 1967). GHB was acutely administered at daytime (a) *per os* at doses of 35–63 mg/kg in the study by Metcalf *et al.* (1966) and (b) intravenously at doses of 10–30 mg/kg in the study by Yamada *et al.* (1967) to healthy subjects. In both studies, treatment with GHB resulted in the induction of a sleep state (Metcalf *et al.* 1966; Yamada *et al.* 1967).

Subsequently, a double-blind study (Lapierre *et al.* 1990) performed to investigate the effect of GHB on REM and NREM phases of sleep demonstrated that a single oral dose of 2.25 g GHB (approximately 30 mg/kg), administered at bedtime, significantly increased time spent in stages 3 and 4 (deep sleep), decreased time spent in stage 1 (light sleep) and did not alter time spent in REM sleep.

Interestingly enough, in contrast to GHB the majority of synthetic hypnotics used for the treatment of sleep disorders suppress REM and delta sleep; furthermore, rebound of REM sleep upon drug withdrawal is usually associated with disturbed, nightmarish sleep (Oswald and Priest 1965; Mamelak *et al.* 1977).

GHB and narcolepsy

In 1979, Broughton and Mamelak tested the ability of GHB in improving night-time sleep in narcoleptic patients in an open-design study. Before treatment with GHB, other drug therapies for narcolepsy had been discontinued for at least 14 days. An initial dose of 1.5–2.25 g GHB was given orally to sixteen narcoleptic subjects at bedtime, followed by further multiple doses of 1–1.5 g during the night at each major reawakening; treatment with GHB was repeated for 20 months. Accordingly, the total quantity of GHB given each night averaged approximately 50 mg/kg, with peaks of approximately 80 mg/kg. GHB efficacy was evaluated by (a) EEG registrations during diurnal and nocturnal sleep and (b) self-reported scores of diurnal alertness.

GHB induced a marked clinical improvement with a reduction of more than 70% in severity of symptoms with respect to baseline. Nocturnal sleep became less restless while nightmares, hallucinations, and attacks of sleep paralysis vanished. All patients found it easier to stay awake during the day. The number of irresistible daytime attacks of sleep and cataplexy diminished without development of tolerance. However, many patients continued to feel tired and drowsy during the day. After addition of 5–10 mg methylphenidate, three times a day, daytime drowsiness and fatigue became minimal. Only a few adverse effects were reported (namely: muscular weakness, urinary urgency and a dream-like confusional state) and these had practically disappeared after 1 week. This study suggested that GHB, given at night and in combination with small doses of methylphenidate, may reduce all major symptoms of narcolepsy with few side effects. The major clinical disadvantage appeared to be the short duration of action and the subsequent need of multiple administrations.

An investigation (Broughton and Mamelak 1980) on the electroencephalographic registration of nocturnal sleep in the same sixteen patients revealed that GHB induced a significant increase of approximately 30% in the duration of slow wave sleep and reduced by approximately 35% the length of stage 1. Furthermore, REM sleep became more efficient and less fragmented. Latency of REM sleep was decreased by approximately 70%. Treatment with GHB also resulted in a significant decrease in the daytime duration of delta sleep (50%) and REM sleep (43%) while it failed to alter diurnal stage 1. The lack of GHB effect on diurnal stage 1 reflects the persistence of subjective daytime drowsiness. Furthermore, the night-time portion of the electroencephalographic recordings showed sleep consolidation, fewer shifts and less sleep fragmentation, consistent with patients' reports of deeper and more restful night sleep and less frequent nightmares and hallucinations.

In a subsequent open study (Scharf *et al.* 1985) including thirty narcoleptic patients, doses of 5–7 g GHB (corresponding to 65–90 mg/kg) were fractioned in two administrations, the first at bedtime and the second 4 h later. Patients were required to stay in bed for 8 h. Stimulant medications (methylphenidate, pemoline and dextroamphetamine) were also administered to all patients to control excessive daytime sleepiness. As in the study by Broughton and Mamelak (1979), the number of awakening episodes decreased by approximately 35%, while stages 3 and 4 increased by approximately 200%. After the first week of treatment, the frequency of cataplexy, sleep paralysis, hypnagogic hallucinations and sleep attacks decreased by 57%, 81%, 80% and 64%, respectively, with respect to baseline values. Lower doses of stimulants were required to control daytime sleepiness. With regard to possible side effects, a single episode of protracted sleep paralysis was reported to occur in three patients, each episode taking place shortly after the initial nightly dose of GHB; moreover, enuresis and increased transient sexual drive were described.

A 9-year-long study that followed (Mamelak *et al.* 1986) demonstrated the lack of development of tolerance to the alleviating effect of GHB (up to about 90 mg/kg) on narcolepsy in forty-eight patients.

In 1989 Scrima *et al.* conducted the first double-blind, placebo-controlled trial on the effectiveness of GHB in narcoleptic patients. Twenty narcoleptic patients (ten females and ten males, aged between 16 and 65 years), unaffected by other major diseases, were admitted to the study. Subjects were allowed to use methylphenidate (up to 30 mg/kg) to counter excessive daytime sleepiness. GHB was administered at the dose of 50 mg/kg (25 mg/kg at bedtime and 25 mg/kg 3 h later, due to its short half-life) for 4 weeks. Diagnostic tests included electroencephalographic registration of sleep and self-evaluation of daytime sleepiness. GHB treatment caused a robust reduction in cataplexy; indeed, the number of daily cataleptic events in GHB-treated patients was approximately 50% and 70% lower on the first and last week of treatment, respectively, in comparison to placebo-dosed patients. The number of arousals from sleep declined by approximately 30%. Finally, GHB decreased the severity of hypnagogic hallucinations. However, GHB treatment did not alter the subjective quality of sleep, the number of sleep attacks, daytime sleepiness, need for methylphenidate, onset of night-time sleep, total sleep time and the number of naps per day. The apparent discrepancy of these results with those of previous open studies (Broughton and Mamelak 1979; Scharf *et al.* 1985) might be due to a non-specific or placebo effect and to the use of higher doses of GHB (up to 80 mg/kg in the study by Broughton and Mamelak (1979) and up to 90 mg/kg in the study by Scharf *et al.* (1985). Total number of adverse reactions observed during GHB treatment was lower than that monitored during placebo treatment.

The overnight electroencephalographic recordings revealed that GHB improved sleep depth and continuity (Scrima *et al.* 1990). Indeed, at the end of treatment period, delta sleep was increased (by 53% with respect to placebo treatment) while stage 1 was decreased (17%), and there were fewer awakenings (21%) and sleep stage shifts (13%). REM sleep latency was significantly decreased in males but not in females. However, stage 0 was significantly increased (25%) during GHB treatment. By dividing night-time into two periods (first 6 and last 2 h), the increase of stage 0 was limited to the last 2 h (47%). These data may be explained by the fact that patients were requested to stay in bed for 8 h, even if they were ready to get up earlier.

In a second double-blind placebo-controlled study (Lammers *et al.* 1993), GHB was orally administered to twenty-four narcoleptic patients (thirteen males and eleven females, aged between 16 and 65 years) for 4 weeks. Two daily doses of 30 mg/kg each were given, the first at bedtime and the second 4 h later. In this study all subjects continued to use their co-medications. The duration of nocturnal sleep was not predefined. The results of this study indicated that GHB induced a reduction, compared to baseline, of all narcoleptic symptoms, namely (a) daily number of hypnagogic hallucinations (by 67%) and (b) daily number and severity of daytime sleep attacks (38% and 21%, respectively). GHB failed to decrease the number of diurnal cataplexy attacks and awakenings at night. Nocturnal electroencephalographic registrations showed that GHB stabilized nocturnal REM sleep, reducing (a) the percentage of wakefulness during REM sleep (by 44%) and (b) the number of awakenings out of REM sleep (by 27%). However, in contrast to the double-blind study by Scrima *et al.* (1989), GHB did not modify (a) the duration of delta sleep and stage 1 and (b) the number of stage shifts and awakenings.

These results, taken together, suggest that GHB is capable of reducing the signs and symptoms of narcolepsy, without suppression of REM sleep; its withdrawal is not associated with REM rebound. Other pharmacotherapies (like stimulants or antidepressants) can induce serious side effects; moreover, they lose efficacy on continuing treatment. In contrast, GHB caused only minor side effects, usually limited to the first days of treatment; they included: gastrointestinal adverse effects, dizziness, weight loss, urinary urgency and sleep walking. Development of tolerance to GHB was not reported even after daily use of GHB for as long as 9 years. The major clinical disadvantage of GHB therapy appears to be its short duration of action.

GHB as an adjuvant in anaesthesia

The ability of GHB to induce hypnosis instigated tests as to whether GHB could be used in anaesthesia. In 1962, twenty-six patients (sixteen male and ten female, aged from 7 months to 83 years) were admitted to a first study on GHB use during surgical procedures (Blumenfeld *et al.* 1962). Both relatively low-risk and long surgical operations (e.g. gastrectomy or pneumonectomy) were selected. Thirty minutes before GHB administration, patients were intravenously injected with preanaesthetic medications; these included analgesic agents such as meperidine, and atropine or scopolamine to facilitate endotracheal intubation. Subsequently, an initial dose of 4–6 g GHB (up to 80 mg/kg) was intravenously administered over a 30-min period (it was observed that an excessively rapid administration induced the occurrence of gross muscular movements). An additional dose of up to 2 g GHB was administered depending upon the length of the intervention. Nitrous oxide–oxygen was used as anaesthetic agent. GHB treatment resulted in (a) profound relaxation of the jaw muscles, permitting extensive manipulations of the upper airways without depression of spontaneous ventilation and of circulatory system, and (b) decreased need for other analgesic agents. No side effects were observed.

In a larger clinical study (Solway and Sadove 1965), GHB was administered to 348 patients, 306 male and 42 female, whose ages ranged from 5 to 87 years. Several types of major surgery were performed, with the exception of those involving extracorporeal circulation, lasting from 30 min to 7 h. Subjects were given the common premedications (meperidine and atropine) and anaesthestics (halotane, diethyl ether and methoxyflurane). GHB dose varied from 1 to 8 g (up to approximately 100 mg/kg). In this study, GHB injection time was shortened to 5 min. Induction and maintenance of anaesthetic state after GHB administration were evaluated. Criteria for satisfactory induction were: (1) smoothness of induction with minimal or no extrapyramidal movements (primarily of the shoulders, arms, and facial muscles, as manifested by jerks, twitching movements and facial grimaces); (2) adequacy of rate and depth of respiration; (3) stability of blood pressure and pulse; and (4) rapid onset of a non-arousable state. Maintenance of anaesthesia was evaluated according to the amount of additional anaesthetic agents required by patients during the intervention. In this study, 88% and 94% of patients were found to have a satisfactory induction and maintenance of anaesthesia, respectively. An abrupt rise in systolic and diastolic pressure at the time of surgical stimulation or incision was observed in a number of patients. Only few patients showed an unsatisfactory maintenance of anaesthesia; these patients required relatively large concentrations of supplementary anaesthetic agents or exhibited temporary instability of blood pressure. GHB administration lowered the amount of anaesthetic agents needed to maintain the desired level of anaesthesia. At this lower dose range, no cardiovascular or respiratory disadvantages, intrinsic in some anaesthetics, were observed. Emergence of delirium and bradycardia were the only side effects reported, occurring in no more than 2% of patients.

More recently, a clinical study (Kleinschmidt *et al.* 1997) compared the effect of GHB on maintenance of anaesthesia with those of a widely used anaesthetic drug, propofol and a benzodiazepine, midazolam. Forty-five patients with coronary disease and maintained left ventricular function, aged 45–75 years, were scheduled for an elective coronary artery bypass grafting. Anaesthesia for this operation requires minimal haemodynamic changes, maintenance of myocardial oxygen balance and adequate suppression of sympathetic reflex responses to surgical stimuli. Usually, a total intravenous anaesthesia is obtained using high-dose opioids (morphine or sufentanil), with good analgesia and haemodynamic stability.

However, hypertension, tachycardia and myocardial ischaemia may occur during sternotomy unless supplementary administration of opioids is provided. Moreover, even high doses of opioids did not prevent intra-operative awareness.

In the study by Kleinschmidt *et al.* (1997), the effect of propofol, midazolam and GHB in reducing the required doses of opioids were compared. The premedication treatment (flunitrazepam) was administered orally; anaesthesia was induced with sufentanil. Patients were randomized to receive either propofol (2 mg/kg/h, $n = 15$), midazolam (0.06 mg/kg/h, $n = 15$) or GHB (20 mg/kg/h, $n = 15$) in addition to sufentanil. Haemodynamic measurements were performed after induction and at various times during the intervention. In patients with coronary disease, treatment with propofol, midazolam and GHB resulted in comparable haemodynamic stability without periods of myocardial ischaemia. When compared to propofol, GHB and midazolam produced a significant lower prevention of hypertension after sternotomy. Hypertension (defined as mean arterial pressure greater than 30% of the baseline value for more than 1 min) was treated with a supplemental bolus of sufentanil. In GHB- and midazolam-dosed groups, the percentage of patients who required supplemental bolus of sufentanil after sternotomy was significantly higher than in propofol-treated group (namely, 86%, 86% and 30%, respectively). These similar haemodynamic responses to sternotomy monitored in GHB- and midazolam-treated groups was likely due to a transiently inadequate anaesthetic depth, associated with the activation of an endocrine response to surgery (e.g. stimulation of the renin-angiotensin-aldosterone system; release of adrenaline and/or noradrenaline). The greater anaesthetic properties of propofol may account for these differences.

The haemodynamic parameters during anaesthesia induced by GHB and sufentanil were investigated in twenty-one patients with unimpaired left ventricular function and seventeen patients with impaired left ventricular function scheduled for coronary artery bypass grafting (Kleinschmidt *et al.* 1998). Anaesthesia was induced by etomidate, sufentanil and pancuronium, and maintained with a priming dose of 40 mg/kg GHB followed by 20 mg/kg/h GHB. As found in the previous study (Kleinschmidt *et al.* 1997), GHB maintained adequate haemodynamic functions with no episode of myocardial ischaemia in either group. A similar percentage of patients in both groups required supplementary administrations of opioids to control temporary hypertension after sternotomy. These results are suggestive of GHB as a suitable anaesthetic adjuvant also in patients with impaired left ventricular function undergoing coronary artery bypass grafting.

Finally, GHB and propofol were found to induce similar effects on haemodynamic, respiratory and endocrinological parameters in patients undergoing spinal anaesthesia (Kleinschmidt *et al.* 1999). These results suggest that GHB might be a useful alternative to propofol in spinal anaesthesia.

However, despite these positive results, the use of GHB in anaesthesia has never gained widespread acceptance, probably because of reports of its lacking analgesic properties and producing seizure-type EEG activity in animals that resembles petit mal epilepsy (see Tunnicliff 1992).

GHB efficacy in schizophrenia

According to the dopaminergic hypothesis of schizophrenia, drugs that decrease dopamine activity, particularly in the mesolimbic dopamine system, could be useful in the treatment of this psychotic disease. Low and high doses of GHB were found to stimulate and inhibit, respectively, the firing rate of dopamine neurons originating in the ventral tegmental area of rats (Diana *et al.* 1991) and to increase intracellular concentrations of dopamine subsequent to stimulation of its synthesis and blockade of release (Gessa *et al.* 1966, 1968). These observations prompted an investigation to test GHB efficacy in schizophrenic patients.

In the first investigation addressing this issue (Tanaka *et al.* 1966), an oral dose of GHB ranging between 10 and 100 mg/kg, divided in three daily administrations, was given to forty-eight schizophrenic patients for approximately 6 months. Subjects were allocated into four groups based on different schizophrenia subtypes. The results of the study showed that GHB treatment reduced insomnia, anxiety and agitation, by 60–70%, in the catatonic and hebephrenic groups (the latter characterized by autism, inactivity and apathy). On the contrary, GHB was ineffective in patients affected by schizophrenia with hallucination and atypical psychosis. GHB-induced comatose and anaesthetic state was observed in three patients. The authors concluded that GHB may have a 'regulating effect on psychosis in neuropsychiatric patients'.

A subsequent small double-blind, placebo-replacement trial (Schulz *et al.* 1981) evaluated the efficacy of GHB in the treatment of schizophrenia. Seven schizophrenic subjects were given an initial oral dose of 2 g/day GHB, divided into five daily administrations (corresponding to approximately 25 mg/kg/day). Every 4 days, the daily dose of GHB was increased until a maximum of 16 g was reached. Drug treatment was repeated for approximately 3 weeks. GHB efficacy was evaluated by means of global psychosis ratings recorded daily by the nursing staff and weekly by physicians. Treatment with GHB did not significantly modify severity of schizophrenia; briefly, two patients became non-psychotic, in three symptoms worsened, while two patients did not respond to GHB therapy. Interestingly enough, the two patients who benefited from GHB treatment did not show a satisfactory response to other antipsychotic drugs. It was found that, before GHB administration, these two patients had levels of homovanillic acid in the cerebrospinal fluid (a reliable indicator of dopamine turnover) higher than in the other five patients. These results led to the hypothesis that GHB may possess some efficacy in schizophrenic patients with a high dopaminergic tone.

This hypothesis was not confirmed by the results of a subsequent double-blind cross-over study (Levy *et al.* 1983), where GHB was administered orally for three consecutive weeks at doses increasing from 6 to 12 g/day, in combination with a low dose (5 mg) of the neuroleptic, fluphenazine, to ten schizophrenic patients. Since GHB and fluphenazine are considered to affect dopamine neurotransmission by means of two different mechanisms of action (inhibition of dopamine release and blockade of dopamine receptors, respectively), this study was aimed at determining whether GHB and fluphenazine effects on schizophrenia were additive. The severity of schizophrenia was scored by the Brief Psychiatric Rating Scale; plasma and cerebrospinal fluid levels of prolactin and homovanillic acid, respectively, were also assessed. The overall results showed that GHB had no significant antipsychotic effect in schizophrenic patients nor did it modify plasma prolactin levels. Moreover, in the few patients responding to GHB treatment, there was no relationship between the degree of clinical improvement on the one hand and prolactin response or cerebrospinal fluid homovanillic acid levels on the other hand.

Other therapeutic uses of GHB

GHB efficacy as an antidepressant

In order to evaluate GHB efficacy in the treatment of patients affected by depressive disorders, thirty hospitalized depressed women were treated with a single daily administration of GHB

(4 g; corresponding to approximately 50 mg/kg), given by means of a slow intravenous infusion, for 3–12 consecutive days (Rinaldi *et al.* 1967). The severity of depressive symptoms was scored using a self-rating scale. Treatment with GHB resulted in a complete and immediate remission of depressive symptoms in twenty-five patients, while three subjects showed a decrease in severity score; in contrast, a worsening of severity score was monitored in two patients. Vomiting was the only side effect reported. Unfortunately, to our knowledge to date no study has attempted to confirm these promising results.

GHB efficacy as an anxiolytic

Recently, a small double-blind cross-over study (Ferrara *et al.* 1999) evaluated GHB effects on performance and subjective feelings in twelve young healthy volunteers. After a single oral administration of 12.5 or 25 mg/kg GHB or 0.03 mg/kg of the benzodiazepine, lorazepam, subjects were required to (a) self-report possible subjective feelings of calmness, alertness and contentedness, (b) perform several psychomotor tests and (c) describe the occurrence of possible side effects. The dose of 12.5 mg/kg GHB was more effective than lorazepam in producing subjective feelings of calmness and was devoid of any impairing effect on psychomotor performances. Transitory dizziness was reported in 50–60% of patients receiving both doses of GHB.

Finally, GHB (50 mg/kg given orally in three daily administrations) was found to decrease anxiety severity score to a similar extent but with a more rapid onset than diazepam (0.75 mg/kg given orally in six daily administrations) in alcoholic patients suffering from alcohol withdrawal syndrome (Addolorato *et al.* 1999b; see above for a more detailed description of the study).

GHB efficacy in fibromyalgia

Fibromyalgia is a disease characterized by the following symptoms: widespread musculoskeletal pain, chronic fatigue, non-restorative sleep, in the absence of other demonstrable pathologies. Recent epidemiologic surveys demonstrate that an estimated 2% of Americans suffer from this disorder (Wolfe *et al.* 1995). Fibromyalgic patients show an alpha electroencephalographic NREM sleep anomaly correlated with musculoskeletal pain and altered mood (Moldofsky 1986).

When GHB was administered to patients with concurrent diagnoses of narcolepsy and fibromyalgia, it was accidentally discovered that GHB, beside exerting an antinarcoleptic effect, decreased the severity of fibromyalgia (Scharf *et al.* 1998). This finding prompted a specific investigation on this unexpected effect. To this aim, a total of 4.5 g GHB (corresponding to approximately 30 mg/kg) was given to eleven fibromyalgic subjects, firstly at bedtime and then 4 h later (Scharf *et al.* 1998). Treatment with GHB improved the subjective estimates of (a) pain levels (by approximately 80%), (b) fatigue (more than 110%) and (c) wellness (80%), compared to baseline. Moreover, electroencephalographic registrations of sleep showed that GHB induced (a) an increase in delta sleep by approximately 60% and (b) a decrease in NREM sleep periods with alpha intrusion by approximately 70%.

GHB use in obstetric procedures

Several drugs have been tested to relieve pain, anxiety and tension during delivery, without interfering with labour progress or adversely affecting mother or foetus; to date, none have been found to be ideal. In 1968, Geldenhuys *et al.* tested GHB because of its capability of inducing sedation and limited toxicity.

In this study, 4 g GHB (corresponding to approximately 50 mg/kg) was intravenously given to 100 healthy patients. GHB was administered once cervix uteri dilatation had started. Treatment with GHB induced (a) a state resembling a deep natural sleep throughout the labour, (b) an initial increase of uterine contractions, followed by a reduction which requested oxytocin administration in half of the patients, and (c) episodes of restlessness during the sleep state. The majority of multiparae patients were able to bear down involuntarily while asleep. However, 'a large number of babies … appeared to be sleepy' and the mean Baby Apgar Score was lower than seven.

GHB and intracranial pressure

In patients with severe head injury, augmentation of intracranial pressure may result in secondary brain damage. GHB (4 g) has been reported to reduce intracranial pressure in three out of four patients with traumatic intradural haematomas or cortical contusion (Strong 1984); the magnitude of the reduction was positively related to pretreatment pressure. These results require confirmation in further studies.

Conclusions

To date, the total number of clinical studies performed to test GHB effects has been relatively small. Apparently, an initial phase of enthusiasm, originating mostly from Henri Laborit's work, led to the testing of GHB in a number of pathologies, at times even with obscure rationales. Most of these studies produced negative results; this outcome, together with occurrence of cases of seizures and coma after ingestion of high doses of the drug, caused GHB to sink into oblivion. Renewed interest on the potential therapeutic use of GHB was elicited by the finding of its efficacy in the treatment of narcolepsy and alcoholism. In these two pathologies, studies conducted to date, although often testing samples of limited size, feature GHB as an effective, well-tolerated and safe medication.

Nowadays, the alleged capability of GHB to induce salient feelings of euphoria, disinhibition, anxiolysis and relaxation at low doses and sedation and hypnosis at higher doses has provoked an increasing number of illicit and dangerous episodes of self-administration, mostly outside therapeutic protocols (see Tunnicliff 1997; Galloway *et al.* 2000). The majority of cases of illicit consumption of GHB have been reported in the US and Britain. As a consequence, in the US GHB is not approved and its use outside FDA-approved physician-supervised protocols is illegal.

However, analysis of data on GHB abuse should distinguish between the illicit use of GHB for recreational purposes and possible episodes occurring in narcoleptic and alcoholic patients under medical control. While the recreational abuse of GHB clearly constitutes a medical and social hazard, data available at present indicate that no case of GHB abuse has been observed in narcoleptic patients (even in a 9-year study (Mamelak *et al.* 1986)). Moreover, self-directed intake of GHB among alcoholics undergoing therapy with GHB appears to be a limited phenomenon (10–15% of patients (Addolorato *et al.* 1996; Gallimberti *et al.* 2000)) that, to our understanding, should not undermine its medical use.

Acknowledgement

The authors are grateful to Mrs Anne Farmer for language editing of the manuscript.

References

Addolorato G., Castelli E., Stefanini G. F., Casella G., Caputo F., Marsigli L., Bernardi M. and Gasbarrini G. (1996) An open multicentric study evaluating 4-hydroxybutyric acid sodium salt in the medium-term treatment of 179 alcohol dependent subjects. *Alcohol Alcohol.* 31, 341–45.

Addolorato G., Cibin M., Capristo E., Beghè F., Gessa G. L., Stefanini G. F. and Gasbarrini G. (1998a) Maintaining abstinence from alcohol with γ-hydroxybutyric acid. *Lancet* 351, 38.

Addolorato G., Cibin M., Caputo F., Capristo E., Gessa G. L., Stefanini G. F. and Gasbarrini G. (1998b) γ-Hydroxybutyric acid in the treatment of alcoholism: dosage fractioning utility in non-responder alcoholic patients. *Drug Alcohol Depend.* 53, 7–10.

Addolorato G., Balducci G., Capristo E., Attilia M. L., Taggi G., Gasbarrini G. and Ceccanti M. (1999a) Gamma-hydroxybutyric acid (GHB) in the treatment of alcohol withdrawal syndrome: a randomized comparative study versus benzodiazepine. *Alcohol. Clin. Exp. Res.* 23, 1596–1604.

Addolorato G., Caputo F., Capristo E., Bernardi M., Stefanini G. F. and Gasbarrini G. (1999b) A case of gamma-hydroxybutyric acid withdrawal syndrome during alcohol addiction treatment: utility of diazepam administration. *Clin. Neuropharmacol.* 22, 60–62.

Addolorato G., Caputo F., Capristo E., Colombo G., Gessa G. L. and Gasbarrini G. (2000a) Ability of baclofen in reducing alcohol craving and intake: II – preliminary clinical evidence. *Alcohol. Clin. Exp. Res.* 24, 67–71.

Addolorato G., Caputo F., Capristo E., Stefanini G. F. and Gasbarrini G. (2000b) Gamma-hydroxybutyric acid: efficacy, potential abuse and dependence in the treatment of alcohol addiction. *Alcohol* 20, 217–22.

Agabio R., Colombo G., Loche A., Lobina C., Pani M. L., Reali R. and Gessa G. L. (1998) γ-Hydroxybutyric acid (GHB) reducing effect on ethanol intake: evidence in favour of a substitution mechanism. *Alcohol Alcohol.* 33, 465–74.

Association of Sleep Disorder Centers and the Association for the Psychophysiological Study of Sleep (1979) Diagnostic classification of sleep and arousal disorders – first edition. *Sleep* 2, 1–154.

Babor T. F., Hofmann M., DelBoca F. K., Hesselbrock V., Meyer R. E., Dolinsky Z. S. and Rounsaville B. (1992) Types of alcoholics, I, evidence for an empirically derived typology based on indicators of vulnerability and severity. *Arch. Gen. Psychiat.* 49, 599–608.

Beghè F. and Carpanini M. T. (2000) Safety and tolerability of gamma-hydroxybutyric acid in the treatment of alcohol dependent patients. *Alcohol* 20, 223–25.

Blumenfeld M., Suntay R. G. and Harmel M. H. (1962) Sodium gamma-hydroxybutyric acid: a new anaesthetic adjuvant. *Anesth. Analg.* 41, 721–26.

Broughton R. (1971) Neurology and sleep research. *Can. Psychiatr. Assoc. J.* 16, 283–92.

Broughton R. and Mamelak M. (1979) The treatment of narcolepsy–cataplexy with nocturnal gamma-hydroxybutyrate. *Can. J. Neurol. Sci.* 6, 1–6.

Broughton R. and Mamelak M. (1980) Effects of gamma-hydroxybutyrate on sleep/waking patterns in narcolepsy–cataplexy. *Can. J. Neurol. Sci.* 7, 23–31.

Cloninger C. R. (1987) Neurogenetic adaptive mechanisms in alcoholism. *Science* 236, 410–16.

Diana M., Mereu G., Mura A., Fadda F., Passino N. and Gessa G. L. (1991) Low doses of γ-hydroxybutyric acid stimulate the firing of dopaminergic neurons in unanesthetized rats. *Brain Res.* 566, 208–11.

Eckardt M. J., File S. E., Gessa G. L., Hoffman P. L., Grant K. A., Guerri C., Kalant H., Koob G. F., Li T. K. and Tabakoff B. (1998) Effects of moderate alcohol consumption on the central nervous system. *Alcohol. Clin. Exp. Res.* 22, 998–1040.

Edwards G. (1990) Withdrawal symptoms and alcohol dependence: fruitful mysteries. *Br. J. Addict.* 85, 447–61.

Ferrara S. D., Zotti S., Tedeschi L., Frison G., Castagna F., Gallimberti G., Gessa G. L. and Palatini P. (1992) Pharmacokinetics of γ-hydroxybutyric acid in alcohol patients after single and repeated oral doses. *Br. J. Clin. Pharmacol.* 34, 231–35.

Ferrara S. D., Giorgetti R., Zancaner S., Orlando R., Tagliabracci A., Cavarzeran F. and Palatini P. (1999) Effects of single dose of gamma-hydroxybutyric acid and lorazepam on psychomotor performance and subjective feelings in healthy volunteers. *Eur. J. Clin. Pharmacol.* 54, 821–27.

Gallimberti L., Canton G., Gentile N., Ferri M., Cibin M., Ferrara S. D., Fadda F. and Gessa G. L. (1989) Gamma-hydroxybutyric acid for treatment of alcohol withdrawal syndrome. *Lancet* 2, 787–89.

Gallimberti L., Ferri M., Ferrara S. D., Fadda F. and Gessa G. L. (1992) Gamma-hydroxybutyric acid in the treatment of alcohol dependence: a double-blind study. *Alcohol. Clin. Exp. Res.* 16, 673–76.

Gallimberti L., Cibin M., Pagnin P., Sabbion R., Pani P. P., Pirastu R. and Ferrara S. D. (1993) Gamma-hydroxybutyric acid for treatment of opiate withdrawal syndrome. *Neuropsychopharmacology* 9, 77–81.

Gallimberti L., Schifano F., Forza G., Miconi L. and Ferrara S. D. (1994) Clinical efficacy of gamma-hydroxybutyric acid in treatment of opiate withdrawal. *Eur. Arch. Psych. Clin. Neurosci.* 244, 113–14.

Gallimberti L., Spella M. R., Soncini C. A. and Gessa G. L. (2000) Gamma-hydroxybutyric acid (GHB) in the treatment of alcohol and heroin dependence. *Alcohol* 20, 257–62.

Galloway G. P., Frederick S. L., Seymour R., Contini S. E. and Smith D. E. (2000) Abuse and therapeutic potential of gamma-hydroxybutyrate. *Alcohol* 20, 263–69.

Garbutt J. C., West S. L., Carey T. S., Lohr K. N. and Crews F. T. (1999) Pharmacological treatment of alcohol dependence: a review of the evidence. *J. Am. Med. Assoc.* 281, 1318–25.

Geldenhuys F. G., Sonnendecker E. W. W. and De Klerk M. C. C. (1968) Experience with sodium-gamma-4-hydroxybutyric acid (gamma-OH) in obstetrics. *J. Obstet. Gynaec. Cwlth.* 75, 405–13.

Gerra G., Fertonani-Affini G., Caccavari R., Zaimovic A., Tagliavini P., Riva M. and Delsignore R. (1994) Gamma-hydroxybutyric acid in the treatment of heroin addiction. *New Trends Clin. Neuropharmacol.* 8, 357–62.

Gessa G. L., Vargiu L., Crabai F., Boero G. C., Caboni F. and Camba R. (1966) Selective increase induced by gamma-hydroxybutyrate. *Life Sci.* 5, 1921–30.

Gessa G. L., Crabai F., Vargiu L. and Spano P. F. (1968) Selective increase of brain dopamine induced by γ-hydroxybutyrate: study of the mechanism of action. *J. Neurochem.* 15, 377–81.

Gessa G. L., Agabio R., Carai M. A. M., Lobina C., Pani M., Reali R. and Colombo G. (2000) Mechanism of the anti-alcohol effect of gamma-hydroxybutyric acid (GHB). *Alcohol* 20, 271–76.

Gremmelt A. and Braun U. (1995) Analgesia and sedation in patients with head–brain trauma. *Anaesthesist* 44, S559–S565.

Guilleminault C. (1986) Narcolepsy 1985. *Sleep* 9, 99–101.

June H. L., Williams J. A., Cason C. R., Devaraju S., Lin M., Murphy J. M., Lewis M. J., Lumeng L. and Li T.-K. (1995) Low doses of gamma-hydroxybutyric acid (GHB) attenuate ethanol intake in alcohol-preferring (P) rats. *Alcohol. Clin. Exp. Res.* 19, 14A.

Kleinschmidt S., Grundmann U., Janneck U., Kreienmeyer J., Kulosa R. and Larsen R. (1997) Total intravenous anaesthesia using propofol, gamma-hydroxybutyrate or midazolam in combination with sufentanil for patients undergoing coronary artery bypass surgery. *Eur. J. Anaesthesiol.* 14, 590–99.

Kleinschmidt S., Grundmann U., Knocke T., Silomon M., Bach F. and Larsen R. (1998) Total intravenous anaesthesia with gamma-hydroxybutyrate (GHB) and sufentanil in patients undergoing coronary artery bypass graft surgery: a comparison in patients with unimpaired and impaired left ventricular function. *Eur. J. Anaesthesiol.* 15, 559–64.

Kleinschmidt S., Schellhase C. and Mertzufft F. (1999) Continuous sedation during spinal anaesthesia: gamma-hydroxybutyrate vs propofol. *Eur. J. Anaesthesiol.* 16, 23–30.

Laborit H. (1964) Sodium 4-hydroxybutyrate. *Int. J. Neuropharmacol.* 3, 433–52.

Lammers G. J., Arends J., Declerck A. C., Ferrari M. D., Schouwink G. and Troost J. (1993) Gammahydroxybutyrate and narcolepsy: a double-blind placebo-controlled study. *Sleep* 16, 216–20.

Lapierre O., Montplaisair J., Lamarre M. and Bedard M. A. (1990) The effect of gamma-hydroxybutyrate on nocturnal and diurnal sleep of normal subjects: further considerations on REM sleep-triggering mechanisms. *Sleep* 13, 24–30.

Lesch O. M. and Walter H. (1996) Subtypes of alcoholism and their role in therapy. *Alcohol Alcohol.* 31, 63–67.

Levy M. I., Davis B. M., Mohs R. C., Trigos G. C., Mathé A. A. and Davis K. L. (1983) Gamma-hydroxybutyrate in the treatment of schizophrenia. *Psychiatry Res.* 9, 1–8.

Litten R. Z. and Allen J. P. (1998) Advances in development of medications for alcoholism treatment. *Psychopharmacology* 139, 20–33.

Mamelak M., Escriu J. M. and Stokan O. (1977) The effects of gamma-hydroxybutyrate on sleep. *Biol. Psychiat.* 12, 273–88.

Mamelak M., Scharf M. B. and Woods M. (1986) Treatment of narcolepsy with γ-hydroxybutyrate. A review of clinical and sleep laboratory findings. *Sleep* 9, 285–89.

Metcalf D. R., Emde R. N. and Stripe J. T. (1966) An EEG-behavioral study of sodium hydroxy-butyrate in humans. *Electroenceph. Clin. Neurophysiol.* 20, 506–12.

Mitler M. M., Shafor R., Hajdukovich R., Timms R. M. and Browman C. P. (1986) Treatment of narcolepsy: objective studies on methylphenidate, pemoline, and protryptyline. *Sleep* 9, 260–64.

Moldofsky H. (1986) Sleep and musculoskeletal pain. *Am. J. Med.* 8, 85–89.

Moncini M., Masini E., Gambassi F. and Mannaioni P. F. (2000) Gamma-hydroxybutyrate (GHB) and alcohol-related syndromes. *Alcohol* 20, 285–91.

Montplaisir J., Billiard M., Takahashi S., Bell I., Guilleminault C. and Dement W. C. (1978) 24-hours polygraphic recording in narcoleptics with special reference to nocturnal sleep disturbance. *Biol. Psychiat.* 13, 73–89.

Morse R. M. and Flavin D. K. (1992) The definition of alcoholism. *J. Am. Med. Assoc.* 268, 1012–14.

Morton A., Laird L. K., Crane D. F., Partovi N. and Frye L. H. (1994) A prediction model for identifying alcohol withdrawal seizures. *Am. J. Drug Alcohol Abuse* 20, 75–86.

Nimmerrichter A. A., Walter H., Gutierrez-Lobos K. E. and Lesch O. M. (2002) Double-blind con-trolled trial of gamma-hydroxybutyrate and clomethiazole in the treatment of alcohol withdrawal. *Alcohol Alcohol.* 37, 67–73.

O'Brien C. P., Volpicelli L. A. and Volpicelli J. R. (1996) Naltrexone in the treatment of alcoholism: a clinical review. *Alcohol* 13, 35–39.

O'Malley S. S., Jaffe A. J., Chang G., Rode S., Schottenfeld R., Meyer R. E. and Rounnsaville B. (1996) Six-month follow-up of naltrexone and psychotherapy for alcohol dependence. *Arch. Gen. Psychiat.* 53, 217–24.

Oswald I. and Priest R. G. (1965) Five weeks to escape the sleeping-pill habit. *Brit. Med. J.* 2, 1093.

Palatini P., Tedeschi L., Frison G., Padrini R., Zordan R., Orlando R., Gallimberti L., Gessa G. L. and Ferrara S. D. (1993) Dose-dependent absorption and elimination of gamma-hydroxybutyric acid in healthy volunteers. *Eur. J. Clin. Pharmacol.* 45, 353–56.

Parkers J. (1976) Amphetamines and alermess. In *Narcolepsy* (Guilleminault C., Dement W. and Passouant P., eds), pp. 643–58. New York, Spectrum.

Passouant P., Popoviciu L., Velok G. and Baldy-Moulinier M. (1968) Etude polygraphique des narcolepsies au cours du nycthemere. *Rev. Neur.* 118, 431–41.

Pelc I., Verbanck P., Le Bon O., Gavriloc M., Lion K. and Lehert P. (1997) Efficacy and safety of acamprosate in the treatment of detoxified alcohol-dependent patients. *Br. J. Psychiat.* 171, 73–77.

Pieninkeroinen I. P., Telakivi T. M. and Hillbom M. E. (1992) Outcome in subjects with alcohol-provoked seizures. *Alcohol. Clin. Exp. Res.* 16, 955–59.

Rinaldi F., Puca F. M., Mastrosimone F. and Memoli G. (1967) Sull'impiego del gamma-idrossibutirrato di sodio in terapia psichiatrica. *Acta Neurol.* 22, 21–41.

Rosen M. I., Pearsall H. R., Woods S. W. and Kosten T. R. (1996) The effect of gamma-hydroxybu-tyric acid on naloxone-precipitated opiate withdrawal. *Neuropsychopharmacology* 14, 187–93.

Sass H., Soyka M., Mann K. and Zieglgansberger W. (1996) Relapse prevention by acamprosate. *Arch. Gen. Psychiat.* 53, 673–80.

Scharf M. B. and Fletcher K. A. (1989) GHB-new hope for narcoleptics? *Biol. Psychiat.* 26, 329–30.

Scharf M. B., Brown D., Woods M., Brown L. and Hirschowitz J. (1985) The effects and effectiveness of γ-hydroxybutyrate in patients with narcolepsy. *J. Clin. Psychiat.* 46, 222–25.

Scharf M. B., Fletcher K. A. and Jennings S. W. (1988) Current pharmacologic management of narcolepsy. *Am. Fam. Physician* 38, 143–48.

Scharf M. B., Hauck M., Stover R., McDannold M. and Berkowitz D. (1998) Effect of gamma-hydroxybutyrate on pain, fatigue, and alpha sleep anomaly in patients with fibromyalgia. Preliminary report. *J. Rheumatol.* 25, 1986–90.

Schulz S. Ch., van Kammen D. P., Buchsbaum M. S., Royh R. H., Alexander P. and Bunney W. E. (1981) Gamma-hydroxybutyrate treatment of schizophrenia: a pilot study. *Pharmacopsychiatry* 14, 129–34.

Scrima L., Hartman P. G., Johnson F. H. and Hiller F. C. (1989) Efficacy of gamma-hydroxybutyrate versus placebo in treating narcolepsy–cataplexy: double-blind subjective measures. *Biol. Psychiat.* 26, 331–43.

Scrima L., Hartman P. G., Johnson F. H., Thomas E. E. and Hiller F. C. (1990) The effect of γ-hydroxybutyrate on the sleep of narcolepsy patients: a double-blind study. *Sleep* 3, 479–90.

Sinton C. M. and McCarley R. W. (2000) Neuroanatomical and neurophysiological aspects of sleep: basic science and clinical relevance. *Semin. Clin. Neuropsych.* 5, 6–19.

Solway J. and Sadove M. S. (1965) 4-hydroxybutyrate: … a clinical study. *Anesth. Analg.* 44, 532–41.

Strong A. J. (1984) γ-Hydroxybutyric acid and intracranial pressure. *Lancet* 1, 1304.

Swift R. M. (1999) Drug therapy for alcohol dependence. *New Eng. J. Med.* 340, 1482–90.

Tanaka Z., Mukai. A., Takayanagi Y., Muto A., Mikami Y., Miyakoshi T., Araya M., Ohdaira T. and Aizawa H. (1966) Clinical application of 4-hydroxybutyrate sodium and 4-butyrolactone in neuropsychiatric patients. *Folia Psychiatr. Neurol. Jpn.* 20, 9–17.

Tunnicliff G. (1992) Significance of γ-hydroxybutyric acid in the brain. *Gen. Pharmacol.* 23, 1027–34.

Tunnicliff G. (1997) Sites of action of gamma-hydroxybutyrate (GHB) – a neuroactive drug with abuse potential. *Clin. Toxicol.* 35, 581–90.

Weinrieb R. M. and O'Brien C. P. (1997) Naltrexone in the treatment of alcoholism. *Ann. Rev. Med.* 48, 477–87.

Wolfe F., Ross K., Anderson J., Russell I. J. and Hebert L. (1995) The prevalence and characteristics of fibromyalgia in the general population. *Arthritis Rheum.* 38, 19–28.

Woodward J. J. (1999) Overview of the effects of alcohol on the cerebral nervous system. *Neurochem. Int.* 35, 93–94.

Yamada Y., Yamamoto J., Fujiki A., Hishikawa Y. and Kaneko Z. (1967) Effect of butyrolactone and gamma-hydroxybutyrate on the EEG and sleep in man. *Electroenceph. Clin. Neurophysiol.* 22, 558–62.

Zarcone V. (1973) Narcolepsy. *New Engl. J. Med.* 288, 1156–66.

Zorick F., Roehrs T., Witting R., Lamphere J., Sicklesteel J. and Roth T. (1986) Sleep–wake abnormalities in narcolepsy. *Sleep* 9, 189–93.

11 Abuse potential and toxicology of γ-hydroxybutyrate

B. U. Raess and G. Tunnicliff

Introduction

Several decades ago the sodium salt of *n*-butyric acid was reported to produce a sleep-like state after intravenous administration to rabbits (White and Samson 1956). EEG activity recorded at the same time was consistent with the behavioral observations. This hypnotic action of sodium butyrate was confirmed by Jouany *et al.* (1960). On the heels of these observations was the accumulation of evidence that suggested that γ-aminobutyric acid (GABA) was an inhibitory substance in the central nervous system (CNS) (Krnjevic 1991). Owing to its hydrophilic and net charge properties, however, GABA was to all intents and purposes unable to enter the brain from the bloodstream upon oral or parenteral treatment in animals.

Since γ-hydroxybutyrate (GHB) is a structural analogue of both *n*-butyric acid and GABA, Laborit (1964) investigated its possible role as a GABA-mimetic which would have easy access to the brain. He, too, reported that his animals fell into a somnolent state soon after administration of the drug. Later this property was exploited for clinical gain when it was used to induce general anesthesia (Blumenfeld *et al.* 1962; Solway and Sadove 1965). Since GHB possessed no analgesic activity, patients were usually additionally administered nitrous oxide to control surgical pain (Vickers 1969). GHB exhibits many features of a CNS depressant, including depression of respiration. On the other hand, the drug also has been observed to induce neuronal excitation. This is manifested in the form of epileptic activity in experimental animals, especially in the form of petit mal seizures (Godschalk *et al.* 1977). In patients this excitation might take the form of random clonic movements of the face or limbs (Vickers 1969). There are also a few reports of individuals exhibiting generalized seizures after an overdose of this drug (Dyer 1991; Chin *et al.* 1992; Steele and Watson 1995).

After its overt pharmacological effects had been documented, endogenous GHB was discovered in brain, and an uneven distribution was reported (Bessman and Fishbein 1963; Vayer *et al.* 1988). Certain cortical regions had the greatest concentrations whereas the hindbrain and spinal cord possessed the lowest levels. Within the neuron, GABA is the precursor of GHB formation. In the mitochondria, succinic semialdehyde is synthesized from GABA by the action of GABA aminotransferase. Succinic semialdehyde is evidently transported into the cytosol where a reductase converts it to GHB (Kaufman and Nelson 1991; Tunnicliff 1992). GHB can be released from nerve endings upon depolarization (Maitre *et al.* 1983; Kemmel *et al.* 1998). Evidence for the existence of a plasma membrane receptor specific for GHB is quite compelling (Benavides *et al.* 1982a; Snead 2000). However, GHB is an agonist at both GHB receptors and GABA$_B$ receptors. Occupation of either receptor leads to G-protein activation (Mathivet *et al.* 1997; Snead 2000) and a resulting hyperpolarization of the nerve cell. These events are associated with facilitation of K^+ efflux in the case of $GABA_B$ receptor activation and with the modulation of Ca^{2+} conductances in the case of GHB receptor activation (Lorente *et al.* 2000; Harris *et al.* 1989; Kemmel *et al.* 1998). A sodium- and chloride-dependent transport system removes GHB from the synaptic gap (Benavides *et al.* 1982b; McCormick and Tunnicliff 1998). It is then oxidized by GHB dehydrogenase to succinic semialdehyde, and the carbon skeleton ends up in the citric acid cycle (Kaufman and Nelson 1991). These accumulated data strongly suggest that GHB can modulate neuronal activity, perhaps as an inhibitory neurotransmitter.

Besides its use in general anesthesia, other potential clinical uses of GHB include the treatment of both alcohol and opiate withdrawal (Gallimberti *et al.* 1992, 1993). Further, GHB has been investigated as a sleep-promoting drug. According to Mamelak *et al.* (1977), it seems to produce a type of sleep similar to physiological sleep. GHB also has been studied as a possible treatment for certain sleep disorders, including narcolepsy (Mamelak *et al.* 1986). In addition, it has been reported to have some potential use in fibromyalgia therapy (Scharf *et al.* 1998), as well as in the reduction of intracranial pressure and cerebral edema (Strong 1984).

There is some evidence that GHB can reduce the oxygen needs of certain tissues. In a review by Li *et al.* (1998b), work is cited claiming that GHB is beneficial in reducing damage from myocardial infarction in experimental animals. Furthermore, isoproterenol-induced myocardial damage in gerbils was markedly reduced in the presence of anesthetic doses of GHB (Kolin *et al.* 1993). Evidently GHB also can protect against stroke and several types of ischemia in animals. To date, however, it is not clear that such benefits extend to humans. Yet if it transpires that these observations have utility in patients, then the clinical uses of GHB are likely to expand.

Pharmacokinetics of GHB

If taken orally, GHB is rapidly and well absorbed and has an onset of action within 15 min (Vickers 1969). The bioavailability in rat is in the range of 52–65% (Lettieri and Fung 1976) and reaches peak plasma levels of 0.87–1.15 mmol/l 1.5–2 h following oral doses of 75–200 mg/kg in humans (Hoes *et al.* 1980). Its duration of action is approximately 3 h. Its volume of distribution (V_d) is between 0.4 and 0.58 l/kg (Helrich *et al.* 1964), suggesting total body water distribution similar to ethanol and other small water-soluble substances. The half-life of GHB varies considerably (35 min to 2.68 h, appears to be species- and dose-dependent, and non-linear in humans (Addolorato *et al.* 1998; Kohrs and Porter 1999; Couper and Logan 2000; Doherty *et al.* 1975; Lettieri and Fung 1976; Van der Pol *et al.* 1975; Snead 1977). Approximately 1% is excreted unchanged in the urine over a period of a few hours (Ferrara *et al.* 1992, 1993; Kavanagh *et al.* 2001). Reported GHB concentrations determined in human fatalities vary over a large range of postmortem blood levels (27 to 330 mg/l) (Kalasinsky *et al.* 2001). This suggests either co-administration and potentiation of other drugs or idiosynchratic responses to GHB that result in death from lower doses.

GHB as a drug with abuse potential

During the late 1980s and early 1990s, reports began to trickle in that suggested that some people were consuming GHB with the intention of experiencing a euphoric effect. At the same time this chemical was readily available in health food stores as a supposed aid to body building. By 1990 some states had banned the sale of GHB (Chin *et al.* 1992). At about the same time, a report was released by the Centers for Disease Control (CDC 1990) in which about fifty cases of acute poisonings attributed to GHB ingestion were documented. Apparently GHB is one of several drugs used by revelers at all-night dance parties. Another disturbing use of GHB is as a 'date rape' drug where it has been reportedly added to alcoholic drinks to rapidly intoxicate victims (Ropero-Miller and Goldberger 1998). In the US, severe restrictions have been imposed on the availability of this drug by the FDA which classified it as a Schedule I substance in 2000. Since 1994 the FDA has designated GHB investigational orphan drug status for the treatment of narcolepsy with marketing approval pending (FDA 2001).

While GHB was readily available as an over-the-counter drug or food supplement, the recommended dose for body building, on account of its known growth hormone releasing effect (Takahara *et al.* 1977), was around 35 mg/kg (Galloway *et al.* 2000). Apparently even a dose as low as 10 mg/kg effects muscle relaxation and has a measurable action on CNS activity (Dyer 1991). However, a dose of 40–50 mg/kg produces extreme drowsiness and sleep. Higher doses (60–70 mg/kg) can lead to a coma lasting several hours (Galloway *et al.* 1997). Lower doses of GHB will produce the same effects in the presence of alcohol (McCabe *et al.* 1971). If large enough amounts of GHB are ingested, especially accompanied by alcohol and other drugs, the CNS depressant effects can be substantial and the user probably will require emergency medical intervention. The report by Li *et al.* (1998a) illustrates a significant medical problem. Seven cases are described where patients were admitted to the emergency department of a New Orleans hospital. These individuals were treated over a period of 3 months. Each person was unconscious upon arrival and exhibited marked respiratory depression, the signs of which ranged from shallow breathing to apnea. It is often difficult to ascribe adverse symptoms to GHB intake alone. Typically, subjects co-ingest a number of drugs. For instance, Li *et al.* (1998a) report that they detected a minimum of three drugs in the urine of their seven patients and two individuals had six drugs present (Table 11.1). All patients required physical restraint for the safety of all concerned. The widespread combative behavior of the subjects made attempts at intubation extremely difficult. Of the six attempted intubations, four were eventually successful. All patients recovered without any overt or apparent lasting side effects or toxicities.

Other reports document patients being treated by hospital emergency departments. Ross (1995) reported that two patients became unresponsive and comatose after GHB intake.

Table 11.1 Clinical course of GHB intoxication

Case	Ingestion	Co-ingestants	Condition
1	Single dose	GHB, alcohol, cocaine, caffeine, cocaethylene diphenhydramine	Unconscious, apneic
2	Single dose	GHB, alcohol, cocaine, cocaethylene, nicotine	Unconscious
3	Single dose	GHB, alcohol, caffeine, nicotine	Unresponsive
4	Two doses	GHB, alcohol, cocaine, cocaethylene, nicotine ibuprophen	Euphoria then unconscious
5	Single dose	GHB, ephedrine, marijuana	Diaphoretic, unresponsive
6	Single dose	GHB, alcohol, cocaine	Unconscious, apneic
7	Unknown amount	GHB, alcohol, cocaine, cocaethylene, nicotine fluconazole	Unconscious, apneic

Source: Data taken from Li *et al.* (1998a).

Summary of seven case reports presented to the emergency department of Charity Hospital, New Orleans between September and November 1995.

Chin *et al.* (1992) described a further five subjects who required treatment. Typically these patients volunteered that they had ingested GHB, but no urinalysis was used to confirm this. At various stages these individuals tended to have respiratory problems, mental confusion, euphoria, coma and perhaps seizures. All recovered. The existence of many other reports suggests that the consumption of GHB in doses that lead to emergency department admissions is widespread (Dyer 1991; Steele and Watson 1995; Thomas *et al.* 1997; Boyce *et al.* 2000). To date reliable reports have documented at least seven deaths that might be linked to GHB toxicity. The first death was reported by Ferrara *et al.* (1995) although the victim reportedly had concomitantly self-administered heroin. In 1996, a 17-year-old female with no history of drug and alcohol use attended a local dance club and died the following day of cardiac arrest and a serum level of 27 mg/l GHB (CDC 1997). Two other fatalities occurred in Sweden and were reported by Timby *et al.* (2000). These victims had consumed other drugs (including alcohol) in addition to the GHB. Other deaths have been reported by Dyer *et al.* (2001) and by Karch *et al.* (2001), although the latter authors cautioned that in one of the victims they examined, myocardial fibrosis might have been the primary cause. There are two major concerns regarding the availability and accuracy of data on deaths caused by GHB toxicity. First, the vast majority of individuals consume other CNS depressants and/or stimulants along with GHB (e.g. Table 11.1). Second, fatalities due to GHB overdose might not appear in the scientific or medical literature for years after the event, if at all. Thus, although at least five reports on GHB-related deaths are readily accessible, this might be far from the actual situation. A similar problem exists in attempting to document the epidemiology of GHB abuse in the US. According to Nicholson and Balster (2001), there was a steady increase in the number of emergency department cases involving GHB over the period 1992–8. These numbers are tabulated by the Drug Abuse Warning Network (DAWN) system and show that the frequency of cases rose from 20 in 1992 to 1,282 in 1998. As pointed out by Nicholson and Balster (2001), the fact that GHB usage is linked to a patient treated in an emergency setting does not necessarily mean that GHB was the sole intoxicant or, indeed, that GHB even was involved. Many times a urinalysis might not screen for GHB, or might not be performed at all. Undoubtedly information from patients, or from accompanying persons, can often be the sole source of knowledge in the absence of a drug screen. It is axiomatic that such information is unreliable. Interestingly, in 1996 the majority of cases seen by emergency department personnel occurred in the south, midwest or west, with eastern cities reporting relatively few cases.

It would be useful for hospital emergency department personnel to have an agent to reverse the GHB-induced coma. Naloxone, which is widely used to reverse opiate toxicity, has not been shown to possess an antidotal effect in patients intoxicated with GHB. A drug that might offer some benefit in treating GHB-related coma is physostigmine. This drug acts as a competitive inhibitor of acetylcholinesterase and thus potentiates cholinergic neurotransmission. Henderson and Holmes (1976) have reported on the use of physostigmine to accelerate recovery from GHB anesthesia. Twenty-five patients were given 2 mg of the drug intravenously. The subjects regained consciousness within 2–10 min, the mean recovery time being 6 min. These results might have importance in reviving individuals who are in a coma as a result of GHB intoxication. Indeed, two cases of GHB overdose have been reported (Yates and Viera 2000) in which both subjects were brought into the emergency department in a state of unconsciousness after what witnesses reported was GHB ingestion. One patient was administered 2 mg physostigmine intravenously and awoke 5 min later. A second comatose patient who was given the same dose of physostigmine had a similar outcome. In neither case were adverse effects apparent. Since the presence of GHB in the urine was not measured in either patient, some doubt must remain as to whether GHB was involved. It is obviously important to see if physostigmine can successfully reverse the effects of GHB in other subjects before we can be confident that this is a useful treatment option for GHB poisoning. In the above cases the mechanism of the physostigmine effect is not clear except to note that if GHB is a CNS depressant, activating central cholinergic system should increase excitatory pathway activity and presumably compensate for the excessive inhibition brought about by GHB.

From animal studies there is little evidence pointing to GHB being addictive or inducing tolerance or physical dependence. Self-administration experiments concluded that GHB has no, or only weak, reinforcing effects (Beardsley *et al.* 1996; Woolverton *et al.* 1999). Usually, such results indicate a drug has limited potential to be abused. Some tolerance was noted after chronic treatment of rats with GHB but this was much weaker than that produced by alcohol (Colombo *et al.* 1995). Cases of physical dependence in animals have not been documented. In humans, though, incidences of physical dependence are beginning to be reported (Addolorato *et al.* 1999; Friedman *et al.* 1996; Galloway *et al.* 1994; Price 2000). In a communication by Galloway *et al.* (1997), a case involving the development of tolerance was reported. An additional case was described by Craig *et al.* (2000) where a patient had apparently abused GHB daily for over two years and an abrupt ending of the drug intake resulted in severe agitation, increased arterial pressure and tachycardia. More recently, Dyer *et al.* (2001) have documented a GHB withdrawal syndrome. Eight patients were studied after being treated in hospital emergency departments for the effects of discontinuing GHB use. Each subject experienced psychosis and agitation and had to be sedated. The withdrawal syndrome described included anxiety, insomnia and tremors, which started from 1 to 6 h after the last dose and continued for 5–15 days. One patient died on day 13 of admission. The other subjects eventually recovered. Another consequence of chronic GHB intake is Wernicke–Korsakoff syndrome (Friedman *et al.* 1996). The case described involved a patient who evidently consumed GHB on a daily basis for at least 18 months, without concomitant intake of alcohol. Several times she tried to detoxify herself but invariably she started to develop tremor and marked insomnia and paranoia. The patient was admitted to hospital and was observed to exhibit signs of Wernicke–Korsakoff syndrome. The symptoms were eventually resolved after treatment that included thiamine supplementation.

Management of the overdosed patient

The primary approach to the management of GHB overdose is to supply supportive measures. Both Li *et al.* (1998b) and O'Connell *et al.* (2000) have made certain recommendations concerning the incapacitated patient brought to the hospital emergency department. They include the advice that if the subject exhibits spontaneous breathing, ensurance of clear airways is important. This could involve the use of aspiration. Pulse oximetry should be employed, and oxygen supplementation might be necessary.

Sometimes atropine administration is useful in patients with persistent bradycardia (Virtue *et al.* 1966). As discussed above, in a case-controlled study and in emergency department situations, intravenous physostigmine administration was used successfully to induce a relatively rapid (2–10 min) recovery from GHB-induced unconsciousness when no other CNS depressants were present (Henderson and Holmes 1976; Yates and Viera 2000). In a more complicated situation after GHB ingestion, rapid-sequence intubation together with succinylcholine administration might be appropriate. Li *et al.* (1998b) emphasized that at this stage sedation is already provided by the GHB. If it is possible that other psychotropic drugs are present, gastric lavage and treatment with activated charcoal and sorbitol might be in order. Li *et al.* (1998b) suggest that in straightforward cases, patients who recover and are in a stable condition can be released after 6 h. Typically, complicated cases are admitted for observation.

Despite the above reports of untoward effects of GHB ingestion, this drug is remarkably inert by many criteria. In rats it has an LD_{50} of 1.7 g/kg although all the animals died at the slightly higher dose of 2 g/kg (Jouany *et al.* 1960). The animals succumbed to respiratory depression. This is an exceedingly high dose and by most measures GHB cannot be described as a potent substance when administered on its own. When patients are poisoned by GHB, invariably they have consumed great quantities in the presence of other psychotropic drugs.

Conclusions

GHB is gradually gaining widespread use as one of a number of recreational club scene drugs among the 18–25-year-old age bracket. This has led to declaring the formerly freely available drug a DEA schedule I substance. The CNS depressant effects of GHB have potential licit medical uses in narcolepsy, anesthesia, sleep disorders, ischemic conditions, and alcohol and opiate withdrawal. However, because of its other apparent desirable effects which include euphoria and heightened sexual desire, by itself or in combination with other recreational drugs, its misuse associated with social predatory practices make it a devastatingly dangerous and potentially lethal drug. Although this naturally occurring metabolite of GABA was synthesized over forty years ago, much still remains in understanding its pharmacological mechanism of action. Based on its pharmacological and toxicological profile, dopaminergic, serotonergic, cholinergic and GABAergic mechanisms have been proposed to mediate a myriad of desirable and undesirable effects. Acute CNS depressant symptoms of GHB, which may persist for several hours, can be readily reversed by intravenous administration of physostigmine. This again suggests a cholinergic involvement in the production of the hypnotic, amnesic, anesthetic and respiratory depressant effects. Unfortunately, it seems to be the case that when misused, other mood altering drugs, both CNS excitatory and depressant, often are ingested concomitantly. This makes treatment of the intoxicated patient particularly challenging. Clearly, basic research leading to a well-defined mechanism of GHB action is long overdue. Moreover, a greater emphasis should be placed on educating potential users of the inherent short- and long-term consequences associated with situations and environments conducive to intentional and unintentional ingestion of GHB.

References

Addolorato G., Cibin M., Caprista E., Beghe F., Gessa G., Stefanini G. F., Gasbarrini G. and Capristo E. (1998) Maintaining abstinence from alcohol with gamma-hydroxybutyric acid. *Lancet* 351 (9095), 38.

Addolorato G., Caputo F., Capristo E., Bernardi M., Stefanini G. F. and Gasb G. (1999) A case of gamma-hydroxybutyric acid withdrawal syndrome during alcohol addiction treatment: utility of diazepam administration. *Clin. Neuropharmacol.* 22, 60–62.

Beardsley P. M., Balster R. L. and Harris L. S. (1996) Evaluation of the discriminative stimulus and reinforcing effects of gammahydroxybutyrate (GHB). *Psychopharmacology* 127, 315–22.

Benavides J., Rumigny J. F., Bourguignon J. J., Cash C., Wermuth C. G., Mandel P., Vincendon G. and Maitre M. (1982a) High affinity binding site for γ-hydroxybutyric acid in rat brain. *Life Sci.* 30, 953–61.

Benavides J., Rumigny J. F., Bourguignon J. J., Wermuth C. G., Mandel P. and Maitre M. (1982b) A high-affinity, Na⁺-dependent uptake system for γ-hydroxybutyrate in membrane vesicles prepared from rat brain. *J. Neurochem.* 38, 1570–75.

Bessman S. P. and Fishbein W. M. (1963) Gamma-hydroxybutyrate, a normal brain metabolite. *Nature* 200, 1207–08.

Blumenfeld M., Suntay R. G. and Harmel W. H. (1962) Sodium gamma-hydroxybutyric acid: a new anaesthetic adjuvant. *Anesth. Analg. Curr. Res.* 41, 721–26.

Boyce S. H., Padgham K., Miller L. D. and Stevenson J. (2000) Gamma hydroxybutyric acid (GHB): an increasing trend in drug abuse. *Eur. J. Emerg. Med.* 7, 177–81.

CDC (1990) Multistate outbreak of poisonings with illicit use of gamma hydroxybutyrate. *Morb. Mortal. Wkly. Rep.* 39, 861–63.

CDC (1997) Gamma hydroxybutyrate use – New York and Texas, 1995–1996. *Morb. Mortal. Wkly. Rep.* 46, 281–83.

Chin M.-Y., Kreutzer R. A. and Dyer J. E. (1992) Acute poisoning from γ-hydroxybutyrate in California. *West. J. Med.* 156, 380–84.

Colombo G., Agabio R., Lobina C., Reali R., Fadda F. and Gessa G. L. (1995) Cross-tolerance to ethanol and gamma-hydroxybutyric acid. *Eur. J. Pharmacol.* 273, 235–38.

Couper F. J. and Logan B. K. (2000) Determination of gamma-hydroxybutyrate (GHB) in biological specimens by gas chromatography-mass spectrophotometry. *J. Analyt. Toxicol.* 24, 1–7.

Craig K., Gomez H. F., McManus J. L. and Bania T. C. (2000) Severe gamma-hydroxybutyrate withdrawal: a case report and literature review. *J. Emerg. Med.* 18, 65–70.

Doherty J. D., Stout R. W. and Roth R. H. (1975) Metabolism of (1-¹⁴C)gamma hydroxybutyric acid by rat brain after intraventricular injection. *Biochem. Pharmacol.* 24, 469–74.

Dyer J. E. (1991) γ-Hydroxybutyrate: a health-food product producing coma and seizurelike activity. *Am. J. Emerg. Med.* 9, 321–24.

Dyer J. E., Roth B. and Hyma B. A. (2001) Gamma-hydroxybutyrate withdrawal syndrome. *Ann. Emerg. Med.* 37, 147–53.

Food and Drug Administration (FDA) (2001) http://www.fda.gov/orphan/designat/list.htm (accessed 26 June, 2001).

Ferrara S. D., Zotti S., Tedeschi L., Frison G., Castagna F., Gallimberti L., Gessa G. L. and Palatini P. (1992) Pharmacokinetics of gamma-hydroxybutyric acid in alcohol dependent patients after single and repeated oral doses. *Br. J. Clin. Pharmacol.* 34, 231–35.

Ferrara S. D., Tedeschi L., Frison G., Castagna F., Gallimberti L., Giorgetti R., Gessa G. L. and Palatini P. (1993) Therapeutic gamma-hydroxybutyric acid monitoring in plasma and urine by gas chromatography–mass spectrometry. *J. Pharm. Biomed. Anal.* 11, 483–87.

Ferrara S. D., Tedeschi L., Frison G. and Rossi A. (1995) Fatality due to gamma-hydroxybutyric acid (GHB) and heroin intoxication. *J. For. Sci.* 40, 501–04.

Friedman J., Westlake R. and Furman M. (1996) 'Grievous bodily harm:' gamma hydroxybutyrate abuse leading to a Wernicke–Korsakoff syndrome. *Neurology* 46, 469–71.

Gallimberti L., Ferri M., Ferrara S. D., Fadda F. and Gessa G. L. (1992) Gamma-hydroxybutyric acid in the treatment of alcohol dependence: a double-blind study. *Alcoholism: Clin. Exp. Res.* 16, 673–76.

Gallimberti L., Cibin M., Pagnin P., Sabbion R., Pani P. P., Pirastu R., Ferrara S. D. and Gessa G.L. (1993) Gamma-hydroxybutyric acid for the treatment of opiate withdrawal syndrome. *Neuropsychopharmacology* 16, 673–76.

Galloway G. P., Frederick S. L. and Staggers F. (1994) Physical dependence on sodium oxybate. *Lancet* 343, 57.

Galloway G. P., Frederick S. L., Staggers F. E., Gonzales M., Stalcup S. A. and Smith D. E. (1997) Gamma-hydroxybutyrate: an emerging drug of abuse that causes physical dependence. *Addiction* 92, 89–96.

Galloway G. P., Frederick-Osborne S. L., Seymour R., Contini S. E. and Smith D. (2000) Abuse and therapeutic potential of gamma-hydroxybutyric acid. *Alcohol* 20, 263–69.

Godschalk M., Dzoljic M. R. and Bonta I. L. (1977) Slow wave sleep and a state resembling absence epilepsy induced in the rat by γ-hydroxybutyrate. *Eur. J. Pharmacol.* 44, 105–11.

Harris N. C., Webb C. and Greenfield S. A. (1989) The effects of gamma-hydroxybutyrate on the membrane properties of guinea-pig pars compacts neurons in the substantia nigra *in vitro*. *Neuroscience* 31, 363–70.

Helrich M., McAslan T. C., Skolnik S. *et al.* (1964) Correlation of blood levels of 4-hydroxybutyrate with state of consciousness. *Anesthesiol.* 25, 771–75.

Henderson R. S. and Holmes C. M. (1976) Reversal of the anaesthetic action of sodium gamma-hydroxybutyrate. *Anaesth. Intensive Care* 4, 351–54.

Hoes M. J. A. J. M., Vree T. B. and Guelen P. J. M. (1980) Gamma-hydroxybutyric acid as hypnotic. Clinical and pharmacokinetic evaluation of gamma-hydroxybutyric acid as hypnotic in man. *L'Encéphale* VI, 93–99.

Jouany J. M., Gerard J., Broussolle B., Reynier M., Orsetti A., Vermuth C. and Baron C. (1960) Pharmacologie comparée des sels de l'acide butyrique et de l'acide 4-hydroxybutyrique. *Agressologie* 1, 417–27.

Kalasinsky K. S., Dixon M. M., Schmunk G. A. and Kish S. J. (2001) Blood, brain and hair GHB concentrations following fatal ingestion. *J. Forensic Sci.* 46, 728–30.

Karch S. B., Stephens B. G. and Nazareno G. V. (2001) GHB. Club drug or confusing artifact? *Am. J. Forensic Med. Pathol.* 22, 266–69.

Kaufman E. E. and Nelson T. (1991) An overview of γ-hydroxybutyrate catabolism: the role of the cytosolic $NADP^+$-dependent oxidoreductase EC 1.1.1.19 and a mitochondrial hydroxyacid-oxoacid transhydrogenase in the initial, rate-limiting step in this pathway. *Neurochem. Res.* 16, 965–74.

Kavanagh P. V., Kenny P. and Feely J. (2001) The urinary excretion of gamma-hydroxybutyric acid in man. *J. Pharm. Pharmacol.* 53, 399–402.

Kemmel V., Taleb O., Perard A., Andriamampandy C., Siffert J. C., Mark J. and Maitre M. (1998) Neurochemical and electrophysiological evidence for the existence of a functional γ-hydroxybutyrate system in NCB-20 neurons. *Neuroscience* 86, 989–1000.

Kohrs F. P. and Porter W. H. (1999) Gamma-hydroxybutyrate intoxication and overdose. *Ann. Emerg. Med.* 33, 475–76.

Kolin A., Brezina A., Mamelak M. and Pandula E. (1993) Cardioprotective action of sodium gamma-hydroxybutyrate against isoproterenol induced myocardial damage. *Int. J. Exp. Path.* 74, 275–81.

Krnjevic K. (1991) Significance of GABA in brain function. In *GABA Mechanisms in Epilepsy* (Tunnicliff G. and Raess B. U., eds), pp. 47–87. Wiley-Liss, New York.

Laborit H. (1964) Sodium 4-hydroxybutyrate. *Int. J. Neuropharmacol.* 3, 433–52.

Lettieri J. and Fung H. L. (1976) Absorption and first-pass metabolism of ^{14}C-gamma-hydroxybutyric acid. *Res. Commun. Chem. Pathol. Pharmacol.* 13, 425–37.

Li J., Stokes S. A. and Woeckener A. (1998a) A tale of novel intoxication: seven cases of γ-hydroxybutyric acid overdose. *Ann. Emerg. Med.* 31, 723–28.

Li J., Stokes S. A. and Woeckener A. (1998b) A tale of novel intoxication: a review of the effects of γ-hydroxybutyric acid with recommendations for management. *Ann. Emerg. Med.* 31, 729–36.

Lorente P., Lacampagne A., Pouzeratte Y., Richards S., Malitschek B., Kuhn R., Bettler B. and Vassort G. (2000) γ-Aminobutyric acid type B receptors are expressed and functional in mammalian cardiomyocytes. *Proc. Nat. Acad. Sci. USA* 97, 8664–69.

McCabe E., Layne E., Sayler D. and Slusher N. (1971) Synergy of ethanol and a natural soporific – gamma hydroxybutyrate. *Science* 171, 404–06.

McCormick S. J. and Tunnicliff G. (1998) Inhibitors of synaptosomal γ-hydroxybutyrate transport. *Pharmacology* 57, 124–31.

Maitre M., Cash C. D., Weissmann-Nanopoulos D. and Mandel P. (1983) Depolarization-evoked release of γ-hydroxybutyrate from rat brain slices. *J. Neurochem.* 41, 287–90.

Mamelak M., Escriu J. M. and Stokan O. (1977) Sleep-inducing effects of gammahydroxybutyrate. *Lancet* 2, 328–29.

Mamelak M., Scharf M. and Woods M. (1986) Treatment of narcolepsy with gamma-hydroxybutyrate: a review of clinical and sleep laboratory findings. *Sleep* 9, 285–89.

Mathivet P., Bernesconi R., De Barry J., Marescaux C. and Bittiger H. (1997) Binding characteristics of gamma-hydroxybutyric acid as a weak but selective GABA$_B$ receptor agonist. *Eur. J. Pharmacol.* 321, 67–75.

Nicholson K. L. and Balster R. L. (2001) GHB: a new and novel drug of abuse. *Drug Alcohol Dep.* 63, 1–22.

O'Connell T., Kaye L. and Plosay J. J. (2000) Gamma-hydroxybutyrate (GHB): a newer drug of abuse. *Am. Fam. Physician* 62, 2478–82.

Price G. (2000) In-patient detoxification after GHB dependence. *Br. J. Psychiatr.* 177, 181.

Ropero-Miller J. D. and Goldberger B. A. (1998) Recreational drugs. Current trends in the 90s. *Clin. Lab. Med.* 18, 727–46.

Ross T. M. (1995) Gamma hydroxybutyrate overdose: two cases illustrate the unique aspects of this dangerous recreational drug. *J. Emerg. Nurs.* 21, 374–76.

Scharf M. B., Hauck M., Stover R., McDannold M. and Berkowitz D. (1998) Effect of gamma-hydroxybutyrate on pain, fatigue, and the alpha sleep anomaly in patients with fibromyalgia. Preliminary report. *J. Rheumatol.* 25, 1986–90.

Snead O. C. (1977) Gamma hydroxybutyrate. *Life Sci.* 20, 1935–44.

Snead O. C. (2000) Evidence for a G-protein-coupled γ-hydroxybutyric acid receptor. *J. Neurochem.* 75, 1986–96.

Solway J. and Sadove M. S. (1965) 4-Hydroxybutyrate: a clinical study. *Anesth. Analg. Curr. Res.* 44, 532–39.

Steele M. T. and Watson W. A. (1995) Acute poisoning from gamma hydroxybutyrate (GHB). *Miss. Med.* 92, 354–57.

Strong A. J. (1984) Gamma-hydroxybutyric acid and intracranial pressure. *Lancet* 1 (8389), 1304.

Takahara J., Yunoki S., Yakushiji W., Yamauchi J., Yamane Y. and Ofuji T. (1977) Stimulatory effects of gamma-hydroxybutyric acid on growth hormone and prolactin release in humans. *J. Clin. Endocrinol. Metab.* 44, 1014–17.

Thomas G., Bonner S. and Gascoigne A. (1997) Coma induced by abuse of γ-hydroxybutyrate (GBH or liquid ecstasy): a case report. *Bri. Med. J.* 314, 35–36.

Timby N., Eriksson A. and Boström K. (2000) Gamma-hydroxybutyrate-associated deaths. *Am. J. Med.* 108, 518–19.

Tunnicliff G. (1992) Significance of γ-hydroxybutyric acid in the brain. *Gen. Pharmacol.* 23, 1027–34.

Van der Pol W., Van der Kleijn E. and Lauw M. (1975) Gas chromatic determination and pharmacokinetics of 4-hydroxybutyrate in dog and mouse. *J. Pharmacokinet. Biopharm.* 3, 99–113.

Vayer P., Ehrhardt J.-D., Gobaille S., Mandel P. and Maitre M. (1988) Gamma hydroxybutyrate distribution and turnover rates in discrete brain regions of the rat. *Neurochem. Int.* 12, 53–59.

Vickers M. D. (1969) Gammahydroxybutyric acid. *Int. Anaesth. Clin.* 7, 75–89.

Virtue R. W., Lund L. O., Beckwitt H. J. and Vogel J. H. (1966) Cardiovascular reactions to gamma hydroxybutyrate in man. *Can. Anaesth. Soc. J.* 13, 119–23.

White R. P. and Samson F. E. (1956) Effects of fatty acid anions on electroencephalogram of unanesthetized rabbits. *Am. J. Physiol.* 186, 271–78.

Woolverton W. L., Rolett J. K., Winger G., Woods J. H., Gerak L. R. and France C. P. (1999) Evaluation of the reinforcing and discriminative stimulus effects of gamma-hydroxybutyrate in rhesus monkeys. *Drug Alcohol Depend.* 54, 137–43.

Yates S. W. and Viera A. J. (2000) Physostigmine in the treatment of γ-hydroxybutyric acid overdose. *Mayo Clin. Proc.* 75, 401–02.

12 4-Hydroxybutyric aciduria

K. M. Gibson

Introduction and historical perspective

In 1981, Dr Cornelis Jakobs (then located at the Free University of Berlin) identified a mentally retarded boy who had very little language development (Rating *et al.* 1982). When Dr Jakobs analyzed organic acids in the urine of this boy, he found massively increased quantities of 4-hydroxybutyric acid (or γ-hydroxybutyric acid (GHB)), a compound which had not been identified in human urine (Fig. 12.1). Dr Jakobs thought that this child, who was born to related Turkish parents, had an inherited defect in the metabolism of GABA. Normally, the enzyme succinic semialdehyde dehydrogenase (SSADH) (E.C.1.2.1.24; McKusick 271980) works in conjunction with GABA transaminase to convert GABA to succinic acid, which can ultimately be used for energy production within the tricarboxylic acid cycle. Jakobs hypothesized that, as the result of the enzyme defect, succinic semialdehyde would accumulate and be converted to 4-hydroxybutyric acid, which is a compound with unusual neuropharmacologic and neurophysiologic properties. As it turned out, his hypothesis was correct.

Next, Dr Jakobs needed to confirm that this enzyme was indeed defective. He attempted to interest several colleagues in the disorder, but without success. Dr Jakobs then arrived at the University of California, San Diego, where he was going to complete his Doctoral research in the laboratories of Dr Lawrence Sweetman and Dr William L. Nyhan. At that time I was a graduate student in the same laboratory, working on a different project. Dr Jakobs was a persuasive individual, and managed to interest me in the patients. I decided to try and identify the enzyme defect, but the concern was that SSADH was an enzyme occurring only in central nervous system. We might be unable to identify the enzyme abnormality without a brain biopsy from our first patients, which we knew would be difficult to do.

In Dr Sweetman and Nyhan's laboratories, we developed preliminary enzyme assays to estimate SSADH activity in cells, and used these methods in control skin fibroblasts derived from healthy donors (Gibson *et al.* 1983, 1985a; Gibson and Sweetman, 1983). The results were disappointing, and we found no enzyme activity. I, of course, was disappointed but Dr Jakobs urged me not to abandon the project. As a last resort, I had one of the technicians in the laboratory draw my own blood, in order to obtain white blood cells to try our assay in a different cell type. Luck was on our side, and the assay worked in white blood cells. This allowed us a method and a source of cells in which to show the enzyme defect in patients with 4-hydroxybutyric aciduria. This historical perspective serves to remind us that in science, there is nothing more important than persistence.

It is widely accepted that GABA is a major central nervous system inhibitory neurotransmitter (Tillakaratne *et al.* 1995). The role of the GABA metabolite, GHB (Fig. 12.1), in

198   *K. M. Gibson*

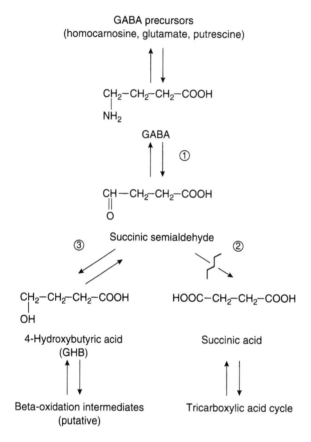

Figure 12.1 Metabolic interconversions of GABA. Numbered enzymes include: (1) GABA-transaminase; (2) SSADH; and (3) 4-hydroxybutyrate dehydrogenase (potentially carried out by one or more enzymes). The site of the block in patients with SSADH deficiency is indicated by a zig-zag pattern. Abbreviations: GABA – 4-aminobutyric acid.

central nervous system function, however, is not completely clear. Although it has been established that GHB possesses a variety of neuromodulatory features, additional experimental work is needed to establish if GHB is a neurotransmitter (Gibson *et al.* 1998a). GHB, developed in the early 1960s as an analogue of GABA which is capable of traversing the blood–brain barrier, is well known for its sedative and anesthetic properties. GHB was initially employed as an agent for induction of anesthesia in children, but interest in its use waned due to a lack of analgesic effects and the presence of epileptogenic side effects (Snead 1977).

Recently, the clinical application of GHB has undergone a major renaissance. GHB is believed to be an important component of energy regulation and a tissue protectant during hypoxia and/or excessive metabolic demand. GHB has been used in the treatment of narcolepsy, alcohol and opiate withdrawal, fatal familial insomnia, difficult cases of labor and delivery of infants, and in the preservation of tissues for transplantation (Gallimberti *et al.* 2000). More recently, GHB has been used to provide adequate sedation for patients under controlled ventilation or for surgical patients experiencing difficulties with mechanical respiration (Gibson *et al.* 1998a). GHB has even evolved into a popular recreational drug in the US, with reports of toxicity. Because of its putative number of multiorgan effects, model systems in which GHB either naturally accumulates, or is artificially depressed, would provide important insight into the mode of action of this molecule.

Our interest in GHB has evolved from the study of patients with inherited deficiency of SSADH, the final enzyme of the GABA degradative pathway (Fig. 12.1) (Gibson *et al.* 1984a,b,c; Divry *et al.* 1983; Jakobs *et al.* 1984; Rating *et al.* 1984). This rare and unusual disorder is of considerable interest to neuroscientists and geneticists in representing an inherited human disease in which two neuroactive compounds, GABA and GHB, accumulate (Fig. 12.1). In response to the inherited defect, succinic semialdehyde accumulates and is reduced to GHB. Elevation of GHB is the biochemical hallmark of the disease, detected using routine urine organic acid analysis.

One of the goals of our research is to impart to practicing pediatricians the need for accurate organic acid analysis in their patients with idiopathic mental retardation. There is little information concerning the pathogenesis of SSADH deficiency. In particular, we do not know whether neurologic findings are a result of 'toxicity' of accumulated GHB, perturbations in GABA pools, or if these phenomena are acting synergistically. A clear picture of pathomechanisms in SSADH deficiency would provide the important groundwork for developing better therapeutic approaches in patients. The objective of this chapter is to provide the reader an overview of 4-hydroxybutyric aciduria, and a summary of future research objectives and directions.

4-Hydroxybutyric aciduria

Clinical phenotype

The author is aware of >300 patients with 4-hydroxybutyric aciduria. While several case reports have been presented, the disease has been reviewed substantially only infrequently (Gibson 1984, 1996; Gibson *et al.* 1986, 1990b, 1997a, 2002; Jaeken *et al.* 1990; Jakobs *et al.* 1993a,b; Hoffmann *et al.* 1994; Hoffmann and Gibson 1996; Gibson and Jakobs, 2001). The clinical features of twenty-three patients (twenty families) with SSADH deficiency (4-hydroxybutyric aciduria) were presented in 1997. The age at diagnosis ranged from 3 months to 25 years in the eleven male and twelve female patients; consanguinity was noted in 39% of families (Gibson *et al.* 1997a).

The following abnormalities were observed (frequency in twenty-three patients): motor delay, including fine motor skills, 78%; language delay, 78%; hypotonia, 74%; mental delay, 74%; seizures, 48%; decreased or absent reflexes, 39%; ataxia, 30%; behavioral problems, 30%; hyperkinesis, 30%; neonatal problems, 26%; and electroencephalographic abnormalities, 26%. Associated findings included psychoses, cranial magnetic resonance or computed tomographic abnormalities, and ocular problems in 22% or less of patients. Therapy with vigabatrin proved beneficial to varying degrees in 35% of the patients. Normal early development was noted in 30% of patients.

The presenting problem varied among the patients studied; however, in general, the presenting symptoms were neurologic. The following presenting signs were observed: motor delay, fourteen patients; hypotonia, ten patients; mental delay, nine patients; language delay, six patients; ataxia, three patients; seizures, two patients; and neonatal problems, one patient. All of the patients presented with one (or more) of these symptoms for which the parents sought help. There was a wide range in phenotypic presentation. Even in sibships, the disease presentation was not completely consistent. For example, patient JIY had hyperkinesis, which was not observed in the younger sibling, whereas the younger sibling, JEY, had evidence of hypotonia not seen in her older sibling. In a sibship of three patients, however, the phenotype was very consistent. The lack of a clear-cut presenting phenotype and the significant variability in overall presentation further underscores the variable nature of SSADH deficiency and the relative likelihood that the disorder is frequently undiagnosed or even misdiagnosed.

In the twenty-three patients, the following nationalities were noted: Turkish, four; American Caucasian, four; Northern European (German, Dutch, Bulgarian), three; Indian, three; Palestinian (Lebanese), two; Korean, two; Syrian, one; Saudi, one; Inuit (Greenland), one; Pakistani, one; and Chinese, one. To date, we know of no patients detected with 4-hydroxybutyric aciduria who are in South America or Africa. One of the first patients in whom SSADH deficiency was diagnosed was thought to have originated from Algeria (Divry *et al.* 1983), but it is unclear whether he was French or Algerian by birth. Thus, SSADH deficiency appears to be primarily a disorder of the Eurasian region. Whether or not a founder effect can be linked to the disease remains to be determined. On the other hand, Eurasian prevalence may simply reflect more efficient metabolic screening of the pediatric population for inherited disease in the Northern European region. Approximately 40% of probands were offspring of consanguineous parents, a very high incidence of consanguinity. This may account for a Middle Eastern predominance in geographic distribution of patients because of cultural habits. On the other hand, it may imply that disease-related alleles are rare in the human population. A high incidence of consanguinity is also consistent with multiple loci, but this will be addressed through molecular genetic studies.

One intriguing finding in the tabulation of our twenty-three patients was the distinction of early- and late-onset forms of SSADH deficiency, or at least forms of the disease differentiated by development during infancy. Remarkably, seven of twenty-three patients had normal early development. This is not to say that the ensuing clinical course in these putative 'late-onset' patients was mild. Moreover, it is difficult to identify any one factor as the cause of the late-onset disease form. Molecular genetic analyses eventually may provide answers to these questions. Another interesting observation was the fact that therapeutic intervention with the GABA transaminase inhibitor, vigabatrin, was only clinically beneficial in about one-third of patients. The antiepileptic vigabatrin (γ-vinyl GABA; 4-amino-5-hexenoic acid; Sabril®) is an irreversible inhibitor of GABA-transaminase, the enzyme responsible for the first step in the degradation of GABA.

The review of twenty-three patients demonstrated that the clinical phenotype is extremely nonspecific, which suggests that this disease may be significantly underdiagnosed. It is likely that all patients with 4-hydroxybutyric aciduria will manifest some degree of psychomotor deficit within their childhood years; however, speech and language delay and hypotonia are not necessary concomitants of the disorder, as previously thought. Although the findings varied, delayed psychomotor and language development and hypotonia were observed in approximately three-fourths of the patients. Notable exceptions were two patients, IC and AF (the latter an adult patient who attended primary school), both of whom had normal development of speech and language. The detection of seizures in half of the patients was surprising, because seizures were not previously believed to be prevalent in 4-hydroxybutyric aciduria. On the other hand, the detection of ataxia in only one-third of patients was surprising, because ataxia was previously believed to be common in SSADH deficiency.

Two adult patients have been detected with 4-hydroxybutyric aciduria, suggesting the potential importance of screening for metabolic disease in adults diagnosed with idiopathic mental retardation (Jakobs *et al.* 1990). The first patient, a 23-year-old female, was institutionalized in a Dutch center for the mentally retarded at the time of the initial investigations. She was the firstborn daughter of healthy consanguineous Turkish parents. Two sisters and two brothers were healthy. Although details about early development were not available, in the first decade mild psychomotor retardation with severe learning difficulties were apparent. The second decade was complicated by some generalized epileptic events and increasingly aggressive behavior. After psychiatric evaluation, the girl was institutionalized in a center for the mentally disabled. Examination at age 23 showed a dysmorphic, shy and thick-set woman without dysmorphic features. Eye movements were normal. Eye grounds showed a normal picture. There was a mild dysarthria with a saccadic speech and little developed language. She walked with a propulsive gait but coordination was grossly intact. Fine motor skills were poorly developed. Muscle tone was elevated; muscle strength was intact. Deep tendon reflexes were absent. There were no abnormalities of sensory function.

The second adult patient was a 25-year-old Chinese man. The pregnancy, birth and neonatal periods were normal. The parents were not related. Motor milestones were delayed with unassisted walking at 2 years. Early cognitive and language development were normal, with first words at 10 months and full sentences at 2 years. At age 3–5 years, he was described as lacking coordination, motor control and social skills. He entered primary school but repeated the first grade. Special education placement was made at 9 years because of 'minimal brain dysfunction' found on psychologic evaluation in the USA. During adolescence he developed his first grand mal seizure. He is now seizure-free on carbamazepine and mysoline. At age 22 he completed a special education program in the USA and returned to Hong Kong, where he developed hallucinations and psychotic symptomatology, successfully treated with haloperidol. Genetics evaluation at age 25 revealed an uncooperative, psychotic, bilingual man with tall stature, macrocephaly, long triangular face and an unremarkable neurologic examination. Head MRI was normal. Organic acid analysis revealed 4-hydroxybutyric aciduria (119 mmol/mol creatinine; control < 4). SSADH activity was undetectable in extracts of isolated and cultured leukocytes.

Heterogeneity in clinical phenotype

Patient CA, a 15-year-old male, was the fourth child in a sibship of seven (Gibson *et al.* 1997b). Three brothers and two sisters (ages 20, 19, 17, 6 and 1.5 years) were clinically well. His 13-year-old sister is described below. The parents of CA are first cousins; the paternal grandmother and maternal grandfather are siblings. Pregnancy, delivery and the neonatal period were uneventful. The mother noted that the baby was too quiet and did not cry. For the first few months of life, he was suspected of being blind. Milestones were achieved slowly, and hypotonia was apparent. He was able to crawl and support his head at 10 months. He sat unaided and smiled at 1 year, walked at 2 years and spoke his first words at 3 years, which was followed by simple phrases. He was toilet trained by age 2–3 years, but has never attended school.

At age 15 years, he manifested global psychomotor and intellectual retardation, functioning at the developmental level of 4 years. He can comprehend simple commands, eats and sleeps normally, and is capable of dressing himself and bathing. He is described as having a good control of the vernacular language, and can perform simple, routine daily tasks. His sight is felt to be poor, and growth is subnormal. His tongue is cracked, but it is unclear whether this represents self-mutilation. He has nystagmus and hyporeflexia. There is no pyramidal, extrapyramidal or cerebellar syndrome and no ataxia. Genitalia are underdeveloped and there are no secondary sexual characteristics. He can speak simple sentences, count to 15 and roughly distinguish the time of day, but often has difficulty in focusing on individuals or objects. He is beginning to show the capacity to interact with other individuals.

Patient CJM, the younger female sibling of CA, is more severely affected. At 1 year of age she has hypotonic and apathetic. She smiled at one and one-half years and sat unaided at the same time. Developmental milestones were progressively delayed, with walking at 4 years and daytime toilet training achieved at 7 years of age. Nocturnal bladder control has not been achieved. She spoke only two words, and used them infrequently. Like her brother, her eyesight was thought to be poor. At age 11, she developed tonic/clonic seizures, which were only partially controlled by depakote. Treatment with vigabatrin aggravated the convulsive disorder. The family voluntarily ceased all anticonvulsant therapies, and she continues to have one to two convulsions per month. There was no ataxia, and she is described as timid and afraid.

From ages 6 to 9 years, CJM attended an institution for the developmentally delayed on the island of Fiji, at which time some developmental achievements were made. Further progress in development ceased when she returned to live with her family. At age 13 years, CJM was described as an easily agitated adolescent who frequently kept her hands in front of her eyes. She cannot maintain eye contact. She had difficulty in comprehending the presence of other people, and frequently rocks back and forth. Her language consists of five words. Although she can follow simple commands, she responds to frustration with anger, even though there is no true aggressive behavior. She cannot write, and spends the majority of her time gathering fallen leaves, feeding animals and playing with blocks and dolls. Psychiatric examination indicates consistent signs of autism and stereotypical psychotic aspects. Further evidence for increasing clinical heterogeneity in 4-hydroxybutyric aciduria is evidenced in a number of individual patient case reports (Gibson *et al.* 1985b, 1988; Haan *et al.* 1985; De Vivo *et al.* 1988; Hodson *et al.* 1989, 1990; Onkenhout *et al.* 1989; Daly *et al.* 1991; Rahbeeni *et al.* 1994; Worthen *et al.* 1994; Opp *et al.* 1996; Asai and Wada 1998; Peters *et al.* 1999; Al-Essa *et al.* 2000; Yalcinkaya *et al.* 2000).

Metabolic phenotype

An important step in our understanding of 4-hydroxybutyric aciduria was the development of isotope-dilution mass spectrometric methods for quantitation of GHB, and different forms of GABA, in cerebrospinal fluid (CSF) samples derived from patients (Gibson *et al.* 1990a; Howells *et al.* 1992). These techniques have provided an important tool with which to investigate the metabolic anomalies in patients. While it has been our experience that urinary concentrations of GHB can vary significantly in the same patient from day-to-day (and within the same 24 h period), the concentration of GHB in CSF samples from affected patients remains relatively constant (Table 12.1). Moreover, CSF is the most representative fluid from which to analyze brain-specific metabolites.

A quantitative assay for 4-hydroxybutyric acid was developed using D_6-4-hydroxybutyric acid as an internal standard. 4-Hydroxybutyric acid was isolated by liquid chromatography and the amount quantified by selected ion monitoring, ammonia chemical ionization gas chromatography/mass spectrometry of the trimethylsilyl derivatives. The concentrations of 4-hydroxybutyric acid in control physiologic fluids were: 0–7 mmol/mol creatinine in urine, 0–4 umol/l in plasma, 0–2.1 umol/l in cerebrospinal fluid and 0–2.3 umol/l in amniotic fluid. The stable isotope dilution assay of 4-hydroxybutyric acid in physiologic fluid samples is sensitive and accurate for quantification, and a valuable technique for the prenatal diagnosis of 4-hydroxybutyric aciduria (Table 12.2).

Table 12.1 Concentrations of 4-hydroxybutyric acid and different forms of GABA in CSF of patients with SSADH deficiency[a]

Patient	Age (yr)	GHB (µM)	GABA (µM)		Clinical findings
			Free	Total	
2	2	450	0.14	21.4	Severe psychomotor retardation, language delay, febrile seizures, hypotonia
3	2	525	0.178	22.0	Moderate retardation, delayed language development
	5	263	0.183	36.7	Hypotonia, lesions in globus pallidus, thalamus and brain stem
4	3	775	—	18.0	Severe psychomotor retardation, vegetative
5	4	420	0.227	17.4	Delayed speech development, seizures
6	7	512	—	19.8	Psychomotor and language retardation, hyperkinesis
7	8	830	0.38	—	Mental retardation, epilepsy, hypotonia, seizures
8	8	431	—	—	Retardation, hypotonia, ataxia, seizures, delayed language development
9	10	410	—	—	Mild retardation, hypotonia, hyperkinesis
10	10	535, 660	—	—	Retardation, delayed speech, hypotonia
11	12	565	0.189	—	Retardation, seizures, hyperkinesis, aggressive behavior
(Control range)		<3.0	<0.15	<14	

Note

a GHB and GABA concentrations in CSF samples determined as described (Gibson *et al.* 1990; Howells *et al.* 1992).

The mean concentration of GHB in CSF of patients with SSADH deficiency was 526 ± 43 µM (±SEM), more than 200-fold higher than the upper limit of the control range (Table 12.1). Mean values for free and total GABA were 0.22 ± 0.03 µM ($n = 6$) and 22.2 ± 2.5 µM ($n = 7$), respectively, less than twofold elevated in comparison to the upper limit of control (Table 12.1). Is accumulated GHB in CSF (and thereby brain) responsible for the neurologic insult occurring in patients? It could be argued that the data of Table 12.1 reflect a rapid outflow of GHB from brain into the CSF. Rapid efflux of GHB was observed

Table 12.2 Concentrations of 4-hydroxybutyric acid in physiological fluids from controls and patients with 4-hydroxybutyric aciduria

Subject	Urine (mmol/mol creatinine)	Plasma (μmol/l)	CSF (μmol/l)
Control	<10 (n = 18)[a]	<8 (n = 16)	<3 (n = 12)
Patient MS	208	98	410
Patient BS	663–7,600 (n = 9)[b]	1,500	3,100
Patient KM	402	—	—
Patient MM	1,450	—	—
Patient DW	116–135 (n = 3)	144–326 (n = 7)	431
Patient JMW	939–2024 (n = 5)	885–1,096 (n = 3)	—

Notes
a Values in parentheses following control means indicate the number of subjects studied.
b Where analyses were performed on multiple specimens from one patient, the data are presented as the range of values.

in dogs given an intravenous injection of GHB capable of inducing anesthesia (Gibson *et al.* 1998a). There does not appear to be an age-dependent decrease in GHB levels in CSF, as levels detected in patients at age 2 and 12 years are comparable, despite a significant change in the ratio of brain to body mass. This would suggest that GHB is produced in brain in fairly constant amounts. Because age is highly correlated with length, this might suggest an absence of rostro-caudal gradient of GHB, perhaps consistent with local production of GHB by the cord or blood. The quantities of GABA and GABA conjugates in our patients may be misleading, since the transport of these charged species into CSF is likely very inefficient.

The data of Table 12.1 present additional paradoxical features of SSADH deficiency. Of the eleven patients shown, four were hyperactive and one demonstrated aggressive behavior. This does not correlate with studies in the cat, in which intravenous administration of a GHB bolus which induced anesthesia (approximately 3.5 mmol/kg) yielded a brain tissue concentration of <1 μM GHB (Gibson *et al.* 1998a). These data may indicate that retention of GHB by mammalian brain is transient, or perhaps excitatory receptors are being targeted in these patients. Only five of eleven patients manifested seizures or epileptiform activity on EEG (Table 12.1). At the concentrations detected in CSF, it is perhaps remarkable that seizure activity was not observed in all patients. On the other hand, the majority of data on seizure induction by GHB administration is from animal studies.

Elevated 4-hydroxybutyric acid excretion has shown a strong correlation with enzyme deficiency. However, in three male patients this correlation was absent (Gibson *et al.* 1998a). Patient SD presented with mild mental retardation, ataxia, morbid obesity and microgenitalia. Routine screening revealed increased 4-hydroxybutyric acid in two samples, with normal GABA and 4-hydroxybutyric acid concentrations in plasma and CSF. Patient MT died shortly after birth with multiorgan, circulatory and respiratory failure. GHB was elevated in three urine samples (47–103 mmol/mol creatinine; normal <1). We, and others, have noted increased GHB in autopsied organs. Patient WG presented in early childhood with seizures, delayed mental and language development, and behavioral disturbances. CSF total GABA (350–396 nmol/l; normal 40–150) and GHB (140 μM; normal <3) levels were elevated with only trace amounts of GHB in urine. Vigabatrin therapy improved behavior and concentration, and decreased CSF 4-hydroxybutyric acid (36 μM). SSADH activity in cell extracts was normal in all. The clinical phenotype of patients SD (except obesity and microgenitalia) and WG were consistent with SSADH deficiency. All metabolic and clinical features in patient WG suggested SSADH deficiency, except for minor GHB excretion in urine. Patient WG may represent a kinetic variant of SSADH deficiency, or possibly a tissue-specific abnormality. These three patients illustrate important points: (1) that GHB does not always correlate with SSADH deficiency; (2) that non-GABA sources of GHB may exist in humans, perhaps through omega-oxidation of fatty acids or other compounds; and (3) the importance of SSADH determination in patients with increased GHB in physiologic fluids (Brown *et al.* 1987; Roesel *et al.* 1987; Gibson *et al.* 1989b; Gibson and Nyhan 1989; Shih *et al.* 1990; Bonham *et al.* 1994; Pitt *et al.* 1997).

Enzyme phenotype

The method of Cash and co-workers was adapted to the determination of SSADH activity in extracts of cultured human lymphoblasts, leucocytes, fibroblasts, amniocytes and chorionic villus tissue (Pattarelli *et al.* 1988; Gibson *et al.* 1991, 1994; Thorburn *et al.* 1993; Chambliss *et al.* 1995b). SSADH activity in the fluorometric assay (determining NADH production in stoichiometry with succinic semialdehyde consumption; excitation 340 nm, emission 470 nm) was linear, with incubation time up to 150 min in assays containing 0.37 mg protein, and with protein up to 0.37 mg for incubation periods of 100 min. Mean lymphoblast SSADH activities for parents, siblings and patients were 37%, 59% and 2%, respectively, of the mean control value. The mean values for all three groups were significantly different from control ($P < 0.001$ for parents and patients; $P < 0.05$ for siblings). We compared SSADH activities for controls with respect to age and sex. The results were: children ($n = 10$), 2.18 ± 0.93 nmol/min/mg protein (range 1.37–4.56); adults ($n = 17$), 2.44 ± 1.04 (range 0.69–4.59); female ($n = 9$), 1.89 ± 1.13 (range 0.69–4.56); male ($n = 6$), 1.96 ± 0.55 (range 1.19–2.60). For both parameters, there was no statistical difference ($P > 0.05$). For control, the day-to-day variability of the assay was $58\% \pm 15\%$ (± 1 SD, range 30–107%, $n = 27$), calculated as the ratio of the SD to the mean SSADH activity for each cell line. The within-day precision of the assay was $1.3\% \pm 1.3\%$ (± 1 SD, range 0–8%, $n = 27$) for control. Activity of control leucocyte SSADH activity was lower than the same activity in transformed cells at 56–461 pmol/min/mg protein (range; $n = 53$ subjects), with patients invariably showing nil activity. SSADH activity was readily estimated in amniocytes from cultured and biopsied chorionic villi, but was much lower in fibroblast extracts (Table 12.3).

Over the last several years, five patients have been identified whose metabolic profiles were consistent with diagnosis of SSADH deficiency, including increased excretion of GHB and/or 3,4-dihydroxybutyric acid, a postulated metabolite of GHB (Gibson *et al.* 1998b). Determination of SSADH activity in cultured cells and/or isolated peripheral leucocytes from these patients revealed partial SSADH deficiency in two and normal activity in the remaining three.

Patient RJ, a five-year-old male, was the first child of healthy nonconsanguineous parents. A younger brother is healthy. RJ was born at term, suffered transient postnatal hypoglycemia, and experienced ongoing feeding problems. Growth parameters remained below the third centile with psychomotor retardation. There were mild dysmorphic features, no ocular abnormalities and computed tomography of the brain were normal. Routine urine organic acid analysis revealed normal GHB concentration with increased 3,4-dihydroxybutyric and glycolic acids, both compounds believed to be derived from further metabolism of GHB. Total GABA was only slightly increased in CSF, whereas the free GABA concentration was normal.

Table 12.3 SSADH activities in extracts of cultured fibroblasts and/or isolated leucocytes from five patients

Subject	Fibroblasts	Leucocytes
RJ	0.14 $(n = 2)^a$	33 $(n = 1)$
Parallel controls[b]	33 ± 6^c $(n = 8$, range 18–67)	100 ± 23 $(n = 4$, range 41–146)
SN	—	0, 20 $(n = 2)$
Parallel controls	—	76 ± 17 $(n = 5$, range 40–140)
SD	—	285 $(n = 1)$
Parallel controls	—	95 ± 21 $(n = 6$, range 56–192)
MT	46 $(n = 1)$	—
Parallel controls	26 ± 5 $(n = 4$, range 18–40)	—
WG	40, 67 $(n = 2)$	199 $(n = 1)$
Parallel controls	26 ± 4 $(n = 6$, range 18–46)	157 $(n = 1)$

Notes

a n = number of determinations.

b For comparison, the following activities (pmol/min/mg protein) were obtained from SSADH-deficient cells during these analyses: leucocytes, 4; fibroblasts, 2; lymphoblasts, 4.

c Mean ± SD of the mean (SEM).

Patient SN, an 11-year-old female, was the second child born to nonconsanguineous parents. The first child was healthy at 19 years of age. The pregnancy was complicated by bleeding in the first and last trimesters. In the newborn period, SN vomited repeatedly and failed to thrive. At 3 weeks of life she underwent surgery for duodenal duplication. At 1 month she was admitted to the hospital in Brno, Czech Republic, because of abnormal phenylketonuria screening results. Hyperphenylalaninemia was detected (phenylalanine 2,660 μmol/l; normal 29–88), and the diagnosis of classical phenylketonuria (PKU) was established. Dietary treatment was started, but compliance was poor. In the first year of life, SN was living in an infant asylum as the family refused to care for her. At 1 year of age, she was profoundly retarded and hypotonic, with ventricular and subarachnoid hydrocephalus revealed on a CT scan, accompanied by photophobia, nystagmus and retinal abnormality. In the second year, repeated episodes of vomiting occurred with intermittently green to brown–green urine and positive urobilinogen reaction. During episodes, there was increased porphyrin excretion and lead in urine. At 4 years of age, automutilation appeared, with finger biting and skin scratching. At 7 years of age, metabolites corresponding to phenylketonuria were detected in urine accompanied by increased excretion of GHB (22–392 mmol/mol creatinine; normal < 10) during episodic decompensation. In some of the subsequent urine samples, no GHB was detected. An association of SSADH deficiency with PKU has not been reported and may represent a coincidental finding or possibly translocation of chromosomes 12 (PKU) and 6 (SSADH).

Patient SD, a 4-year-old male, manifested mild mental retardation and ataxia. Major problems were polyphagia with morbid obesity, microgenitalism and surgically corrected cryptochordism. Prader–Willi syndrome was ruled out via molecular genetic analysis. Routine screening revealed increased GHB in two urine samples; in CSF, free total GABA concentration was normal. CSF and serum GHB levels were also normal.

Patient MT, a male, developed circulatory and respiratory failure shortly after birth, eventually leading to multiorgan failure. The baby expired after a week on extracorporeal membrane oxygenation (ECMO). A single blood lactate concentration was 6.7 mmol/l during ECMO. Urine organic acid analysis in three urine samples revealed lactic acidosis (335–766 mmol/mol creatinine; normal < 121), increased excretion of GHB (47–103 mmol/mol creatinine; normal < 10), and elevated excretion of 3-hydroxyisobutyric acid (71–281 mmol/mol creatinine; normal < 44). Work-up for defects of the respiratory chain was not pursued.

Patient WG, a 9-year-old male, presented with convulsions in early childhood, mental retardation, significantly delayed speech development and severe behavioral disturbances. CSF free GABA (350–396 nmol/l; normal 40–150) and GHB (140 μmol/l; normal < 10) concentrations were elevated and were verified with subsequent CSF samples. Vigabatrin improved behavior and concentration, resulting in a significant decrease in the concentration of CSF GHB (36 μmol/l). In urine, only trace amounts of GHB were detected. Pertinent enzyme results in these five 'variant' forms of 4-hydroxybutyric aciduria are depicted in Table 12.3.

Prenatal diagnosis

Representative findings for the prenatal diagnosis of SSADH deficiency in seven 'at-risk' pregnancies from four unrelated families are depicted in Table 12.4 (Sweetman *et al.* 1986; Jakobs *et al.* 1993a,b). Prenatal diagnosis was performed by determination of GHB concentration in amniotic fluid using isotope-dilution gas chromatography–mass spectrometry in conjunction with assay of SSADH activity in biopsied chorionic villus and/or cultured amniocytes. In three of four pregnancies predicted as affected, confirmation was obtained by demonstration of deficient SSADH activity in fetal tissues.

In three pregnancies in which amniotic fluid GHB was increased, SSADH activity was significantly decreased in biopsied chorionic villus and/or cultured amniocytes (Table 12.4). In three of the pregnancies that were terminated on the basis of our prenatal studies, the enzyme defect documented in biopsied chroionic villus tissue and/or cultured amniocytes was verified by the demonstration of SSADH deficiency in fetal tissues. Because matched control fetal tissues were unavailable for enzyme studies in Family 2 (fifth pregnancy), and Family 3, control tissues were samples of autopsied brain, liver and kidney obtained from neonates or adolescents, submitted to our laboratory for diagnosis of inborn errors unrelated to 4-hydroxybutyric aciduria.

The range of control SSADH activity in neonatal and adolescent tissues was 2.1–4.1 nmol/min/mg protein ($n = 5$ total for all three tissue types, including brain, liver and kidney). The values for Family 2 (fifth pregnancy) and Family 3 were 0–0.7 nmol/min/mg protein ($n = 5$ for all three tissue types). Absolute comparison of these values is questionable, of course, because of age differences in the specimens, time intervals between death and autopsy, length of storage time, and other uncontrollable variables. Further, limited evidence suggests that SSADH activities in adult tissues are higher than activities in fetal tissues, especially in brain (Chambliss *et al.* 1995b).

In our experience with six of seven pregnancies at-risk for SSADH deficiency, the concentration of GHB in amniotic fluid and the SSADH activity in biopsied chorionic villi and/or cultured amniocytes was predictive of pregnancy outcome. Amniocyte SSADH activity was lower than the same activity in cultured lymphoblasts, lymphocytes or biopsied chorionic villi, as noted by other investigators (Thorburn *et al.* 1993). However, we did not differentiate between fibroblast-like and nonfibroblast-like amniocytes as did earlier investigators. Until nucleic acid-based methods for the prenatal detection of 4-hydroxybutyric aciduria become widely employed, the prenatal diagnosis of 4-hydroxybutyric aciduria should be investigated using both metabolite and enzymatic procedures to ensure accuracy and reliability.

Table 12.4 Prenatal diagnosis of 4-hydroxybutyric aciduria

Pregnancy/ Family[b]	GHB[c]	SSADH activity[a]		Outcome	
		Amniocytes (nmol/h/mg *protein*)	Biopsied chorionic villus[d] (nmol/min/mg *protein*)	Projected	Actual
2nd/Family 1	2.3	Normal[e]	NS[f]	Unaffected	Clinically normal child
3rd/Family 2	5.5	0.05 (control 3.7)	NS	Affected	SSADH deficiency in fetal tissues (brain, liver and kidney)[g]
4th/Family 2	1.2	0.85 (no control)	7.8 (control 6.9)	Unaffected	Clinically normal child
5th/Family 2	5.1	0[h]	0.02 (control 0.4)	Affected	SSADH deficiency in fetal tissues (brain and liver)
6th/Family 2	6.1	NS	NS	Affected	Fetal tissues unavailable
3rd/Family 3	8.5	0[h]	0.01 (control 0.5)	Affected	SSADH deficiency in fetal tissues (brain, liver and kidney)
6th/Family 4	2.1	NS	NS	Unaffected	Clinically normal child

Notes

a As determined by fluorimetric enzyme assay (Gibson *et al.* 1991).

b In Family 1, the affected female proband preceded this pregnancy; in Family 2, the affected male and a clinically unaffected male preceded these pregnancies; in Family 3, the affected female proband and a clinically unaffected female preceded this pregnancy; in Family 4, an affected male and four clinically unaffected siblings preceded this pregnancy.

c GHB determined by stable-isotope dilution assay (Gibson *et al.*, 1990a; control range $< 2.3\ \mu$mol/l, $n = 30$).

d Internal controls were cultured lymphoblasts (previous range 0.7–4.6 nmol/min/mg protein, $n = 27$) (Gibson *et al.* 1991).

e Normal activity as determined by whole cell conversion of [U-^{14}C] succinic semialdehyde to $^{14}CO_2$ (Pattarelli *et al.* 1988).

f NS, not studied.

g SSADH deficiency detected in fetal tissues using polyclonal antibodies raised against rat brain SSADH (Chambliss *et al.* 1995).

h The activity of a control mitochondrial enzyme, propionyl-CoA carboxylase, was within the range of controls.

Therapeutic intervention

Vigabatrin has been used therapeutically in numerous patients with 4-hydroxybutyric aciduria (Gibson *et al.* 1989a, 1995; Jakobs *et al.* 1992; Uziel *et al.* 1993; Dietz *et al.* 1996; Matern *et al.* 1996). The results of vigabatrin intervention in six patients are shown in Table 12.5.

Inhibition of GABA-transaminase by vigabatrin is predicted to decrease accumulation of GHB in patient physiologic fluids and result in potential clinical improvement. In five patients, an anticipated increase in the CSF GABA/GHB ratio was indeed observed.

Vigabatrin dosages were 40–100 mg/kg/day for periods of 1 month to 1 year. Five of six patients are still undergoing therapy; for a sixth, treatment with vigabatrin was stopped because of apparent induction of seizures (Patient DS, Table 12.5).

Therapeutic intervention with vigabatrin (in five of six patients tolerating the drug) resulted in variable improvement of cerebellar signs, concentration, hyperactivity, agility and cognitive functions. Significant improvements in psychomotor and language deficits, and hypotonia (when present) were not observed. For three patients the CSF GABA/GHB ratio increased, as expected (Table 12.5). However, the fall of CSF GHB levels was never as

Table 12.5 Clinical and metabolic response to vigabatrin in six patients with 4-hydroxybutyric aciduria

Patient	Clinical observations	Cerebrospinal fluid analyses[a] (μmol/l)			
		Metabolite	Pretreatment	Posttreatment	Change (%)
SM	Decreased ataxia	GHB	116	62	−47
ON	Improved concentration; reduced hyperactivity	Total GABA	13.6	31.2	+229
OI	Improved agility and able to stand briefly	GHB	1,110	708	−36
		Total GABA	22.4	32.1	+143
ED	Decreased level of frustration; improvements in concentration, speech and behavioral disturbances	GHB	615	457	−26
		Total GABA	18.3	38.8	+212
KS	No clinical information	GHB	525	384	−27
		Total GABA	22.1	33.6	+152
DS	Initially improved alertness; later, absence-like seizures with normal EEG; further intervention with vigabatrin resulted in generalized tonic/clonic seizures accompanied by slowing on the EEG without signs of epileptiform activity	—	—	—	—

Note
a Control CSF GHB < 3 μmol/l (*n* = 18); control total GABA < 12 μmol/l (*n* = 10).

great as the increase in total CSF GABA. As noted by Howells *et al.* (1992), vigabatrin may not effectively inactivate liver GABA-transaminase to the same extent as the brain enzyme, leading to peripheral resupply of GHB to the brain. This hypothesis would be supported by our observations in treated patients, for whom we detected only modest decreases in GHB levels in CSF.

SSADH – protein structure

SSADH has been partially purified and studied in rat, pig, monkey and human brain. The subunit molecular weights have been reported for the rat and human brain enzyme. To better characterize SSADH of different species and to develop tools for future molecular analyses to elucidate genetic defects in patients with 4-hydroxybutyric aciduria, we purified both rat and human SSADH, produced antiserum to the rat enzyme, and examined the SSADH protein of different species by immunoblots (Chambliss and Gibson 1992).

NAD$^+$-dependent SSADH was purified to apparent homogeneity from rat brain and highly purified from human brain. Molecular exclusion chromatography of the purified enzymes on Sephadex G-150 and G-200 revealed M_r values of 203,000 and 191,000 for rat and human enzymes, respectively. Electrophoresis on sodium dodecylsulfate polyacrylamide gels revealed a single subunit of M_r 54,000 for rat and 58,000 for human. Isoelectric focusing of the purified rat enzyme yielded a pI of 6.1. For both proteins, K_m values for short-chain aldehydes acetaldehyde and propionaldehyde ranged from 0.33 to 2.5 mM; K_m values for succinic semialdehyde were in the 2–4 μM range. The subunit structure of both enzymes was investigated in brain extracts and purified preparations by immunoblotting, using polyclonal rabbit antiserum against the purified rat brain enzyme. For rat and human extracts, single bands were detected at M_r 54,000 and 58,000, comparable to findings in the purified preparation. Immunoblotting analyses in other species (guinea pig, hamster, mouse and rabbit) revealed single subunits of M_r 54,000–56,500, revealing size conservation throughout evolution.

SSADH – molecular cloning

Although the nucleotide sequence of bacterial $NADP^+$-linked SSADH (E.C.1.2.1.16) was known in the early 1990s, a cDNA encoding mammalian NAD^+-dependent SSADH had not been reported. We isolated cDNAs encoding SSADH from rat and human and expressed a rat SSADH cDNA composite in bacteria (Chambliss *et al.* 1995a; Trettel *et al.* 1997). SSADH cDNAs from rat and human recognize two differentially expressed mRNAs of approximately 2.0 and 6.0 kb that are transcribed from a single gene.

During our preliminary cloning work, three rat brain cDNA clones of ~3,500, 1,465 and 1,135 bp in length encoding SSADH were isolated from two cDNA libraries using a polymerase chain reaction derived probe. Restriction mapping and DNA sequencing revealed that the 3.5 kb clone contained an 84 bp (twenty-eight amino acid) insert in the coding region. Composite clones encoding mature SSADH predicted proteins with 488 amino acids ($M_r = 52,188$) when including the insert and 460 amino acids ($M_r = 48,854$) without the insert. The cDNA clones were confirmed by expression of enzyme activity in bacteria and protein sequence data obtained from sequencing purified rat brain SSADH. Two human liver SSADH cDNA clones of 1,091 and 899 bp were also isolated. Northern blot analysis revealed two differentially expressed SSADH transcripts of approximately 2.0 and 6.0 kb in both rat and human tissues. Human genomic Southern blots indicated that the two SSADH transcripts are encoded by a greater than 20 kb single copy gene. Mammalian SSADH contained significant homology to bacterial $NADP^+$-SSADH (E.C.1.2.1.16) and conserved regions of general aldehyde dehydrogenases, suggesting it is a member of the aldehyde dehydrogenase superfamily of proteins.

While it was of value to utilize the rat cDNAs as a springboard from which to isolate the human cDNAs, we found isolation of a full-length human cDNA encoding SSADH impossible. Alignment of all human cDNA sequences obtained in our laboratories generated a 2,323 bp composite clone with an open-reading frame encoding 467 amino acids (Fig. 12.2). The deduced amino acid sequence included a peptide with the sequence DRRALVLK (amino acids 138–145), which had previously been identified from sequencing purified human brain SSADH but was in a region of the protein for which no previous nucleotide sequence had been known (Chambliss *et al.* 1995a). When aligned with the twenty-four N-terminal amino acids known from sequencing human brain SSADH, the mature SSADH polypeptide contained 488 amino acids with a deduced M_r of 52,329 (rat deduced $M_r = 52,186$).

The polypeptide depicted in Fig. 12.2. was a composite of deduced amino acid sequence, amino acid sequences obtained from protein sequencing and amino acids inferred from the deduced rat SSADH polypeptide sequence (Chambliss *et al.* 1995a; Trettel *et al.* 1997). Amino acids deduced from cDNAs began at residue 22 (Trp; Fig. 12.2), while the first twenty-one amino acids were obtained from sequencing of purified human brain SSADH. The nucleotide sequence encoding these amino acids, and the mitochondrial leader sequence, have only been recently identified (P. Malaspina and A. Novelletto, personal communication). Ambiguity in amino acid identification during sequence analysis resulted in Gly_{19} and Gly_{20} being inferred from the corresponding rat SSADH amino acid sequence (Fig. 12.2). Genomic sequencing should eventually clarify these sequences.

MATCIWLRSCGARRLGSTFPGCRLRPRAGGLVPASGPAPGPAQLRCYAGRLAGLSA

ALLRTDSFVGGRWLPAAATFPVQDPASGAALGMVADCGVREARAAVRAAYEAFCRWR

EVSAKE<u>R</u>SSL<u>L</u>RKWYNLMIQNKDDLARIITAES<u>GK</u>PLKEAHGEILYSAFFLEWFSEEARRV

YGDIIHTPAKDRRALVLKQ<u>PIGV</u>AAV<u>IT</u>PW<u>NFP</u>SAMITRKVGA<u>A</u>LAA<u>G</u>CTVVV<u>K</u>PAEDTP

FSALALAELASQAGIPSGVYNVIP**C**SRKNAKEVGEAICTDPLVSKIS<u>FTGS</u>TTT<u>G</u>KILLHHA

ANSVKR<u>V</u>SM<u>E</u>L<u>CGL</u>APFIVFDSANVDQAVAGAMASK<u>F</u>RNT<u>GQTC</u>VCSNQFLVQRGIHD

AFVKAFAEAMKKNLRVGNGFEEGTTQ<u>G</u>PLINEKAVEKVEKQVNDAVSKGATVVT<u>GGK</u>

RHQLGKNFFE<u>PT</u>LLCNVTQDMLCTH<u>EET</u><u>FGP</u>LAPVIKFDTEEEAIAIA<u>N</u>AADVGL<u>A</u>GYFY

SQDPAQIWRVAEQLEVG<u>M</u><u>V</u>GV<u>N</u>EGLISSVECP<u>FGG</u>VKQ<u>SGL</u>GREGSKYGIDEYLELKYV

CYGGL

Figure 12.2 Amino acid sequence of the human SSADH polypeptide. The mitochondrial import signal peptide is depicted in bold lettering. Underlined residues represent those maintained at high conservation in other aldehyde (and semialdehyde) dehydrogenase proteins. Underlined and bold residues (two) represent those which are divergent in human SSADH from other aldehyde dehydrogenase (and semialdehyde dehydrogenase) proteins.

The subunit M_r for human SSADH was consistent with the estimated native M_r of 191,000 (homotetramer) for the human protein using molecular exclusion chromatography of the purified human brain SSADH. The M_r of the deduced polypeptide sequence, however, was somewhat lower than the expected subunit M_r of 58,000, which was determined from SDS–PAGE analysis. We had estimated that the rat SSADH subunit migrated with an estimated M_r of 54,000 (Chambliss and Gibson 1992). Other investigators (Ryzlak and Pietruszko 1988) provided results which indicated that the human brain SSADH protein was a tetramer comprised of weight nonidentical subunits ($M_r = 61,000$ and 63,000). Our results in rat and human, however, were consistent with a homotetrameric structure. Human and rat cDNAs, and their corresponding amino acid sequence alignment in overlapping regions, revealed 83% and 92% homology, respectively, between the two species.

Alignment of the human SSADH amino acid sequence with sequences from sixteen aldehyde dehydrogenase proteins revealed that SSADH maintains the majority of conserved residues (Trettel *et al.* 1997). This alignment included strict consensus residues and those maintained as 87–93% identity. In Fig. 12.2, these residues are underlined, with alignment using ALIGN. As reported (Trettel *et al.* 1997), twenty-three invariant residues, including eleven glycines and three prolines, have suggested evolutionary resistance to alteration of peptide chain-bending points. Human SSADH retained twenty-two/twenty-three invariant residues with one exception (Fig. 12.2). Gly$_{212}$ was conserved in all aldehyde dehydrogenases, whereas in human SSADH this residue was a Cys. This Cys$_{212}$ was maintained in rat SSADH, but the conserved Gly$_{212}$ was detected in bacterial SSADH. In addition, an almost invariant Lys$_{263}$ was Leu in human and rat SSADH, and Asn in bacterial SSADH (see bolded and underlined residues in Fig. 12.2). Invariant Cys and Glu residues, believed important in catalysis, were maintained in human SSADH as residues 259 and 293, respectively (Fig. 12.2). Human SSADH maintains an EiFGP sequence at residues 391–395 (Fig. 12.2), with Thr filling the 392 position (generally Ile in other aldehyde dehydrogenases) in bacterial, rat and human SSADH.

The chromosomal location of the SSADH gene was determined by hybridizing the radiolabeled 1091 bp human liver cDNA to a Southern blot containing DNA purified from twenty different hamster-human somatic cell hybrids. Comparison of the content of human chromosomes in these hybrid cell lines against cell lines that did not hybridize to the radiolabeled probe, and analysis of concordance, revealed that human SSADH segregated with human chromosome 6. Partial sequencing of YACs and cosmids included in a contig encompassing human chromosome 6p22 verified the presence of SSADH coding sequences, and human SSADH was oriented from telomere to centromere (Trettel *et al.* 1997).

Molecular defects in patients

Despite fairly consistent enzymatic and metabolic findings, the clinical phenotype of 4-hydroxybutyric aciduria, featuring mainly nonspecific neurologic findings, is highly heterogeneous. Although it is highly likely that defects in the SSADH gene are the cause of SSADH deficiency, until recently there was no evidence to support this hypothesis. As a first step in beginning an investigation of the molecular genetics of 4-hydroxybutyric aciduria, we utilized genomic cosmid clones to derive the remainder of the cDNA structure of human SSADH, including the putative mitochondrial leader sequence. A composite human cDNA expressed enzymatically active SSADH protein. These sequences were employed to identify two point mutations in the SSADH genes derived from four patients. These mutations, identified by standard methods of reverse transcription, PCR, dideoxy-chain termination and cycle sequencing, alter highly conserved sequences in intron/exon boundaries and prevent the RNA-splicing apparatus from properly recognizing the normal splice junction (GenBank database locations: NM_001080; Y11192; AL031230) (Chambliss *et al.* 1998; Hogema *et al.* 1999).

Each family segregated a mutation in a different splice site, resulting in exon skipping and, in one case, a frameshift and premature termination and, in the other case, an in-frame deletion in the resulting protein. We documented that these mutant alleles existed at the cDNA and genomic levels. Family members, including parents and siblings of these patients (all offspring of consanguineous matings), were shown to be heterozygotes for the splicing abnormality, providing additional evidence for autosomal recessive inheritance. These alleles have been labeled 'ALDH4A1*1' (G → T transversion, first base of intron 9) and 'ALDH4A1*2' (G → A transition, first base of intron 5), according to preferred nomenclature (Chambliss *et al.* 1998). These results provided the first evidence that 4-hydroxybutyric aciduria, resulting from SSADH deficiency, was the result of genetic defects in the human SSADH gene.

To extend these preliminary studies, and attempt a preliminary assessment of genotype/phenotype correlation in 4-hydroxybutyric aciduria, we performed mutation screening in samples obtained from twenty-five patients (twenty-one families) using combined RT–PCR/genomic PCR and sequence analysis. The results were: three RNA spicing mutations (loss of exon 8, loss of exon 9, and a splicing mutation resulting in skipping of the first 14 bp of exon 4); one insertion (AG insertion at nt 457); one deletion (14 bp following nt 460); four nonsense mutations (Y128X, W204X, R412X and R514 X); and eight missense mutations (P182L, C223Y, T233M, A237S, G268E, P382L, G409D and G533R). This represents twenty-one alleles/eighteen mutations and adds substantively to our previous report of two exon-skipping mutations (Hogema *et al.* 1999). With the exception of the P182L, C223Y and G533R alleles, the mutations represented highly conserved residues in other (putative) SSADH proteins and other human aldehyde dehydrogenases. Thus far our analysis has provided no clear insight into genotype/phenotype correlations, and indicates that most mutations are private within the families studied. We continue extensive sequence analysis in patients with 4-hydroxybutyric aciduria, and have devised an expression system in which to assess the functional significance of missense mutations.

Summary and conclusions

Naturally occurring systems in which GHB concentrations are either depressed (not currently known) or elevated (as in 4-hydroxybutyric aciduria) provide a valuable tool with which to study the neuropharmacology of GHB, an unusual compound which may be a nonamine mammalian neurotransmitter (Gibson *et al.* 1998a). Nature has given us an important, albeit unfortunate, model in which to study the effects of accumulated GHB in the form of patients with 4-hydroxybutyric aciduria. It is probable that GHB plays an important role in the development and progression of pathophysiology associated with 4-hydroxybutyric aciduria. It remains unknown if GHB acts alone, synergistically with other metabolites and pathways, or if conversion of GHB to discrete regional pools of GABA has a role. Further, GHB and GABA may exert their effects independently, or in synergy, and possibly in conjunction with other neurogenics whose natural metabolic processes are perturbed by accumulated GHB.

It remains important to understand the exact molecular mechanisms which lead to 4-hydroxybutyric aciduria. These studies may provide a more complete understanding of the clinical heterogeneity which is observed in patients. However, molecular analyses are not likely to provide improved therapeutic approaches, and they will generate minimal new insight into the pharmacologic effects of GHB at the pathologic levels detected in patients. A more useful approach toward understanding the 4-hydroxybutyric aciduria disease process will be the development of an animal model of the disease. Such a model will enable a wide array of experimental approaches. First and foremost, accurate determination of GHB concentrations in brain could be obtained. Alterations in receptor biosynthesis could be evaluated in response to chronic GHB exposure. The use of GHB and $GABA_B$ receptor antagonists would be evaluated, along with their effects on the GABAergic and dopaminergic pathways. In conjunction with *in vivo* microdialysis, the effect of chronic GHB accumulation in discrete regions of brain could be assessed on the concentration of other neurogenic compounds, such as dopamine, serotonin, glutamate and GABA. The response of SSADH-deficient animals to other convulsants, with the capacity to induce absence and generalized seizure activity, would be of interest. With a complete picture of the gene structure, it should be possible to isolate the murine SSADH gene, which will provide the foundation from which to develop a murine knockout for 4-hydroxybutyric aciduria. Eventually, retroviral-mediated transfer of the normal SSADH gene should be feasible in the murine model. An animal model of SSADH deficiency will add greatly to our understanding of the role of GHB and SSADH in cerebral metabolism, and provide new information on the role of GHB in the development of the clinical phenotype of patients with 4-hydroxybutyric aciduria.

Acknowledgments

Limited space precludes acknowledgement of the numerous metabolic specialists, physicians, collaborators and families who have been involved in the above work over the span of some 20 years. None of this work would have been possible, however, without the collaboration and support of Drs C. Jakobs, A. Novelletto, P. Malaspina, K. L. Chambliss, L. Sweetman, D. Rating, W. L. Nyhan, Boris Hogema and Patrick F. Lauer. I gratefully acknowledge the financial support of the GHB Research Fund at OHSU, and Research Grants No. 9813 from the Oregon Health Sciences Foundation, #1-FY00–352 from the March of Dimes Birth Defects Foundation, and NS 40270 from the National Institute of Neurological Disorders and Stroke, National Institutes of Health.

References

Al-Essa M. A., Bakheet S. M., Patay Z. J., Powe J. E. and Ozand P. T. (2000) Clinical, fluorine-18 labeled 2-fluoro-2-deoxyglucose positron emission tomography (FDG PET), MRI of the brain and biochemical observations in a patient with 4-hydroxybutyric aciduria: a progressive neurometabolic disease. *Brain Dev.* 22, 127–31.

Asai M. and Wada Y. (1998) 4-Hydroxybutyric aciduria [Japanese]. *Ryoikibetsu Shokogun Shirizu* 18 (Pt 1), 250–51.

Bonham J. R., Downing M., Pollitt R. J., Manning N. J., Carpenter K. H., Olpin S. E., Allen J. C. and Worthy E. (1994) Quality assessment of urinary organic acid analysis. *Ann. Clin. Biochem.* 31 (Pt 2), 129–33.

Brown G. K., Cromby C. H., Manning N. J. and Pollitt R. J. (1987) Urinary organic acids in succinic semialdehyde dehydrogenase deficiency: evidence of alpha-oxidation of 4-hydroxybutyric acid, interaction of succinic semialdehyde with pyruvate dehydrogenase and possible secondary inhibition of mitochondrial beta-oxidation. *J. Inherit. Metab. Dis.* 10, 367–75.

Chambliss K. L. and Gibson K. M. (1992) Succinic semialdehyde dehydrogenase from mammalian brain: subunit analysis using polyclonal antiserum. *Int. J. Biochem.* 24, 1493–99.

Chambliss K. L., Caudle D. L., Hinson D. D., Moomaw C. R., Slaughter C. A., Jakobs C. and Gibson K. M. (1995a) Molecular cloning of the mature NAD$^+$-dependent succinic semialdehyde dehydrogenase from rat and human. cDNA isolation, evolutionary homology, and tissue expression. *J. Biol. Chem.* 270, 461–67.

Chambliss K. L., Zhang Y. A., Rossier E., Vollmer B. and Gibson K. M. (1995b) Enzymatic and immunologic identification of succinic semialdehyde dehydrogenase in rat and human neural and nonneural tissues. *J. Neurochem.* 65, 851–55.

Chambliss K. L., Hinson D. D., Trettel F., Malaspina P., Novelletto A., Jakobs C. and Gibson K. M. (1998) Two exon-skipping mutations as the molecular basis of succinic semialdehyde dehydrogenase deficiency (4-hydroxybutyric aciduria). *Am. J. Hum. Genet.* 63, 399–408.

Daly D. M., Hodson A. and Gibson K. M. (1991) Central auditory processing in a patient with SSADH deficiency. *Soc. Neurosci. Abs.* 17 (Pt 1), 892.

De Vivo D. C., Gibson K. M., Resor L. D., Steinschneider M., Aramaki S. and Cote L. (1988) 4-Hydroxybutyric acidemia: clinical features, pathogenetic mechanisms, and treatment strategies. *Ann. Neurol.* 24, 304.

Dietz B., Aksu F., Aguigah G., Witting W., Aygen S., Lehnert W. and Jakobs C. (1996) Vigabatrin therapy is a 7 years old boy with succinic semialdehyde dehydrogenase deficiency [German]. *Monatsschr. Kinderheikd.* 144, 797–802.

Divry P., Baltassat P., Rolland M. O., Cotte J., Hermier M., Duran M. and Wadman S. K. (1983) A new patient with 4-hydroxybutyric aciduria, a possible defect of 4-aminobutyrate metabolism. *Clin. Chim. Acta* 129, 303–09.

Gallimberti L., Spella M. R., Soncini C. A. and Gessa G. L. (2000) Gamma-hydroxybutyric acid in the treatment of alcohol and heroin dependence. *Alcohol* 20, 257–62.

Gibson K. M. (1984) 4-Hydroxybutyric aciduria. *Bioessays* 1, 110–13.

Gibson K. M. (1996) Succinic semialdehyde dehydrogenase deficiency. In *Neurobase*, Vol. 4.2 (Gilman S., Goldstein G. W. and Waxman S. G., eds). Arbor Publishing, LaJolla.

Gibson K. M. and Jakobs C. (2001) Disorders of β- and γ-amino acids in free and peptide-linked forms. In *The Metabolic and Molecular Bases of Inherited Disease*, 8th edn (Scriver C. R., Beaudet A. L., Sly W. S. and Valle D., eds), Chapter 91, pp. 2079–105. McGraw-Hill, New York.

Gibson K. M. and Nyhan W. L. (1989) Metabolism of [U-^{14}C]-4-hydroxybutyric acid to intermediates of the tricarboxylic acid cycle in extracts of rat liver and kidney mitochondria. *Eur. J. Drug Metab. Pharmacokinet.* 14, 61–70.

Gibson K. M. and Sweetman L. (1983) Enzymatic preparation of radiolabeled succinic semialdehyde. *Anal. Biochem.* 135, 436–37.

Gibson K. M., Sweetman L., Nyhan W. L., Jakobs C., Rating D., Siemes H. and Hanefeld F. (1983) Succinic semialdehyde dehydrogenase deficiency: an inborn error of gamma-aminobutyric acid metabolism. *Clin. Chim. Acta* 133, 33–42.

Gibson K. M., Jansen I., Sweetman L., Nyhan W. L., Rating D., Jakobs C. and Divry P. (1984a) 4-Hydroxybutyric aciduria: a new inborn error of metabolism. III. Enzymology and inheritance. *J. Inherit. Metab. Dis.* 7 (Suppl. 1), 95–96.

Gibson K. M., Sweetman L., Nyhan W. L. and Rating D. (1984b) Succinic semialdehyde dehydrogenase deficiency. *J. Neurogenet.* 1, 213–18.

Gibson K. M., Sweetman L., Nyhan W. L., Lenoir G. and Divry P. (1984c) Defective succinic semialdehyde dehydrogenase activity in 4-hydroxybutyric aciduria. *Eur. J. Pediatr.* 142, 257–59.

Gibson K. M., Sweetman L., Jansen I., Brown G. K., Haan E. A., Danks D. M. and Nyhan W. L. (1985a) Properties of succinic semialdehyde dehydrogenase in cultured human lymphoblasts. *J. Neurogenet.* 2, 111–22.

Gibson K. M., Sweetman L., Nyhan W. L. and Bowen P. (1985b) Clinical correlation of dysequilibrium syndrome and 4-hydroxybutyric aciduria. *J. Inherit. Metab. Dis.* 8, 58.

Gibson K. M., Nyhan W. L. and Jaeken J. (1986) Inborn errors of GABA metabolism. *Bioessays* 4, 24–27.

Gibson K. M., Hoffmann G., Nyhan W. L., Aramaki S., Thompson J. A., Goodman S. I., Johnson D. A. and Fife R. R. (1988) 4-Hydroxybutyric aciduria in a patient without ataxia or convulsions. *Eur. J. Pediatr.* 147, 529–31.

Gibson K. M., DeVivo D. C. and Jakobs C. (1989a) Vigabatrin therapy in patient with succinic semialdehyde dehydrogenase deficiency. *Lancet* 2, 1105–06.

Gibson K. M., Goodman S. I., Frerman F. E. and Glasgow A. M. (1989b) Succinic semialdehyde dehydrogenase deficiency associated with combined 4-hydroxybutyric and dicarboxylic acidurias: potential for clinical misdiagnosis based on urinary organic acid profiling. *J. Pediatr.* 114 (Pt 1), 607–10.

Gibson K. M., Aramaki S., Sweetman L., Nyhan W. L., DeVivo D. C., Hodson A. K. and Jakobs C. (1990a) Stable isotope dilution analysis of 4-hydroxybutyric acid: an accurate method for quantification in physiological fluids and the prenatal diagnosis of 4-hydroxybutyric aciduria. *Biomed. Environ. Mass. Spectrom.* 19, 89–93.

Gibson K. M., Hoffmann G. and Rating D. (1990b) Acidemia, gamma-hydroxybutyric. In *Birth Defects Encyclopedia* (Buyse M. L., ed.), pp. 17–18. Blackwell Scientific Publications, Cambridge, Massachusetts.

Gibson K. M., Lee C. F., Chambliss K. L., Kamali V., Francois B., Jaeken J. and Jakobs C. (1991) 4-Hydroxybutyric aciduria: application of a fluorometric assay to the determination of succinic semialdehyde dehydrogenase activity in extracts of cultured human lymphoblasts. *Clin. Chim. Acta* 196, 219–21.

Gibson K. M., Baumann C., Ogier H., Rossier E., Vollmer B. and Jakobs C. (1994) Pre- and postnatal diagnosis of succinic semialdehyde dehydrogenase deficiency using enzyme and metabolite assays. *J. Inherit. Metab. Dis.* 17, 732–33.

Gibson K. M., Jakobs C., Ogier H., Hagenfeldt L., Eeg-Olofsson K. E., Eeg-Olofsson O., Aksu F., Weber H.-P., Rossier E., Vollmer B. and Lehnert W. (1995) Vigabatrin therapy in six patients with succinic semialdehyde dehydrogenase deficiency. *J. Inherit. Metab. Dis.* 18, 143–46.

Gibson K. M., Christensen E., Jakobs C., Fowler B., Clarke M. A., Hammersen G., Raab K., Kobori J., Moosa A., Vollmer B., Rossier E., Iafolla A. K., Matern D., Brouwer O. F., Finkelstein J., Aksu F., Weber H.-P., Bakkeren J. A. J. M., Gabreels F., Bluestone D., Barron T. F., Beauvais P., Rabier D., Santos C., Umansky R. and Lehnert W. (1997a) The clinical phenotype of succinic semialdehyde dehydrogenase deficiency (4-hydroxybutyric aciduria): case reports of 23 new patients. *Pediatrics* 99, 567–74.

Gibson K. M., Doskey A. E., Rabier D., Jakobs C. and Morlat C. (1997b) Differing clinical presentation of succinic semialdehyde dehydrogenase deficiency in adolescent siblings from Lifu Island, New Caledonia. *J. Inherit. Metab. Dis.* 20, 370–74.

Gibson K. M., Hoffmann G. F., Hodson A. K., Bottiglieri T. and Jakobs C. (1998a) 4-Hydroxybutyric acid and the clinical phenotype of succinic semialdehyde dehydrogenase deficiency, an inborn error of GABA metabolism. *Neuropediatrics* 29, 14–22.

Gibson K. M., Sweetman L., Kozich V., Pijackova A., Tscharre A., Cortez A., Eyskens F., Jakobs C., Duran M. and Poll-The B.T. (1998b) Unusual enzyme findings in five patients with metabolic profiles suggestive of succinic semialdehyde dehydrogenase deficiency (4-hydroxybutyric aciduria). *J. Inherit. Metab. Dis.* 21, 255–61.

Gibson K. M., Novelettto A., Malaspina P. and Trettel F. (2002) Human succinic semialdehyde dehydrogenase. In *Encyclopedia of Molecular Medicine* (Creighton T. E., ed.), pp. 3031–35. John Wiley, New York.

Haan E. A., Brown G. K., Mitchell D. and Danks D. M. (1985) Succinic semialdehyde dehydrogenase deficiency: a further case. *J. Inherit. Metab. Dis.* 8, 99.

Hodson A. K., Hartlage P., Roesel A. and Gibson K. M. (1989) Succinic semialdehyde dehydrogenase deficiency: seizures and ragged-red fibers. *Cleve. Clin. J. Med.* 56 (Suppl. 1, Pt 2), S278–79.

Hodson A. K., Gibson K. M. and Jakobs C. (1990) Developmental resolution of ataxia in succinic semialdehyde dehydrogenase deficiency. *Ann. Neurol.* 28, 438.

Hoffmann G. F. and Gibson K. M. (1996) Disorders of organic acid metabolism. In *Handbook of Clinical Neurology* (Vinken P. J. and Bruyn G. W., eds), *Neurodystrophies and Neurolipidoses*, Vol. 66 (Moser H. W., Vol. ed.), pp. 638–60. Elsevier, Amsterdam.

Hoffmann G. F., Gibson K. M., Trefz F. K., Nyhan W. L., Bremer H. J. and Rating D. (1994) Neurological manifestations of organic acid disorders. *Eur. J. Pediatr.* 153 (7 Suppl. 1), S94–S100.

Hogema B. M., Jakobs C., Oudejans C. B. M., Schutgens R. B. H., Grompe M. and Gibson K. M. (1999) Mutation analysis in succinic semialdehyde dehydrogenase (SSADH) deficiency (4-hydroxybutyric aciduria). *Am. J. Hum. Genet.* 65, A238.

Howells D., Jakobs C., Kok R. M., Wrennall J. and Thompson G. N. (1992) Vigabatrin therapy in succinic semialdehyde dehydrogenase deficiency. *Mol. Neuropharmacol.* 2, 181–84.

Jaeken J., Casaer P., Haegele K. D. and Schechter P. J. (1990) Review: normal and abnormal central nervous system GABA metabolism in childhood. *J. Inherit. Metab. Dis.* 13, 793–801.

Jakobs C., Kneer J., Rating D., Hanefeld F., Divry P. and Hermier M. (1984) 4-Hydroxybutyric aciduria: a new inborn error of metabolism. II. Biochemical findings. *J. Inherit. Metab. Dis.* 7 (Suppl. 1), 92–94.

Jakobs C., Smit L. M., Kneer J., Michael T. and Gibson K. M. (1990) The first adult case with 4-hydroxybutyric aciduria. *J. Inherit. Metab. Dis.* 13, 341–44.

Jakobs C., Michael T., Jaeger E., Jaeken J. and Gibson K. M. (1992) Further evaluation of vigabatrin therapy in 4-hydroxybutyric aciduria. *Eur. J. Pediatr.* 151, 406.

Jakobs C., Jaeken J. and Gibson K. M. (1993a) Inherited disorders of GABA metabolism. *J. Inherit. Metab. Dis.* 16, 704–15.

Jakobs C., Ogier H., Rabier D. and Gibson K. M. (1993b) Prenatal detection of succinic semialdehyde dehydrogenase deficiency (4-hydroxybutyric aciduria). *Prenat. Diagn.* 13, 150.

Matern D., Lehnert W., Gibson K. M. and Korinthenberg R. (1996) Seizures in a boy with succinic semialdehyde dehydrogenase deficiency treated with vigabatrin (gamma-vinyl-GABA). *J. Inherit. Metab. Dis.* 19, 313–18.

Onkenhout W., Maaswinkel-Mooij P. D. and Poorthuis B. J. (1989) 4-Hydroxybutyric aciduria: further clinical heterogeneity in a new case. *Eur. J. Pediatr.* 149, 194–96.

Opp J., Raab K., Jakobs C., Lehnert W. and Gibson K. M. (1996) Succinic semialdehyde dehydrogenase deficiency in two siblings. *Monatsschr. Kinderheikd.* 144, 695–98.

Pattarelli P. P., Nyhan W. L. and Gibson K. M. (1988) Oxidation of [U-^{14}C] succinic semialdehyde in cultured human lymphoblasts: measurement of residual succinic semialdehyde dehydrogenase activity in 11 patients with 4-hydroxybutyric aciduria. *Pediatr. Res.* 24, 455–60.

Peters H., Cleary M. and Boneh A. (1999) Succinic semialdehyde dehydrogenase deficiency in siblings: clinical heterogeneity and response to early treatment. *J. Inherit. Metab. Dis.* 22, 198–99.

Pitt J. J., Hawkins R., Cleary M., Eggington M., Thorburn D. R. and Warwick L. (1997) Succinic semialdehyde dehydrogenase deficiency: low excretion of metabolites in a neonate. *J. Inherit. Metab. Dis.* 20, 39–42.

Rahbeeni Z., Ozand P. T., Rashed M., Gascon G. G., al Nasser M., al Odaib A., Amoudi M., Nester M., al Garawi S. and Brismar J. (1994) 4-Hydroxybutyric aciduria. *Brain Dev.* 16 (Suppl.), 64–71.

Rating D., Siemes H., Hanefeld F., Kneer J., Jakobs C., Gibson K. M., Sweetman L. and Nyhan W. L. (1982) An inborn error of GABA-metabolism in atactic syndromes. *Eur. J. Pediatr.* 139, 317.

Rating D., Hanefeld F., Siemes H., Kneer J., Jakobs C., Hermier M. and Divry P. (1984) 4-Hydroxybutyric aciduria: a new inborn error of metabolism. I. Clinical review. *J. Inherit. Metab. Dis.* 7 (Suppl. 1), 90–92.

Roesel R. A., Hartlage P. L., Carroll J. E., Hommes F. A., Blankenship P. R. and Gibson K. M. (1987) 4-Hydroxybutyric aciduria and glycinuria in two siblings. *Am. J. Hum. Genet.* 41, a16.

Ryzlak M.T. and Pietruszko R. (1988) Human brain 'high Km' aldehyde dehydrogenase: purification, characterization, and identification as NAD$^+$-dependent succinic semialdehyde dehydrogenase. *Arch. Biochem. Biophys.* 266, 386–96.

Shih V. E., Younes M. C., Gotoff J. M., Dooling E. C. and Gibson K. (1990) Transient increase in CSF glycine in a patient with succinic semialdehyde dehydrogenase deficiency (SSADH). *Am. J. Hum. Genet.* 47, a166.

Snead O. C. (1977) Minireview: gamma-hydroxybutyrate. *Life Sci.* 20, 1935–44.

Sweetman F. R., Gibson K. M., Sweetman L., Nyhan W. L., Chin H., Swartz W. and Jones O. W. (1986) Activity of biotin-dependent and GABA metabolizing enzymes in chorionic villus samples: potential for 1st trimester prenatal diagnosis. *Prenat. Diagn.* 6, 187–94.

Thorburn D. R., Thompson G. N. and Howells D. W. (1993) A fluorimetric assay for succinic semialdehyde dehydrogenase activity suitable for prenatal diagnosis of the enzyme deficiency. *J. Inherit. Metab. Dis.* 16, 942–49.

Tillakaratne N. J., Medina-Kauwe L. and Gibson K. M. (1995) Gamma-aminobutyric acid (GABA) metabolism in mammalian neural and nonneural tissues. *Comp. Biochem. Physiol.* A112, 247–63.

Trettel F., Malaspina P., Jodice C., Novelletto A., Slaughter C. A., Caudle D. L., Hinson D. D., Chambliss K. L. and Gibson K. M. (1997) Human succinic semialdehyde dehydrogenase. Molecular cloning and chromosomal localization. *Adv. Exp. Med. Biol.* 414, 253–60.

Uziel G., Bardelli P., Pantaleoni C., Rimoldi M. and Savoiardo M. (1993) 4-Hydroxybutyric aciduria: clinical findings and vigabatrin therapy. *J. Inherit. Metab. Dis.* 16, 520–22.

Worthen H. G., al Ashwal A., Ozand P. T., Garawi S., Rahbeeni Z., al Odaib A., Subramanyam S. B. and Rashed M. (1994) Comparative frequency and severity of hypoglycemia in selected organic acidemias, branched chain amino acidemia, and disorders of fructose metabolism. *Brain Dev.* 16 (Suppl.), 81–85.

Yalcinkaya C., Gibson K. M., Gunduz E., Kocer N., Ficicioglu C. and Kucukercan I. (2000) MRI findings in succinic semialdehyde dehydrogenase deficiency. *Neuropediatrics* 31, 45–46.

13 γ-Hydroxybutyrate and oxidative stress

M. Mamelak and D. Hyndman

Introduction

The extraordinary range of pharmacological effects of γ-hydroxybutyrate (GHB) has attracted scientific attention for more than three decades. Intraperitoneal doses of this simple 4-carbon fatty acid in excess of 200 mg/kg can induce sleep and anaesthetic states in small animals, dramatically raise brain dopamine levels, reduce cerebral deoxyglucose utilization and lower body temperature (Gessa *et al.* 1966; Roth and Suhr 1970; Lin *et al.* 1979; Kuschinsky *et al.* 1985; Mamelak 1989). The behavioural and electroencephalographic changes induced by GHB in small animals also appear to have certain features in common with *petit mal* epilepsy (Snead 1988). In humans, the unique anaesthetic properties of GHB have been used to advantage during labour and perhaps more widely, following head trauma to reduce cerebral edema (Vickers 1969; Strong 1984; Dabadie *et al.* 1986). Grace Kelly, the former Princess of Monaco, was treated with GHB soon after her arrival in hospital following her motor vehicle accident (Balaban-Quine 1989). Lower oral doses of GHB in man, in the range of 25–35 mg/kg, can stimulate the release of prolactin and growth hormone (Van Cauter *et al.* 1997). These low doses also can induce both REM sleep and slow wave sleep, and in contrast to hypnotic agents in common use, the effects of GHB on sleep do not change with time (Mamelak *et al.* 1986). GHB has been shown to inhibit lipid peroxidation and to prevent reperfusion injury and the tissue damage caused by radiation and high oxygen pressure (Laborit 1973; Lavyne *et al.* 1983; Meerson *et al.* 1983; Boyd *et al.* 1990). Indeed, GHB has been found to have tissue protective effects in animals and man in many different organs including brain, liver, lung, heart, kidney, gut and pancreatic beta cells (Dosmagambetova 1983; Lavyne *et al.* 1983; Boyarinov *et al.* 1984; Sholokov *et al.* 1986; Boyd *et al.* 1990; Pierrefiche *et al.* 1991; Sherman *et al.* 1994; Bouix *et al.* 1995; Yamasaki *et al.* 1999).

GHB: tissue source, distribution and turnover

The mechanism of action of these pharmacological doses of GHB is not known. Nor is it known how these actions relate to the natural function of GHB which occurs, widely distributed in very low concentrations, in many organ systems. GHB has been identified in brain, heart, kidney, liver, lung and skeletal muscle in pooled organ concentrations ranging from 1.42 nmol/g in liver to 28.4 nmol/g in kidney. Concentrations as high as 37.4 nmol/g have been found in brown fat (Nelson *et al.* 1981). GHB concentrations in specific regions of adult human brain range from 1.26 nmol/g in the cervical cord to 42.4 nmol/g in the hypothalamus. Foetal brain concentrations are much higher. For example, foetal cerebellar concentrations of GHB are as high as 90 nmol/g at 12–19 weeks of gestation. In rats, whole brain GHB levels decline by about 400% from 14 days before birth until 20 days after (Snead and Morley 1981).

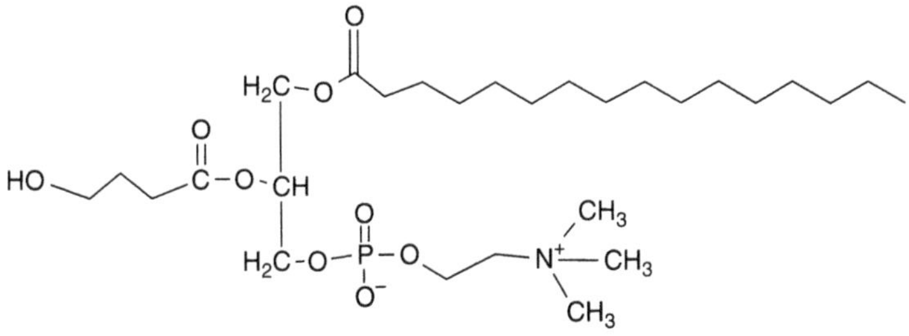

Figure 13.1 Phospholipid containing ester-linked GHB.

The high turnover rate of GHB in the brain, 0.44 h, gives credence to the possibility that GHB has a physiological role. This turnover rate is almost three times more rapid than that of whole brain serotonin, six to eight times more rapid than that of whole brain dopamine and thirteen to nineteen times more rapid than that of whole brain norepinephrine (Gold and Roth 1977). Children with 4-hydroxybutyric aciduria who cannot metabolize succinic semialdehyde, the primary catabolite of GHB, to succinic acid because they lack the enzyme succinic semialdehyde dehydrogenase, excrete as much as 2.5 mmol/l of GHB in their urine as well as smaller amounts of succinic semialdehyde and traces of beta oxidation products (Jakobs *et al.* 1981). It would be surprising for such large quantities of GHB to be produced without a key function in the cellular economy. It has been suggested that GHB acts as a neurotransmitter or neuromodulator but this begs the question of its function in peripheral tissues (Cash 1994; Maitre 1997).

The function of endogenous GHB in all tissues might be more obvious if its source was known. In brain, at least, studies have shown that GHB can be formed from γ-aminobutyric acid (GABA) (Roth and Giarman 1969; Gold and Roth 1977). However, tissue levels of GABA are vanishingly low in other organ systems yet these organs can have high GHB concentrations as well as concentrations of GHB catabolysing enzymes that are ten to twenty times higher than in brain (Barker *et al.* 1985). Even in brain, the ontogeny of GHB differs significantly from that of GABA (Snead and Morley 1981). This suggests that there is an alternative source of GHB in the periphery and perhaps as well in brain (Snead and Morley 1981; Barker *et al.* 1985). Vree *et al.* (1978) found that 1,6 hexanediol and its breakdown product, 6-hydroxyhexanoic acid, are metabolized to GHB in the monkey. This finding requires replication but suggests that GHB can be derived from long chain fatty acids. In this regard, there is also some evidence that lipid peroxidation, a recognized generator of hydroxy-fatty acids, may lead to the formation of a glycerophosphocholine with GHB as a constituent ester bound hydroxy-fatty acid (Freeman and Crapo 1982; Tanaka *et al.* 1993). Hydrolytic cleavage of the ester bond could release GHB (Fig. 13.1). GHB is also a natural constituent of sherry wines but, again, its source in these natural products has never been determined. There are, however, studies suggesting that certain anaerobic bacteria can reduce succinate to GHB (Webb and Kepner 1962; Scherf *et al.* 1994).

GHB catabolism

Recent work on the catabolism of GHB may provide a clue to its function in all tissues and, as well, draws attention back to the original proposal made by Henri Laborit that GHB activates the pentose phosphate pathway (PPP) (Laborit 1973). Kaufman and Nelson (1991) demonstrated that two enzymes, GHB keto-transhydrogenase and GHB dehydrogenase, can oxidize GHB to succinic semialdehyde which, as just described, is then converted to succinic acid and enters the citric acid cycle (Fig. 13.2). GHB keto-transhydrogenase is responsible for most of the oxidation of GHB in the adult animal. It is a mitochondrial enzyme that couples

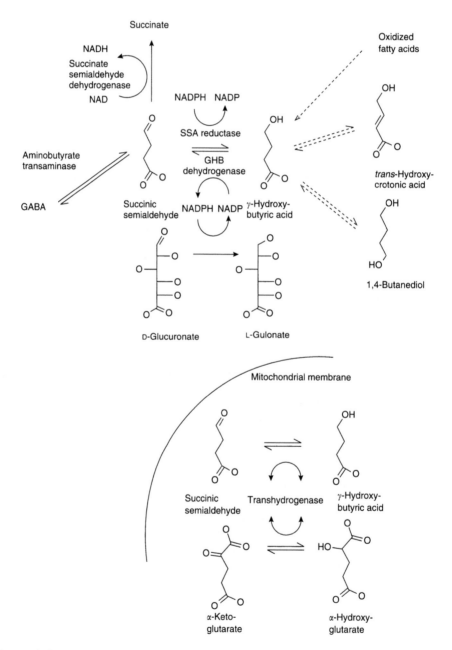

Figure 13.2 Cytosolic and mitochondrial GHB metabolism.

the oxidation of GHB to the reduction of α-ketoglutarate to form α-hydroxyglutarate, a metabolic intermediate in a variety of pathways. However, the full implication of the increase in the rate of production of α-hydroxyglutarate is not known (Van der Knaap *et al.* 1999). GHB keto-transhydrogenase, also, has not yet been well characterized. It may, however, be pertinent to note here that certain transhydrogenases may contribute to NADPH formation (Jackson *et al.* 1999). GHB dehydrogenase, on the other hand, is a cytosolic NADP-dependent oxido-reductase that is responsible for most of the GHB breakdown in

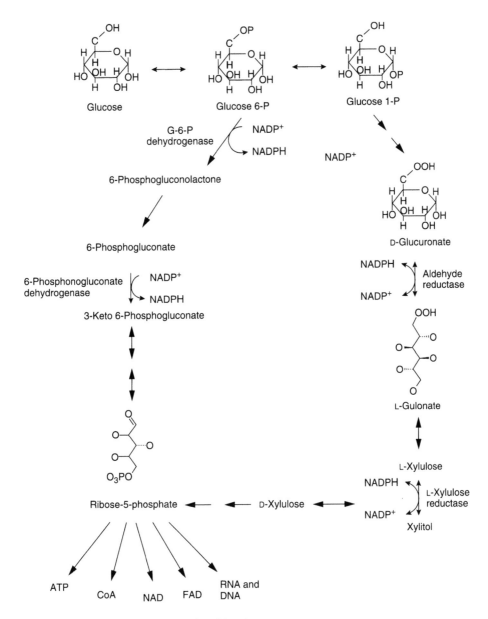

Figure 13.3 PPP and the auxiliary uronic acid pathway.

the foetus and young animal (Nelson and Kaufman 1994). It couples the NADPH generated by the oxidation of GHB to the reduction of glucuronic acid to gulonic acid, a precursor of ascorbic acid and of xylulose and other 5-carbon sugars. The formation of 5-carbon sugars from glucuronic acid serves as an auxiliary pentose shunt (Fig. 13.3) (Tulsiani and Touster 1979).

GHB dehydrogenase, known also as aldehyde reductase (ALR1), is a member of the NADPH-dependent family of aldo-keto reductases found in mammals, plant, yeast, protozoa and bacteria that metabolize a wide variety of carbonyl containing compounds. They are considered to be highly evolved detoxification catalysts (Cromlish and Flynn 1985; Flynn and Kubiseski 1997; Jez *et al.* 1997). GHB dehydrogenase is generally considered to act as a high K_m aldehyde reductase but its K_m for the oxidation of GHB falls four- to fivefold when this oxidation is coupled to the reduction of glucuronic acid (Fig. 13.2) (Kaufman and Nelson 1991). Aldose reductase (ALR2), another member of the aldo-ketose reductase family, functions as a low K_m aldehyde reductase. ALR1 has an exclusive requirement for NADPH while ALR2 can use NADH as a cofactor but far prefers NADPH (Morjana and Flynn 1989).

GHB: NADPH generation and tissue detoxification

The oxidation and catabolism of GHB may generate the NADPH required for the reducing and detoxifying actions of the aldehyde and aldose reductases as well as for the function of other detoxification systems like glutathione peroxidase. For example, osones, which are toxic products of cell metabolism, have been found to be good substrates for the NADPH-dependent aldehyde and aldose reductases (Feather *et al.* 1995). Osones, or carbohydrate 2-oxoaldehydes, are derived from glycated proteins and contain reactive ketone and aldehyde groups (Fig. 13.4). Glycation refers to the non-enzymatic attachment of glucose to certain amino acid residues, especially tyrosine and lysine, in proteins. Osones cause damage to cells and tissues by cross-linking proteins to form advanced glycation end products (AGEs). One of the manifestations of ageing, diabetes, Alzheimer's disease and other degenerative disorders is an increase in the tissue prevalance and rate of sugar-induced cross-linking of proteins (Feather *et al.* 1995). ALR2 also utilizes NADPH to eliminate the toxic aldehydes released by the breakdown of brain monoamines by converting them to their corresponding alcohols. The enzyme preferentially utilizes NADPH to reduce the aldehyde form of D-glucose to sorbitol in diabetes when glucose is present in high concentrations (Cromlish and Flynn 1985; Flynn and Kubiseski 1997). NADPH, acting in concert with glutathione peroxidase, is also required for the detoxification of major products of lipid peroxidation such as 4-hydroxynonenal and for the elimination of the reactive oxygen species hydrogen peroxide H_2O_2, an important cause of tissue oxidative stress (Schraufstatter *et al.* 1985; Canuto *et al.* 1994). Indeed, one manifestation of free radical stress is the depletion of NADPH required for the reduction of glutathione disulfide and the regeneration of glutathione (Schraufstatter *et al.* 1985)

Finally, it should be noted that the beta subunit of many voltage-dependent potassium channels is structurally related to the cytosolic aldo-keto reductases and contains $NADP^+$ in the binding cleft at its active site (Gulbis *et al.* 1999). It has been proposed that gating at the potassium channel may be mediated by a conformational change in the enzyme that occurs when $NADP^+$ is reduced to NADPH (Gulbis *et al.* 1999; Perez-Garcia *et al.* 1999). NADPH is known to produce a conformational change in both aldehyde and aldose reductases (Flynn and Kubiseski 1997; Jez *et al.* 1997).

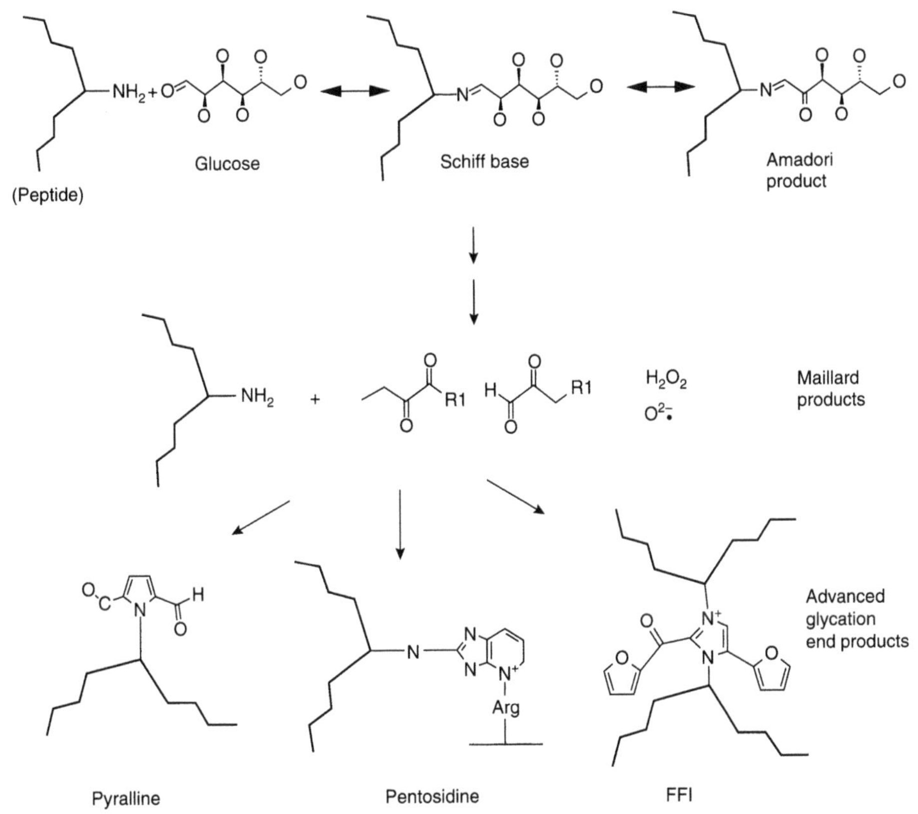

Figure 13.4 Formation of advanced glycation end products.

GHB: a promoter of the pentose PPP

Thus, it is of great interest that Taberner *et al.* (1972) demonstrated that the intraperitoneal administration of 500 mg/kg GHB in mice and rats does increase the rate of NADPH formation in whole brain by about 27% and that the time course of this increase correlates with the sleeping time or duration of anaesthesia. Taberner and colleagues also confirmed Laborit's earlier insight that GHB orients intermediary metabolism towards the PPP. Laborit was originally drawn to this conclusion when he found that GHB had a protective effect against radiation and high-pressure oxygen toxicity and he attributed this tissue protective effect to NADPH derived from the PPP (Laborit 1973). Taberner *et al.* (1972) showed that GHB increased the $1\text{-}^{14}C/6\text{-}^{14}C$ ratio in expired air in mice by 300%. An almost identical result was obtained in rat brain cortical slices in which the $1\text{-}^{14}C/6\text{-}^{14}C$ ratio increased from 1.72 to 3.63. A value of unity for the $1\text{-}^{14}C/6\text{-}^{14}C$ ratio is associated with the symmetric splitting of the glucose molecule in glycolysis. Values greater than one imply the participation of the pentose shunt. GHB did not alter this ratio in either homogenates of the cerebral cortex or slices of kidney and diaphragm. GHB did increase oxygen consumption by cerebral cortical slices as Ornellas and Laborit (1966), but not Taberner and colleagues (Taberner *et al.* 1972; Kerkut *et al.* 1972), had found in liver slices but it was not able to support respiration in cortical slices in the absence of glucose.

GHB has additional effects on carbohydrate metabolism that merit further study. A number of investigators have found that GHB causes a marked rise in brain glucose concentration and that this can occur without the induction of anaesthesia (Leonard and Watkinson 1971; Taberner 1973). Moreover, its anaesthetic actions are not associated with a particularly significant depression of glycolysis in contrast to the effects of other CNS depressants like the barbiturates. GHB does not alter or only slightly decreases brain lactate and fructose diphosphate levels. In addition, GHB does not alter blood glucose levels nor, according to Taberner (1973), does it depress glucose uptake by the brain. It does, however, appear to have powerful inhibitory effects on deoxyglucose utilization (Kuschinsky *et al.* 1985).

GHB and the functions of the pentose PPP

The known functions of the PPP may provide an insight into the metabolic role of GHB (Baquer *et al.* 1988). In this pathway, which is virtually irreversible because of the large change in free energy, NADPH and ribose 5-phosphate are generated when glucose 6-phosphate is oxidized (see Fig. 13.3):

Glucose 6-phosphate $+ 2NADP^+ + H_2O \rightarrow$ Ribose 5-phosphate $+ 2NADPH + 2H^+ + CO_2$.

NADPH, amongst its many roles, is most commonly associated with reductive biosynthesis. Large amounts of NADPH, for example, are consumed by adipose tissue in the reductive synthesis of fatty acids from acetyl-CoA. But NADPH is also involved in cholesterol biosynthesis, in the formation of nitric oxide from arginine and in the formation of glutamate and GABA from α-ketoglutarate (Baquer *et al.* 1988; Moncada *et al.* 1991). NADPH, as described earlier, is also required for the reduction of glutathione and for the detoxification of many potentially tissue damaging metabolites as well as for the elimination of xenobiotics through the microsomal P_{450} system (Schraufstatter *et al.* 1985; Flynn and Kubiseski 1997; Canuto *et al.* 1994; Maser 1995). Ribose 5-phosphate and its derivatives are required for the synthesis of key molecules such as ATP, coenzyme A, NAD^+, FAD, RNA and DNA that are critical intermediates and cofactors in energy metabolism and in protein and nucleic acid synthesis. Thus, the PPP through the formation of NADPH and ribose 5-phosphate appears to function in tissue detoxification, repair and biosynthesis.

Control of the PPP

PPP activity can be increased in fasted animals when hypertonic glucose is infused in the presence of insulin (Laborit 1973). The increase in pentose shunt activity is associated with the induction of sleep and slow waves on the EEG. Fasting decreases the $NADP^+/NADPH$ ratio and shunt activity increases in the presence of electron acceptors such as $NADP^+$ or phenazine methosulfate. Indeed, PPP activity appears to be increased by all conditions that increase NADPH utilization (Baquer *et al.* 1988). Some studies suggest that only 1–2% of brain glucose is metabolized through the PPP under basal conditions but this can be increased to 34.3% with phenazine methosulfate in mixed neuronal glial cell cultures and to 77% in pure glial cultures (Ben-Yoseph *et al.* 1996). Glia appear to possess significantly higher PPP reserve capacity than neurons. Oxidative stress and the formation of hydrogen peroxide stimulate the activity of the PPP and it is of some interest that glutathione peroxide and glutathione reductase are both largely concentrated in neuroglia, close to the source of NADPH they require for their detoxification functions (Ben-Yoseph *et al.* 1996). The utilization of NADPH to reduce the toxic aldehydes and hydrogen peroxide formed by the breakdown brain monoamines may account for the known capacity of these metabolites to drive the activity of the PPP (Hothersall *et al.* 1982). Mild hypothermia and hypoxia also promote glucose metabolism through the PPP (Baquer *et al.* 1988; Kaibara *et al.* 1999). The apparent age-related decrease in the activity of the PPP may be related to a decrease in cell turnover with age or to a decrease in the availability of NADP (Baquer *et al.* 1988; Wood 1986). The concentration and turnover of GHB in each tissue appears to correlate with the activity of the PPP. Hypoxia, for example, increases both tissue GHB levels and shunt activity (Baquer *et al.* 1988; Kaufman and Nelson 1991). GHB levels are high in foetal and adipose tissue where the PPP is also very active (Nelson *et al.* 1981; Snead and Morley 1981; Wood 1986; Baquer *et al.* 1988). A common factor, perhaps lipid peroxidation, if this is a source of GHB, may account for this correlation.

The pentose shunt is naturally more active in some organs than in others and, within organs, in certain cells rather than in others. Under ordinary circumstances, for example, it is more active in adipose tissue than in skeletal muscle (Baquer *et al.* 1988; Wood 1986). Neuroglia, as discussed earlier, have a greater capacity for pentose shunt activity than neuronal cells. Glucose 6-phosphate dehydrogenase activity is the rate-limiting step in the PPP (Zimmer 1996). NADPH inhibits this enzyme but, as discussed earlier, any process that increases NADPH utilization lifts this inhibition. As well, PPP activity can be increased by raising glucose 6-phosphate dehydrogenase levels. Exposure of neuroglial cells in tissue culture to lipopolysaccharides increases the expression of glucose 6-phosphate dehydrogenase mRNA and raises the level of this enzyme and the rate of glucose oxidation through the PPP (Garcia-Nogales *et al.* 1999). This has also been demonstrated in the heart in which the capacity of the PPP is limited by the very low activity of glucose 6-phosphate dehydrogenase. Exposing the heart to catecholamines increases the expression of glucose 6-phosphate dehydrogenase mRNA, raises tissue enzyme levels and increases glucose oxidation through this pathway (Zimmer 1996). The metabolic impact of GHB, then, may depend upon the cell's potential to metabolize glucose through the PPP and specifically upon the availability of $NADP^+$ and the activity of glucose 6-phosphate dehydrogenase.

Comparable actions of GHB and the PPP

The exact mechanism by which GHB activates the PPP remains to be determined. It is not clear how GHB can both increase the rate of NADPH formation and shift glucose metabolism to the PPP when NADPH (or a fall in the $NADP^+/NADPH$ ratio) is known to inhibit glucose 6-phosphate dehydrogenase. The increase in glucose metabolism through the pentose shunt with GHB is not observed in tissue homogenates and, thus, compartmentalization may be required to account for GHB's unexpected actions. Earlier studies on the fate of exogenously administered GHB, in fact, had suggested that it was selectively metabolized in the compartment associated with glial and synaptosomal structures as opposed to the compartment associated with neuronal structures (Mohler *et al.* 1976; Doherty and Roth 1978). More data is needed on the effects of GHB on intermediary metabolism in different tissues and under different circumstances. As well, the significance of GHB oxidation by transhydrogenase needs to be better understood.

Be that as it may, the generation of NADPH from GHB, as demonstrated by Taberner and colleagues, could account for many of the effects of GHB. For example, it would be useful to determine whether the widespread neuronal depression repeatedly demonstrated with GHB in the cortical and specifically in nigrostriatal neurons can be accounted for by an

NADPH linked conformational change in the K^+ channel that alters its conductivity, hyperpolarizes the cell membrane and perhaps especially the more sensitive neuroglial cell membrane and in this way reduces the responsiveness of the cell to external stimulation (Olpe and Koella 1978; Emri *et al.* 1996; Madden and Johnson 1998). This is in keeping with Laborit's proposal that slow wave sleep induced by GHB is caused by glial hyperpolarization, and that this in turn inhibits the propagation of signals along the axodentritic neuronal extensions that the glia envelop (Laborit 1973). Neurophysiologic data do indicate that GHB can hyperpolarize the cell membrane and that it does reduce the responsiveness of the cell to external stimuli (Madden and Johnson 1998). If these effects are mediated by a conformational change in the structural proteins of the K^+ channel, then GHB would at the same time also reduce the energy required to maintain the cell membrane potential and would permit the diversion of cell energy resources to other needs – perhaps to tissue repair and reconstruction (Mamelak 1997). The correlation between the duration of anaesthesia induced in rats by GHB and the rise and fall in the rate of brain NADPH generation suggests that NADPH may be intimately involved with the neuronal depression that mediates slow wave sleep and anaesthesia (Tabener *et al.* 1972). The PPP, too, promotes slow wave sleep. Both GHB and the PPP appear to cocoon the cell, protect it from environmental stimuli while simultaneously facilitating the removal of damaging metabolites and promoting the biosynthesis of ATP and other essential molecules and cofactors necessary for tissue viability.

It is tempting to propose that GHB owes its tissue protective effects to these metabolic capacities. The protective effects of GHB against reperfusion injury, well documented in brain and gut, may be cited as examples (Lavyne *et al.* 1983; Boyd *et al.* 1990). A compelling body of evidence suggests that tissue-damaging reactive oxygen species such as O^-, H_2O_2 and OH^- are generated when molecular oxygen is reintroduced into ischaemic tissues. Ischaemic tissues are not able to maintain requisite ATP levels and this in turn allows calcium to enter the cell and activate the production of free radicals and other reactive oxygen species (Freeman and Crapo 1982; McCord 1985). GHB strikingly protects the hamster gut from reperfusion injury even when it is given immediately before blood flow is restored after 30 min of complete ischaemia (Boyd *et al.* 1990). Similarly, γ-butyrolactone, the congener of GHB and its prodrug, substantially protects the rat brain against neuronal loss after 30 min of vessel occlusion even with intraperitoneal doses as low as 100 mg/kg which are not sufficient to alter behaviour (Lavyne *et al.* 1983). It is noteworthy that H_2O_2 rapidly activates the PPP and the glutathione redox cycle (Schraufstatter *et al.* 1985; Kashiwagi *et al.* 1996). Hypoxia also activates the PPP (Baquer *et al.* 1988). GHB, thus, appears to further enhance the activities of the PPP and strengthen innate cellular defences against tissue damage. In this regard, it has been shown that GHB prevents the fall in brain ATP levels that normally occur with hypoxia (MacMillan 1978).

GHB also has been shown to protect against diabetes in experimental animal models of the disease induced by streptozotocin and alloxan (Pierrefiche *et al.* 1991; Bouix *et al.* 1995). The streptozotocin molecule consists of glucose linked to a reactive nitrosourea moiety and it is internalized by the pancreatic beta cell glucose transporters. Pancreatic beta cells take up glucose more actively than other cells. Once inside the cell, the nitosourea moiety is released and actively damages the cell by free radical alkylation of the DNA. This process activates poly (ADP-ribose) synthetase which excises and repairs the alkylated sites but which, at the same time, decreases cell NAD^+ levels and places cell function in jeopardy. Streptozotocin also appears to inhibit the PPP. A combination of GHB and nicotinamide, the precursor of NAD^+, has been shown to protect against the development of diabetes (Pierrefiche *et al.* 1991). Neither one alone is as effective in the streptozotocin model of diabetes but GHB alone has been shown to protect against the development of alloxan-induced diabetes (Bouix *et al.* 1995). Alloxan is thought to damage pancreatic beta cells by generating reactive oxygen species. High doses of glucose that can activate the PPP also protect beta cells from damage by alloxan (Malaisse 1982). Thus, again, GHB appears to protect against reactive oxygen species damage by enhancing intrinsic tissue protective mechanisms.

Therapeutic applications of GHB

These experimental studies suggest that GHB may be therapeutically efficacious in man in conditions in which the persistent and excessive generation of reactive oxygen species is thought to play a major role. A large literature exists on this topic but, for purposes of discussion, the application of GHB in three disorders of ageing – diabetes, Parkinson's disease and Alzheimer's disease – will be considered.

Diabetes

Chronic hyperglycaemia appears to be the central initiating factor responsible for the development of all diabetes specific complications (Brownlee 1992). It is believed that glucose or glucose-derived metabolites irreversibly modify long-lived extracellular and/or intracellular macromolecules. The best understood example of such irreversible modification by sugars is the formation and accumulation of AGEs. In diabetes, AGE formation begins by the non-enzymatic glycation or attachment of glucose to certain amino acids, especially lysine, in proteins. In model systems, it has been shown that these Schiff base and Amadori products (Fig. 13.2) increase the rate of free radical production nearly fiftyfold compared with non-glycated protein (Mullarkey *et al.* 1990). Molecular rearrangements following the formation of Schiff bases and Amadori products lead to the formation of extremely reactive osone intermediates that react again with the amino groups in proteins to form AGEs. In diabetics, AGE formation alters the functional properties of several important matrix components such as type IV collagen, laminin and vitronectin and leads to the overproduction of basement membrane material. Matrix accumulation of AGEs may accelerate vascular occlusion and this process may be aggravated because nitric oxide, the endothelium-derived relaxing factor and antiproliferative factor, is quenched by AGEs in a dose-dependent fashion. Moreover, macrophages interact with AGE modified proteins and respond by secreting tumour necrosis factor, interleukin-1 and other agents that promote the proliferation of endothelial, mesangial and smooth muscle cells. AGEs also react with endothelial cells to induce procoagulatory factors on the surface of these cells. AGEs may also damage cell nuclear proteins and DNA. For all of these reasons, agents have been sought to inhibit AGE formation. One such agent, aminoguanidine, inhibits AGE formation by reacting with the intermediary osones. However, it is also known that osones can be reduced *in vivo* by the action of NADPH-dependent reducing enzymes (Feather *et al.* 1995). Directed studies are required to determine whether GHB can promote this process and inhibit the formation of AGEs through the generation of NADPH and whether it can as well increase nitric oxide synthesis in diabetes and improve vascular function.

Competition for NADPH in diabetes may be another reason for the increased formation of AGEs and other forms of free radical damage (Cheng and Gonzales 1986; Srivistava *et al.* 1989; Tomlinson *et al.* 1994). Excess glucose is converted to sorbitol by a reductive process mediated by aldose reductase and the preferential utilization of NADPH. This may exhaust cell NADPH and limit its availability for other key functions such as the reduction of glutathione, a vital step in the defence against oxidative stress. Reduced glutathione levels, for example, are low in diabetic lens and it has been proposed that oxidative stress, rather than the accumulation of sorbitol, leads to the development of cataracts in diabetes. However, no studies have been done to determine whether GHB can raise reduced gluta-thione levels and protect against the development of cataracts. Other antioxidants such as butylated hydroxytoluene are able to maintain normal ATP levels in the lens under experi-mental *in vitro* hyperglycaemic conditions designed to lower ATP levels and promote the development of cataracts (Win *et al.* 1994).

Parkinson's disease

Parkinson's disease is characterized by neuronal cell loss in the substantia nigra compacta and other brain stem nuclei coupled with the presence of intracellular neuronal inclusions such as Lewy bodies which accumulate very early in the course of the disease before it becomes symptomatic (Munch *et al.* 1998). There now is histochemical evidence that cross-linking by AGEs contributes to the functional loss and insolubility of the proteins present in these Lewy bodies (Castellani *et al.* 1996). Lewy bodies predominantly contain biochemically altered neurofilament protein. AGEs serve as markers for oxidative stress and provide evidence that this phenomenon is an important factor in the pathogenesis of Parkinson's disease and, as well, that the formation of inclusions from cytoskeletal proteins is linked to oxidative stress. As work on AGEs has progressed, it has become clear that these inclusions are not inert but rather exert multiple deleterious effects on cells particularly by their inter-action with cell surface receptors and intracellular signal transduction. Tissue culture studies demonstrate that the interaction of AGEs with glial cells generates reactive oxygen species with subsequent activation of transcription via NFkB leading to the increased formation of NFkB gene products such as interleukin-1B, interleukin-6 and tumour necrosis factor-alpha all of which are increased in Parkinson's disease (Mogi *et al.* 1994; Blum-Degen *et al.* 1996). NFkB has been shown to be upregulated in Parkinson's disease, and activated glial cells have been detected in the substantia nigra compacta in this condition particularly in regions that have undergone maximal neuronal degeneration (Hunot *et al.* 1997; Jenner and Olanow 1998). In cultured dopaminergic neurons, tumour necrosis factor-alpha receptor stimulation causes nuclear translocation of NFkB, transient free radical formation and apoptotic cell death. Thus, AGE formation appears to set in motion a feedback loop that leads to a vicious cycle of neurodegeneration.

A general increase in the level of oxidized proteins, that is, protein carbonyls, in the brains of patients with Parkinson's disease provides further evidence for oxidative stress (Alam *et al.* 1997). The highest concentration of these carbonyls is found in the substantia nigra compacta. Protein carbonyls are formed when reactive oxygen species attack amino acid residues on proteins to produce carbonyl groups (Yoritaka *et al.* 1996). These carbonyl groups can then react with other agents such as 4-hydroxynonenal to cause cross-linking of proteins and protein aggregation followed by AGE formation with its deleterious conse-quences. In Parkinson's disease, the finding of increased immuno-reactivity for 4-hydroxy-nonenal-linked proteins in the substantia nigra provides evidence for increased lipid peroxidation in this disease as well, and serves as an additional sign of oxidative stress (Yoritaka *et al.* 1996). Again, directed studies are required to determine whether GHB can strengthen tissue defence mechanisms against oxidative stress and retard the progress of Parkinson's disease.

Alzheimer's disease

A very similar proposal can be made for Alzheimer's disease. Indeed, some investigators propose that both Alzheimer's and Parkinson's diseases are really extremes of a spectrum of neurodegeneration and they cite the many clinical and neuropathological features in common (Perl *et al.* 1998). Since these features are also similar to those just described in diabetes, it may be more accurate to posit that tissue deterioration with age, in health and disease, is mediated by similar mechanisms, centrally and peripherally.

Alzheimer's disease is characterized by the progressive accumulation of amyloid plaques formed by the aggregation of the forty-two beta amyloid peptide and of neurofibrillary tangles formed by the aberrant phosphorylation of cell cytoskeletal elements such as the microtubule associated protein tau and the neurofilaments. Amyloid plaques are mainly composed of aggregated copies of the forty-two amino acid beta amyloid peptide derived by the cleavage of a larger amyloid peptide precursor protein. Aggregated but not monomeric beta amyloid peptide is actively neurotoxic and plaque numbers increase in tandem with neuronal degeneration and cognitive decline (Pike *et al.* 1991, 1993). AGE modification increases beta amyloid aggregation (Vitek *et al.* 1994). Recent work shows that purified plaque fractions from the brains of patients with Alzheimer's disease contain about threefold more AGE modifications per mg protein than do parallel fractions prepared from healthy age-matched controls. AGE formation in Alzheimer's disease, therefore, may accelerate plaque formation and increase their resistance to digestion and removal (Vitek *et al.* 1994).

Other work has shown that the paired helical filament tau in neurons is subject to non-enzymatic glycation and AGE formation (Smith *et al.* 1994; Yan *et al.* 1994). These neurons exhibit signs of oxidative stress such as the production of malondialdehyde epitopes and heme oxgenase-1-antigen (Yan *et al.* 1994). In cell tissue culture, paired helical filament tau isolated from post mortem tissue and recombinant AGE-tau each generate oxygen free radicals thereby not only activating transcription via NFkB, but also inducing the release of beta amyloid peptides (Smith *et al.* 1994; Yan *et al.* 1994, 1995).

Increased levels of carbonyl proteins and other products of oxidative stress have also been detected in Alzheimer's disease, the latter particularly in the region of neurofibrillary tangle bearing neurons (Hensley *et al.* 1995). Markers of lipid peroxidation such as malondialdehyde, isoprostanes and 4-hydroxynonenal also are significantly increased (Hensley *et al.* 1995; Markesbery and Ehmann 1999). Increased formation of reactive species in the brains of patients with Alzheimer's disease may account for the elevated levels of heat shock proteins that was reported by Hoyer (1993). Heat shock protein levels rise in response to tissue stress and their concentration is dramatically increased in Alzheimer's disease in regions adjacent to plaques and tangles (Hamos *et al.* 1991). Heat shock proteins appear to be synthesized preferentially by neuroglia rather than neurons. They are believed to promote cell protection and repair but they have also been shown capable of activating the amyloid precursor protein gene and thus may be contributing to the self-perpetuating cycle of plaque formation and tissue damage (Hoyer 1993). High levels of interleukin-1 in Alzheimer brain may also reflect the neuroglial activation in this disease (Mrak *et al.* 1995; Blum-Degen *et al.* 1996). Interleukin-1 can activate the gene that leads to the production of amyloid precursor protein (Rogers *et al.* 1999). An excellent summary of the causes and consequences of oxidative stress on protein, lipid and DNA metabolism in Alzheimer brain has recently been published (Markesbery *et al.* 1999). As in Parkinson's disease and diabetes, oxidative stress and AGE formation appear to be part of an ever accelerating self-perpetuating process of tissue destruction. GHB may help quench oxidative stress in Alzheimer's disease by reducing the rate of formation of reactive oxygen species and AGEs and by hyperpolarizing and deactivating neuroglia. In this way, GHB may restrain the vicious cycle of neurodegeneration.

In all clinical conditions in which oxidative stress is thought to play a major role and, indeed, even in the healthy elderly, GHB can be given at night to induce sleep, rest the brain and strengthen innate defences against oxidative damage. The release of growth hormone at night stimulated by GHB in synchrony with slow wave sleep may also promote lipolysis and produce salutary anabolic effects on muscle and bone (Van Cauter *et al.* 1997). Centrally and systemically, GHB, acting in concert with the pentose phosphate shunt, may provide cells with the reducing power required to remove toxic metabolites and xenobiotics, repair damaged molecules and assemble new structures. In this way, GHB may delay ageing and extend life.

Summary

Despite more than three decades of intensive effort, the tissue source, natural function and mechanism of action of GHB remain largely unknown. In this review, evidence has been presented that GHB may be an ongoing product of lipid peroxidation and that its release signals a metabolic response by pentose phosphate intermediates to contain this self-sustaining process and repair its damage. GHB shifts glucose metabolism towards the PPP and promotes the formation of NADPH, a key factor in reductive detoxification and biosynthesis. These metabolic capacities may account for GHB's anaesthetic, antioxidant and tissue protective properties and suggest that GHB may be used to advantage in conditions like normal ageing, diabetes, and Parkinson's and Alzheimer's diseases in which lipid peroxidation and other forms of oxidative stress are thought to play a major role. GHB may delay ageing and extend life.

Acknowledgement

I thank Susy O'Neill for her expert help with this manuscript.

References

Alam Z. I., Daniel S. E., Leas A. J., Marsden D. C., Jenner P. and Halliwell B. (1997) A generalised increase in protein carbonyls in the brain in Parkinson's disease but not incidental Lewy body disease. *J. Neurochem.* 69, 1326–29.

Balaban-Quine J. (1989) *The Bridesmaids: Grace Kelly, Princess of Monaco and Six Intimate Friends.* Weidenfeld and Nicolson, London.

Baquer N. Z., Hothersall J. S. and McLean P. (1988) Function and regulation of the pentose phosphate pathway in brain. *Current Topics Cell. Reg.* 29, 265–89.

Barker S. A., Snead O. C., Poldrugo F., Liu C.-C., Fish F. P. and Settine R. L. (1985) Identification and quantitation of 1,4 butanediol in mammalian tissues: an alternative biosynthetic pathway for gamma-hydroxybutyric acid. *Biochem. Pharmacol.* 34, 1849–52.

Ben-Yoseph O., Boxer P A. and Ross B. D. (1996) Assessment of the role of the glutathione and pentose phosphate pathways in the protection of primary cerebrocortical cultures from oxidative stress. *J. Neurochem.* 66, 2329–37.

Blum-Degen D., Muller T., Kuhn W., Gerlach M., Przuntek H. and Riederer P. (1996) Interleukin 1 beta and interleukin 6 are elevated in the cerebrospinal fluid of Alzheimer's and *de novo* Parkinson's disease patients. *Neurosci. Lett.* 202, 17–20.

Bouix O., Reynier M., Guintrand-Hugret R. and Orsette A. (1995) Protective effect of gamma-hydroxybutyrate and nicotinamide on low dose streptozotocin-induced diabetes in mice. *Horm. Metab. Res.* 27, 216–20.

Boyarinov G. A., Shvets N. A., Snopova L. B., Khvorov N. V. and Knov S. M. (1984) Effect of sodium hydroxybutyrate on macroergic phosphates, function and ultrastructure of the myocardium under blood loss. *Biull. Eksp. Biol. Med.* 97, 309–12.

Boyd A. J., Sherman I. A., Saibil F. G. and Mamelak M. (1990) The protective effect of gamma-hydroxybutyrate in regional intestinal ischemia in the hamster. *Gastroenterology* 99, 860–62.

Brownlee M. (1992) Glycation products and the pathogenesis of diabetic complications. *Diabetes Care* 15, 1835–43.

Canuto R. A., Ferro M., Muzio G., Bassi A. M., Leonarduzzi G., Maggiora M., Adamo D., Poli G. and Lindahl R. (1994) Role of aldehyde metabolizing enzymes in mediating effects of aldehyde products of lipid peroxidation in liver cells. *Carcinogenesis* 15, 1359–64.

Cash C. D. (1994) Gamma-hydroxybutyrate: an overview of the pros and cons for it being a neuro-transmitter and/or a useful therapeutic agent. *Neurosci. Biobehav. Rev.* 18, 291–304.

Castellani R., Smith M. A., Richey P. L. and Perry G. (1996) Glycoxidation and oxidative stress in Parkinson disease and diffuse Lewy body disease. *Brain Res.* 737, 195–200.

Cheng H. M. and Gonzalez R. G. (1986) The effect of high glucose and oxidative stress on lens metabolism, aldose reductase, and senile cataractogenesis. *Metabolism* 35 (4 Suppl. 1), 10–14.

Cromlish J. A. and Flynn T. G. (1985) Identification of pig brain aldehyde reductases with the high-K_m aldehyde reductase, the low-K_m aldehyde reductase and aldose reductase, carbonyl reductase, and succinic semialdehyde reductase. *J. Neurochem.* 44, 1485–993.

Dabadie P., Maurette P., Brule J. F., Kays C., Destandeau J., and Castel and Erny P. (1986) Control of intracranial pressure in traumatic brain injury. Our experience with gamma-hydroxybutyric acid or thiopental and fentanyl. In *Intracranial Pressure VI* (Miller J. D., Teasdale G. M., Rowan J. O., Galbraith S. L. and Mendelow A. D., eds), pp. 750–78. Springer-Verlag, Berlin, Heidelberg.

Doherty J. D. and Roth R. H. (1978) Metabolism of gamma-hydroxy-[1-^{14}C] butyrate by rat brain: relationship to the Krebs cycle and metabolic compartmentation of amino acids. *J. Neurochem.* 30, 1305–09.

Dosmagambetova R. S. (1983) Prevention of stress-induced disturbances of contractility of non-ischemic regions of the heart in myocardial infarction by gamma-hydroxybutyrate. *Biull. Eksp. Biol. Med.* 96, 28–30.

Emri Z., Antal K. and Crunelli V. (1996) Gamma-hydroxybutyric acid decreases thalamic sensory excitatory postsynaptic potentials by an action on presynaptic GABA$_B$ receptors. *Neurosci. Lett.* 216, 121–24.

Feather M. S., Flynn T. G., Munro K. A., Kubiseski T. J. and Walton D. J. (1995) Catalysis of reduction of carbohydrate 2-oxoaldehydes (osones) by mammalian aldose reductase and aldehyde reductase. *Biochim. Biophys. Acta* 1244, 10–16.

Flynn G. T. and Kubiseski T. J. (1997) Aldo-ketoreductases: structure, mechanism and function. In *Comprehensive Toxicology. Vol 3: Biotransformation.*(S:pes I. G. McQueen C. A. and Gandolfi A. J., eds), pp. 133–47. Elsevier Science, The Netherlands.

Freeman B. A. and Crapo J. D. (1982) Biology of disease: free radicals and tissue injury. *Lab. Invest.* 47, 412–42.

Garcia-Nogales P., Almeida A., Fernandez E., Medina J. M. and Bolanos J. P. (1999) Induction of glucose 6 phosphate dehydrogenase by lipopolysaccharide contributes to preventing nitric oxide-mediated glutathione depletion in cultured rat astrocytes. *J. Neurochem.* 72, 1750–58.

Gessa G., Vargie L., Crabai F., Odero G., Carboni F. and Camba R. (1966) Selective increase of brain dopamine induced by gamma-hydroxybutyrate. *Life Sci.* 5, 1921–30.

Gold B. I. and Roth R. H. (1977) Kinetics of *in vivo* conversion of gamma-[^{3}H]aminobutyric acid to gamma-[^{3}H]hydroxybutyric acid by rat brain. *J. Neurochem.* 28, 1069–73.

Gulbis J. M., Mann S. and MacKinnon R. (1999) Structure of a voltage-dependent K$^+$ channel beta subunit. *Cell* 97, 943–52.

Hamos J. E., Oblas B., Pulaski-Salo D., Welch W. J., Bole D. G. and Drachman D. A. (1991) Expression of heat shock proteins in Alzheimer's disease. *Neurology* 4, 345–50.

Hensley K., Hall N., Subramaniam R., Cole P., Harris M., Aksenov M., Aksenova M., Gabbita S. P., Wu J. F., Carney J. M., Lovell M., Markesbery W. R. and Butterfield D. A. (1995) Brain regional correspondence between Alzheimer's disease histopathology and biomarkers of protein oxidation. *J. Neurochem.* 65, 2146–56.

Hothersall J. S., Greenbaum A. L. and McLean P. (1982) The functional significance of the pentose phosphate pathway in synaptosomes: protection against peroxidative damage by catecholamines and oxidants. *J. Neurochem.* 39, 1325–32.

Hoyer S. (1993) Brain oxidative energy and related metabolism, neuronal stress, and Alzheimer's disease: a speculative synthesis. *J. Ger. Psychiatr. Neurol.* 6, 3–13.

Hunot S., Brugg B., Ricard D., Michel P. P., Muriel M.-P., Ruberg M., Faucheux B. A., Agid Y. and Hirsch E. C. (1997) Nuclear translocation of NFkB is increased in dopaminergic neurons of patients with Parkinson's disease. *Proc. Natl Acad. Sci. USA* 94, 7531–36.

Jackson J. B., Peake S. J. and White S. A. (1999) Structure and mechanism of proton translocating transhydrogenase. *FEBS Lett.* 464, 1–8.

Jakobs C., Bojasch M., Monch E., Rating D., Siemes H. and Haneford F. (1981) Urinary excretion of gamma-hydroxybutyric acid in a patient with neurological abnormalities. The probability of a new born error of metabolism. *Clin. Chim. Acta* 111, 169–78.

Jenner P. and Olanow C. W. (1998) Understanding cell death in Parkinson's disease. *Ann. Neurol.* 44, S72–S84.

Jez J., Bennett M. J., Schlegel B. P., Lewis M. and Penning T. M. (1997) Comparative anatomy of the aldo-keto reductase superfamily. *Biochem. J.* 326, 625–36.

Kaibara T. M., Sutherland G. R., Colburne F. and Tyson R. L. (1999) Hypothermia: depression of tricarboxylic acid cycle flux and evidence for pentose phosphate shunt upregulation. *J. Neurosurg.* 90, 339–47.

Kashiwagi A., Asahina T., Nishio Y., Ikebuchi M., Tanaka Y., Kikkawa R. and Shigeta Y. (1996) Glycation, oxidative stress and scavenger activity: glucose metabolism and radical scavenger dysfunction in endothelial cells. *Diabetes* 45 (Suppl. 3), S84–S86.

Kaufman E. E. and Nelson T. (1991) An overview of gamma-hydroxybutyrate catabolism: the role of the cytosolic $NADP^+$-dependent oxidoreductase EC 1.1.1.19 and of a mitochondrial hydroxy-acid–oxoacid transhydrogenase in the initial, rate limiting step in this pathway. *Neurochem. Res.* 16, 965–74.

Kerkut G. A., Rick J. T. and Taberner P. V. (1972) The effects of gamma-hydroxybutyric acid on brain respiration *in vitro*. *Br. J. Pharmacol.* 45, 174–75.

Kuschinsky W., Suda S. and Sokoloff L. (1985) Influence of the gamma-hydroxybutyrate on the relationship between local cerebral glucose utilization and local cerebral blood flow in the rat brain. *J. Cerebral Blood Flow Met.* 5, 58–64.

Laborit H. (1973) Gamma-hydroxybutyrate, succinic semialdehyde and sleep. *Prog. Neurobiol.* 1, 257–74.

Lavyne M., Hariri R., Tankosic T. and Babiak T. (1983) Effect of low dose gamma-butyrolactone therapy on forebrain neuronal ischemia in the unrestrained awake rat. *Neurosurgery* 12, 430–34.

Leonard B. E. and Watkinson W. D. (1971) Some effects of 4-hydroxybutyric acid on brain carbohydrate metabolism. *Life Sci.* 10 (II), 713–19.

Lin M. T., Chern Y. F., Wang H. S. and Chandra A. (1979) Effects of gamma-hydroxybutyric acid on metabolic, respiratory and vasomotor activities and body temperatures in rats. *J. Pharmacol. Exp. Ther.* 211, 167–70.

McCord J. M. (1985) Oxygen-derived free radicals in postischemic tissue injury. *New Eng. J. Med.* 312, 159–63.

Macmillian V. (1978) The effects of gamma-hydroxybutyrate and gamma-butyrolactone upon the energy metabolism of the normoxic and hypoxic rat brain. *Brain Res.* 146, 177–87.

Madden T. E. and Johnson S. W. (1998) Gamma-hydroxybutyrate is a GABA$_B$ receptor agonist that increases a potassium conductance in rat ventral tegmental dopamine neurons. *J. Pharmacol. Exp. Ther.* 287, 261–65.

Maitre M. (1997) The gamma-hydroxybutyrate signalling system in brain: organization and functional implications. *Progr. Neurobiol.* 51, 337–61.

Malaisse W. J. (1982) Alloxan toxicity to the pancreatic B cell. A new hypothesis. *Biochem. Pharmacol.* 31, 3527–34.

Mamelak M. (1989) Gammahydroxybutyrate: an endogenous regulator of energy metabolism. *Neurosci. Biobehav. Rev.* 13, 187–98.

Mamelak M. (1997) Neurodegeneration sleep and cerebral energy metabolism: a testable hypothesis. *J. Ger. Psychiatr. Neurol.* 10, 29–32.

Mamelak M., Scharf M. and Woods M. (1986) Treatment of narcolepsy with gamma-hydroxy-butyrate: a review of the clinical and sleep laboratory findings. *Sleep* 9, 285–89.

Markesbery W. R. and Ehmann W. D. (1999) Oxidative stress in Alzheimer disease. In *Alzheimer Disease*, 2nd edn (Terry R. D., Katzman R., Bick K. L. and Sisodia S. S., eds), pp. 401–14. Lippincott, Williams and Wilkins, Philadelphia.

Maser E. (1995) Xenobiotic carbonyl reduction and physiological steroid oxidoreduction: the pluripotency of several hydroxysteroid dehydrogenases. *Biochem. Pharmacol.* 49, 421–40.

Meerson F. Z., Kagan V. E., Prilipko L. L. and Rozhitskaya I. I. (1983) *The Failing Heart: Adaptation and Deadaptation*, pp. 67–127. Raven Press, New York.

Mogi M., Harada M., Kondo T., Riederer P., Inagaki H., Minami M. and Nagatsu T. (1994) Interleukin-1B, interleukin 6, epidermal growth factor and transforming growth factor are elevated in the brain of Parkinsonian patients. *Neurosci. Lett.* 180, 147–50.

Mohler H., Patel A. J. and Balazs R. (1976) Gamma-hydroxybutyrate degradation in the brain *in vivo*, negligible direct conversion to GABA. *J. Neurochem.* 27, 253–58.

Moncada S., Palmer R. M. J. and Higgs E. A. (1991) Nitric oxide: physiology, pathophysiology, and pharmacology. *Pharmacol. Rev.* 43, 109–42.

Morjana N. A. and Flynn T. G. (1989) Aldose reductase from human psoas muscle. *J. Biol. Chem.* 264, 2906–11.

Mrak R. E., Sheng J. G. and Griffin W. S. T. (1995) Glial cytokines in Alzheimer's disease: review and pathogenic implications. *Human Pathol.* 26, 816–23.

Mullarkey C. J., Edelstein D. and Brownlee M. (1990) Free radical generation by early glycation products: a mechanism for accelerated atherogenesis in diabetes. *Biochem. Biophys. Res. Commun.* 173, 932–39.

Munch G., Gerlach M., Sian J., Wong A. and Riederer P. (1998) Advanced glycation end products in neurodegeneration: more than early markers of oxidative stress? *Ann. Neurol.* 44 (Suppl. 1), S85–S88.

Nelson T. and Kaufman E. E. (1994) Developmental time courses in the brain and kidney of two enzymes that oxidize gamma-hydroxybutyrate. *Dev. Neurosci.* 16, 352–58.

Nelson T., Kaufman E., Kline J. and Sokoloff L. (1981) The extraneural distribution of gamma-hydroxybutyrate. *J. Neurochem.* 37, 1345–48.

Olpe H.-R. and Koella W. P. (1978) Inhibition of nigral and neocortical cells by gamma-hydroxy-butyrate: a microiontophoretic investigation. *Eur. J. Pharmacol.* 53, 359–64.

Ornellas M. R. and Laborit H. (1966) Influence du metabolisme du 4-hydroxybutyrate de sodium sur la consommation d'oxygene et le rapport lactate-pyruvate de coupes de foie de rat. *Agressologie* 7, 481–86.

Perez-Garcia M. T., Lopez-Lopez J. R. and Gonzalez C. (1999) KvB1.2 subunit coexpression in HEK293 cells confers O_2 sensitivity to Kv4.2 but not to Shaker channels. *J. Gen. Physiol.* 113, 897–907.

Perl D. P., Olanow W. and Calne D. (1998) Alzheimer's disease and Parkinson's disease: distinct entities or extremes of a spectrum of neurodegeneration. *Ann. Neurol.* 44 (Suppl. 1), S19–S31.

Pierrefiche G., Topall G., Henriet I. and Laborit H. (1991) Protective effects of gamma-hydroxy-butyrate on alloxan induced diabetes in mice. *Res. Com. Chem. Path. Pharm.* 71, 309–19.

Pike C., Walencewicz A., Glabe C. and Cotman C. (1991) Aggregation related toxicity of synthetic beta amyloid protein in hippocampal cultures. *Eur. J. Pharmacol.* 207, 367–68.

Pike C., Burdick D., Walencewicz A., Glabe C. and Cotman C. (1993) Neurodegeneration induced by beta amyloid peptides *in vitro*: the role of peptide assembly state. *J. Neurosci.* 13, 1676–87.

Rogers J. T., Leiter L. M., McPhee J., Cahill C. M., Zham S. S., Potter H. and Nilsson L. N. (1999) Translation of the Alzheimer amyloid percursor protein mRNA is upregulated by interleukin-1 through 5′-untranslated region sequences. *J. Biol. Chem.* 274, 6421–31.

Roth R. H. and Giarman N. J. (1969) Conversion *in vivo* of gamma-aminobutyric to gammahydroxy-butyric acid in the rat. *Biochem. Pharmacol.* 18, 247–50.

Roth R. H. and Suhr Y. (1970) Mechanism of the gamma-hydroxy-butyrate-induced increase in brain dopamine and its relationship to 'sleep'. *Biochem. Pharmacol.* 19, 3001–12.

Scherf V., Sohling B., Gottschalk G., Linder D. and Bucke W. (1994) Succinate-ethanol fermentation in *Clostridium kluyveri*: purification and characterisation of 4-hydroxybutyryl-CoA dehydratase/vinylacetyl-CoA Δ^3-Δ^2-isomerase. *Acta Microbiol.* 161, 239–45.

Schraufstatter I. U., Hinshaw D. B., Hyslop P. A., Spragg R. G. and Cochrane C. G. (1985) Glutathione cycle activity and pyridine nucleotide levels in oxidant-induced injury of cells. *J. Clin. Invest.* 76, 131–1139.

Sherman I. A., Saibil F. G. and Janossy T. I. (1994) Gamma-hydroxybutyrate mediated protection of liver function after long term hypothermic storage. *Transplantation* 57, 8–11.

Sholokov V. M., Baraniyazov Kh., Kizhaeva E. V., Lyubimov B. I., Samoilov N. N., Yudin G. V. and Eremin G. A. (1986) Effect of lithium oxybutyrate on the viability of preserved donor kidney. *Pharmakol. I. Toksik.* 49, 69–70.

Smith M. A., Taneda S., Richey, P. L., Miyata S., Yan S.-D., Stern D., Sayre L. M., Alzheimer Monnier V. M. and Perry G. (1994) Advanced Maillard reaction end products are associated with disease pathology. *Proc. Natl Acad. Sci. USA* 91, 5710–14.

Snead O. C. (1988) Gamma-hydroxybutyrate model of generalized absence seizures: further characterization and comparison with other absence models. *Epilepsia* 29, 361–68.

Snead O. C. and Morley B. J. (1981) Ontogeny of gamma-hydroxybutyric acid. I. Regional concentration in developing rat, monkey and human brain. *Dev. Brain Res.* 1, 579–89.

Srivistava S. K., Ansari N. H., Liu S., Izban A., Das B., Szabo G. and Bhatnagar A. (1989) The effects of oxidants on biomembranes and cellular metabolism. *Mol. Cell. Biochem.* 91, 149–57.

Strong A. J. (1984) Gamma-hydroxybutyric acid and intracranial pressure. *Lancet* 1 (8389), 1304.

Stryer L. (1988) *Biochemistry*, 3rd edn, pp. 495–516. W.H. Freeman and Company, New York.

Taberner P. V. (1973) Effects of gamma-hydroxybutyric acid and other hypnotics on glucose uptake *in vivo* and *in vitro*. *J. Neurochem.* 20, 669–80.

Taberner P. V., Rick J. T. and Kerkut G. A. (1972) The action of gamma-hydroxybutyric acid on cerebral glucose metabolism. *J. Neurochem.* 19, 245–54.

Tanaka T., Minamino H., Unezaki S., Tsukatani H. and Tokumura A. (1993) Formation of platelet-activating factor-like phospholipids by Fe^{2+}/ascorbate/EDTA-induced lipid peroxidation. *Biochim. Biophys. Acta* 1166, 264–74.

Tomlinson D. R., Stevens E. J. and Diemel L. T. (1994) Aldose reductase inhibitors and their potential for the treatment of diabetic complications. *Trends Pharmacol. Sci.* 15, 293–97.

Tulsiani D. R. and Touster O. (1979) Studies on dehydrogenases of the glucuronate-xylulose cycle in the livers of diabetic mice and rats. *Diabetes* 28, 793–98.

Van Cauter E., Plat L., Scharf M. B., Leproult R., Cespedes S., L'Hermite-Baleriaux M. and Copinschi G. (1997) Simultaneous stimulation of slow-wave sleep and growth hormone secretion by gamma-hydroxybutyrate in normal young men. *J. Clin. Invest.* 100, 745–53.

Van der Knaap M. S., Jakob C., Hoffman G. F., Nyham W. L., Renier W. O., Smeitink J. A. M., Catsman-Berrevoets C. E., Hjalmarson O., Vallance H., Sugita K., Bowe C. M., Herrin J. T., Craigen W. J., Buist N. R. M., Brookfield D. S. K. and Chalmers R. A. (1999) D-2-Hydroxyglutaric aciduria: biochemical marker or clinical disease entity. *Ann. Neurol.* 45, 111–19.

Vickers M. D. (1969) Gammahydroxybutyric acid. *Int. Anesthesiol. Clin.* 7, 75–89.

Vitek M. P., Bhattacharya K., Glendening J. M., Stopa E., Vlassara H., Bucala R., Manogue K. and Cerami A. (1994) Advanced glycation end products contribute to amyloidosis in Alzheimer disease. *Proc. Natl Acad. Sci. USA* 91, 4766–70.

Vree T. B., Damsma J., Van Den Bogert A. G. and Van Der Kleijn E. (1978) Pharmacokinetics of 4-hydroxybutyric acid in man, rhesus monkey and dog. In *Anaestheiologic and Intensivmedizin*, pp. 21–38. Springer, Berlin, New York, Heidelberg.

Webb A. D. and Kepner R. E. (1962) The aroma of flor sherry. *Am. J. Enol. Viticult.* 13, 1–14.

Win T., Asano G. and Shimizu Y. (1994) The efficacy of aldose reductase inhibitor, antioxidant butylated hydroxytoluene and cysteine protease inhibitor E64 on hyperglycemia-induced metabolic changes in organ cultured bovine lens. *Nippon Ika Daigaku Zasshi. J. Nippon Med. School* 61, 180–89.

Wood T. (1986) Distribution of the pentose phosphate pathway in living organisms. *Cell Biochem. Funct.* 4, 235–40.

Yamasaki N., Oka T., Yamamoto S., Nagayasu T., Akamine S., Takahashi T. and Ayabe H. (1999) Twenty-four-hour preservation in gamma-hydroxybutyrate improves lung function in canine single lung allotransplantation. *Transplantation* 67, 529–33.

Yan S.-D., Chen X., Schmidt A. M., Brett J., Godman G., Zou Y.-S., Scott C. W., Caputo C., Frappier T., Smith M. A., Perry G., Yen S. H. and Stern D. (1994) Glycated tau protein in Alzheimer disease: a mechanism for induction of oxidant stress. *Proc. Natl. Acad. Sci. USA* 91, 7787–91.

Yan S.-D., Yan S.-F., Chen X., Fu J., Chen M., Kuppusamy P., Smith M. A., Perry G., Godman G. C., Nawroth P., Zweier J. L. and Stern D. (1995) Non-enzymatically glycated tau in Alzheimer's disease induces neuronal oxidant stress resulting in cytokine gene expression and release of amyloid beta-peptide. *Natl Med.* 1, 693–99.

Yoritaka A., Hattori N., Uchida K., Tanaka M., Stadtman E. R. and Mizuno Y. (1996) Immunohistochemical detection of 4-hydroxynonenal protein adducts in Parkinson disease. *Proc. Natl. Acad. Sci. USA* 93, 2696–701.

Zimmer H.-G. (1996) Regulation of and intervention into the oxidative pentose phosphate pathway and adenine nucleotide metabolism in the heart. *Mol. Cell. Biochem.* 160–161, 101–09.

14 The role of γ-hydroxybutyrate in brain function

M. Maitre, V. Kemmel, C. Andriamampandry,
S. Gobaille and D. Aunis

Introduction

Originally, γ-hydroxybutyrate (GHB) was synthesized about forty years ago in the hope of obtaining a GABA-like substance, able to pass freely through the blood–brain barrier (Laborit 1964). In initial studies, the pharmacological properties of this compound seemed to mimic a general GABAergic effect, despite the fact that an enormous amount of exogenous GHB (several grams) was necessary to induce sedation or anesthesia in humans (Laborit 1973). Later on, the physiological presence of micromolar concentrations of GHB in the brain of several mammalian species was demonstrated (Roth and Giarman 1970). The problem of the functional role of GHB was not addressed at first because interest was mainly focused on its pharmacological and therapeutic use. Until recently, the mechanisms supporting these effects were poorly understood but data have now been accumulated that favor a neuromodulator/neuroregulator role for endogenous GHB (Vayer *et al.* 1987b). Modifications of the plasticity of this system by massive exogenous administration of GHB have shed some light on the molecular pharmacology of these neuromodulatory pathways.

The physiological presence and distribution of GHB in brain

Like many other substances involved in neurotransmission processes (acetylcholine, GABA, glutamate for example), GHB is present outside the brain (Doherty *et al.* 1978; Nelson *et al.* 1981). But the fact of its not being present exclusively in brain does not preclude its having an important role in neuronal functions. In brain, GHB is the product of the reductive route of GABA metabolism, which is a minor pathway, compared to oxidation to succinic acid (Roth 1970). The control of GHB synthesis is strict because only 1–2% of mitochondrial succinic semialdehyde (SSA) is reduced to GHB by the action of succinic semialdehyde reductase (SSR), a cytosolic enzyme present only in neurons (Doherty *et al.* 1975; Gold and Roth 1977). This organization has several consequences. First, the amount of GHB synthesized in the neuronal cytosol could depend on the intensity of GABA metabolism and of GABAergic synapses activity leading to SSA production in the mitochondria. Second, the transport of SSA between the mitochondria and the cytosol could also be a controlled mechanism. Finally, a regulation of SSR activity most probably exists in the neuron. This protein, which has been purified from rat and human brain, is a Michaelian enzyme with a K_m for SSA of about 20 μM (Cash *et al.* 1979; Rumigny *et al.* 1980, 1981). However, after cloning the first SSR protein from a hippocampal cDNA library (Andriamampandry *et al.* 1998), several pieces of evidence suggested the existence of more than one SSR at the level of both

Western and Northern blot. The cloned SSR is not expressed in kidney or liver and possesses several potential phosphorylation sites for protein kinase C, casein kinase II and tyrosine kinase. Thus a possible regulation of SSR activity in brain could be exerted through different mechanisms.

Regional GHB concentration in brain is heterogeneous, ranging from about 1 to 10 μM (Vayer *et al.* 1988), and at the cellular level, the synaptosomal and the cytosolic compartments appear to be the richest (Snead 1987). A rapid accumulation of [^{3}H]GHB, potentiated by the presence of chloride, sodium and potassium ions, has been demonstrated in both vesicles made from synaptosomal membranes and in synaptosomes (Benavides *et al.* 1982b; McCormick and Tunnicliff 1998). These results favor the existence of a concentration gradient from synaptic endings to extracellular spaces. But no direct measurement of extra-cellular brain concentration of GHB has been made so far and it is difficult to speculate about the actual GHB concentration in the synaptic cleft. Fifty to eighty femto moles of GHB are released from a milligram of brain slices submitted to a strong depolarizing stimu-lus (Maitre *et al.* 1983; Vayer and Maitre 1988), indicating that the GHB concentration in the whole extracellular compartment of the brain rise by about 0.5 μM within a minute under these conditions. About 80% of this amount of GHB are released via a calcium-dependent phenomenon which is inhibited by the presence of tetrodotoxin. However, it is an *in vitro* situation with hypoxic brain slices preloaded with [^{3}H]GHB. Up to now, *in vivo* microdialysis on awake animals have never been performed because of the difficulty of measuring low level of GHB in the dialysis medium. But this approach could constitute a method of deter-mining the real GHB concentration in the extracellular space *in vivo*, after electrical-induced depolarization for example.

GHB receptors and the physiological versus pharmacological GHB-induced neuronal response

Even though there is still sometimes confusion between GHB and GABA receptors, almost all criteria needed to define a family of specific receptors for GHB in brain have now been met. The high affinity GHB binding site existing on neuronal brain membranes fulfills all criteria for representing a true receptor site. Kinetic parameters of binding are usually in favor of the existence of two populations of sites (K_m values of about 30–250 nM and 2–16 μM), but under some conditions (membranes washed with low concentrations of detergent), a single population of site is detected with a K_m of about 1.5 μM (Benavides *et al.* 1982a; Hechler *et al.* 1990; Maitre *et al.* 1994). These K_d values are in the range of endoge-nous concentrations of GHB in brain. In addition, the binding of GHB and of GHB-related analogues are dependent on the protein concentration, on the pH of the incubation medium and are stereoselective.

The GHB binding sites are expressed only in particular structures of the brain and in neuronal cells (neurons in primary cultures and neuronal cell lines like NCB-20 cells) (Kemmel *et al.* 1998). Cells of glial origin appear to be devoid of binding sites (astrocytes in primary cultures, C_6 cell line).

In the rat and human brain, the regions which express a high density of receptors are almost exclusively located in the rostral part of the brain (cortex, hippocampus, striatum, thalamus, olfactory tract, some dopaminergic nuclei) but the cerebellum contains only a few sites which are absent from the pons-medulla (Hechler *et al.* 1992; Snead and Liu 1984). This distribution is quite different from those existing for other neurotransmitter receptors, espe-cially GABA receptors (GABA$_A$ and GABA$_B$). As studied by Snead (1994), the developmen-tal pattern of GHB binding sites in the rat brain differs completely from that of GABA$_B$

receptors. In addition, several GHB structural analogues are high affinity ligands for the GHB binding sites but do not displace GABA or baclofen from GABA$_A$ and/or GABA$_B$ sites. Thus it appears that GHB high affinity sites are specifically defined from the kinetic, developmental and pharmacological points of view and they also possess a specific distribution pattern in brain.

However, identification of specific neuronal responses induced by GHB is hampered by the fact that this substance is rapidly transformed into GABA both *in vivo* and *in vitro* (Vayer *et al.* 1985a; Gobaille *et al.* 1999). This transformation is due to the sequential effect of GHB dehydrogenase (GHB-DH) which converts GHB into SSA and then SSA is transaminated by GABA transaminase (GABA-T) (Vayer *et al.* 1985b). Local applications of GHB to neuronal tissue *in vivo* or to tissue slices or sometimes to neuronal cells in cultures often lead to ambiguous results which are a mixture of both a GHB-specific response mediated by GHB receptors and a GABAergic stimulation due to GABA$_B$ and/or GABA$_A$ receptors. However, for low doses of GHB (micromolar range), a specific signal can usually be recorded which is reduced or blocked by the specific GHB receptor antagonist NCS-382 (Ostrovskaya 1982; Kozhechkin 1980; Olpe and Koella 1979; Osorio and Davidoff 1979; Harris *et al.* 1989; Godbout *et al.* 1995; Maitre *et al.* 1990; Kemmel *et al.* 1998). At higher doses of GHB (100–1,000 µM range), most studies have reported a GABA-type response (mainly GABA$_B$) which is blocked or attenuated by GABA$_B$ antagonists (usually CGP 35348 or CGP 55845)

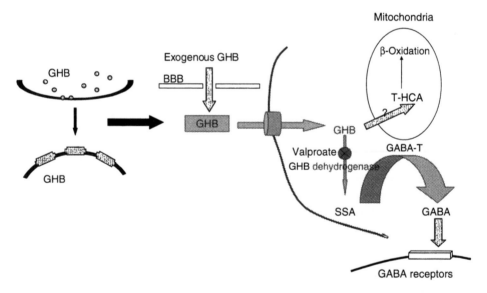

Figure 14.1 GHB catabolism in brain. GHB extracted from the extracellular space or obtained after peripheral administration through the blood–brain barrier is transported by a high-affinity uptake to neighboring cells. These cells contain a cytosolic GHB dehydrogenase which is inhibited by antiepileptic drugs (valproate, ethosuximide, barbiturates) and which transforms GHB into SSA. This SSA is a substrate for mitochondrial GABA-T which produces GABA. This GABA pool derived from the metabolism of GHB can be released to stimulate GABA$_B$ receptors. *In vitro*, the presence of valproate or amino-oxyacetic acid (a GABA-T inhibitor) prevents GHB from displacing [^{3}H]GABA from GABA receptors. An alternative pathway for GHB degradation could be a β-oxidation in the mitochondria leading to the formation of the natural product *trans*-hydroxycrotonate and Krebs cycle intermediates.

but also sometimes by bicuculline (Waldmeier 1991; Xie and Smart 1992; Engberg and Nissbrandt 1993). These results led several authors to propose that GHB is a GABAergic substance, acting mainly via $GABA_B$ receptors. This idea was reinforced by the finding that GHB displaces $GABA_B$ binding *in vitro* but with IC_{50} ranging from 100 to about 800 μM (Bernasconi *et al.* 1992; Ishige *et al.* 1996; Ito *et al.* 1995). This heterogeneity favored the hypothesis that GHB, under the conditions of $GABA_B$ binding *in vitro*, is able to produce variable but significant amounts of GABA which displace $GABA_B$ and/or $GABA_A$ binding.

This mechanism for the GHB effect on GABA receptors *in vitro* and *in vivo* was supported by experiments *in vitro* in which GHB-DH or GABA-T was inhibited, thus blocking the formation of GABA from GHB. Sodium valproate was used as GHB-DH inhibitor and amino-oxyacetic acid as GABA-T inhibitor. In the presence of one of these two inhibitors, the formation of GABA from GHB is greatly reduced. Under these conditions, GHB possesses a very high IC_{50} (millimolar range) for displacing radioactive GABA from its binding sites, indicating that GHB itself is not a ligand for these receptors, at least not *in vitro* (Hechler *et al.* 1997). *In vivo*, the local application or the systemic administration of GHB leads to the rapid formation of GABA whose effects could be blocked by GABA receptor antagonists (CGP 55845 or bicuculline), giving support to the idea that GHB itself is in fact a GABA mimetic. According to *in vitro* experiments, GHB is only the precursor of a GABA pool implicated in GABA receptors stimulation but has no direct affinity for these receptors (Fig. 14.1).

Physiological and biochemical evidence for the existence of specific neuronal events induced by the stimulation of GHB receptors

Kinetic parameters for GHB binding on rat or human brain membranes are in favor of a high-affinity (K_d between 30 and 250 nM) and a low-affinity component (K_d between 1 and 16 μM). These two populations of sites could represent different molecular environments and/or modifications of the same basic population of receptors. In NCB-20 neurons, the cAMP-induced differentiation modifies the K_d for GHB binding sites from 250 to about 1 μM (Kemmel *et al.* 1998). However, these affinities for GHB are in accordance with the endogenous brain GHB concentrations (generally between 1 and 10–20 μM in the rat brain). Thus, the physiological or biochemical experiments which are designed to explore the consequences of GHB receptor stimulation do not need GHB concentrations above 10–20 μM.

Under these conditions, GHB induces cell stimulation of its own, both *in vivo* and *in vitro*, which is selectively reduced by the GHB receptor antagonist NCS-382 and not by $GABA_B$ receptor antagonists or by bicuculline. These effects include modifications of calcium and potassium ion permeability via a mechanism which implicates G proteins. If the neuron in culture or the synaptosome or the nervous tissue slice under study is depolarized by voltage clamp or partial metabolic anoxia, the effect of GHB is mainly a reduction of calcium entry into the cell (Kemmel *et al.* 1998). This result has been confirmed by the use of patch-clamp experiments carried out on NCB-20 cells or by monitoring the calcium-dependent NOS activity (Cash *et al.* 1999). By contrast, if neurons are metabolically active and not stimulated (in cell culture or in tissue slices for example), GHB induces an increase in calcium ion utilization (results to be published). This increase seems to stimulate potassium ion extrusion via the potentiation of calcium-dependent potassium channels, leading to membrane hyperpolarization. These modifications are not inhibited by $GABA_B$ receptor antagonists.

When tested on neuronal cultures or on brain slices, the biochemical modification induced by GHB receptor stimulation suggests a modulation of the Ca^{2+}/NOS/cGMP cascade, in particular in the hippocampus and the frontal cortex (Vayer *et al.* 1987a). The adenylate cyclase/cAMP system seems not to be involved among the second messengers solicited by GHB receptors. Finally, no modification in the inositol phosphate turnover has been found when NCB-20 neurons are stimulated by physiological amounts of GHB.

The global effect of micromolar concentration of GHB in several experimental models is a membrane hyperpolarization, probably via GHB receptors located mainly at the presynaptic level. The consequences of this modification have been studied so far on dopamine and GABA release *in vivo*. Other classical neurotransmitters have not been precisely investigated. Local or peripheral administration of low doses of GHB (not exceeding 250 μM in brain tissue, that is no more than 300 mg/kg GHB i.p.) usually induce a decrease in dopamine and GABA release. The amplitude of this reduction is variable (about 50–100% basal levels) and lasts about 60–120 min. These results have been obtained in striatum or frontal cortex for dopamine and in thalamus and frontal cortex for GABA (Hechler *et al.* 1991; Banerjee and Snead 1995). The problem of whether there is a causal link between the modification of GABA release and that of dopamine remains an open question.

Pharmacological mode of action of GHB

The interaction with GABA$_B$ receptors

Large doses of GHB must be administered in order to observe pharmacological effects following GHB administration in animals and in man. In rats, exogenous GHB penetrates the brain very rapidly and the brain level of GHB peaks a few minutes after i.p. administration (Shumate and Snead 1979; Lettieri and Fung 1979). Doses of 300–1,000 mg/kg are used in rats and in man, 2–3 g GHB must be administered every 4–6 h in order to obtain sedation, sleep and anesthesia (Mamelak *et al.* 1986; Scharf *et al.* 1988). High doses of GHB must also be used in order to induce psychological benefits (Kam and Yoong 1998) or for the treatment of alcohol withdrawal (Biggio *et al.* 1992; Ferrara *et al.* 1992). In rats, these high doses of GHB necessary to induce neuropharmacological effects increase the brain GHB concentration well above 100 μM. These concentrations are considerably higher than the K_d of GHB receptors for their ligand. Thus the question of the role of GHB receptors in mediating the effect of pharmacological doses of GHB has not been resolved.

To answer this question, it can be first postulated that GHB interacts with a low affinity with another class of receptor. Several authors have reported that GHB can displace GABA from the GABA$_B$ receptor with an IC$_{50}$ ranging from 100 to about 800 μM. *In vitro* and *in vivo* studies have shown that the electrophysiological effects of high doses of GHB can be blocked by GABA$_B$ receptor antagonists or sometimes by bicuculline. These results are apparently in favor of an effect of GHB mediated by GABA$_B$ receptors when given at high doses. However, for reasons already discussed above, this effect of GHB is most probably indirect. *In vitro* blockade of GHB conversion into GABA via the inhibition of GHB-DH by valproate or via the inhibition of GABA-T by amino-oxyacetic acid induces the almost complete disappearance of GHB effect on GABA$_B$ binding. *In vivo* and *in vitro*, GHB produces GABA more or less rapidly, depending on the conditions of incubation and the amount of GHB. This variable amount of GABA produced can explain the large range of IC$_{50}$ for GHB on GABA$_B$ binding.

The direct interaction of GHB on GABA$_B$ receptors has been recently re-evaluated after the cloning and expression of GABA$_B$ R$_1$A and GABA$_B$ R$_1$B isoforms (Lingenhoehl *et al.* 1999). Millimolar concentrations of GHB are needed to obtain a significant effect.

Furthermore, GHB has an IC_{50} of about 3 mM in binding experiments with native $GABA_B$ receptors when using an antagonist as radiolabel (Mathivet *et al.* 1997). These millimolar concentrations of GHB in brain are not reached after peripheral administration of GHB at therapeutic or recreative doses.

In addition to these difficulties for explaining the neuropharmacological effects of GHB through a direct interaction at $GABA_B$ receptors, we should recall that the effects of high doses of GHB on dopamine and GABA release are hardly explained via an agonistic effects at $GABA_B$ receptors. When administered at high doses in rats (equal or greater than 500 mg/kg), GHB induces a large increase in dopamine and GABA release. This increase is generally observed with a latency of 10–20 min after GHB administration and has been investigated by *in vivo* microdialysis of awake rats, in frontal cortex and striatum for dopamine and in frontal cortex and thalamus for GABA (Hechler *et al.* 1991; Banerjee and Snead 1995; Gobaille *et al.* 1999). Thus, at low doses of GHB (equal or less than 350 mg/kg), GHB decreases dopamine and GABA release in brain while at higher doses, an increase is observed. These high GHB doses probably represent what happens at pharmacological concentrations of GHB while the low doses reproduce the physiological tone of endogenous GHB levels. The fact that high doses of GHB increase the extracellular levels of dopamine and GABA instead of decreasing it provides some basis for the mechanistic explanation of the recreative and addictive effects of GHB in animals and human. It is difficult to imagine a psychological benefit derived from GHB administration if this drug decreases GABA and dopamine in brain, considering the central role of dopamine and GABA in the mechanism of drug addiction. By contrast, a GHB-induced increase in extracellular GABA and dopamine levels could represent the basis for the mechanism of GHB-induced sedation, anxiolysis, anesthesia and state of well-being which are described by GHB consumers. It is well known that $GABA_B$ agonists like baclofen reduce dopamine and GABA release in several brain regions (Waldmeier and Baumann 1990) and thus an agonistic action of high concentrations of GHB at $GABA_B$ receptors is difficult to reconcile with the GHB-induced increase in dopamine and GABA release.

The desensitization of GHB receptors

A large majority of receptors coupled to G proteins have been reported to be desensitized and down-regulated upon hyperstimulation. GHB receptors belong to this family of receptors (Ratomponirina *et al.* 1995) although no regulatory process by phosphorylation/dephosphorylation has been described so far for them. However, modulation of Ca^{2+} conductance induced by the presence of 5–10 μM GHB in cortical slices has been shown to be dissipated after prolonged incubation with 50–100 μM GHB (Cash *et al.* 1999). In NCB-20 cell culture, micromolar amounts of GHB modify potassium ion permeability through GHB receptors, which is blocked by the antagonist NCS-382. This effect disappears after a long pre-incubation of the cells with micromolar amounts of GHB or after a short but massive stimulation with 100–200 μM GHB (results to be published). *In vivo*, the chronic treatment of rats by intracerebroventricular microinfusion of GHB induces a down-regulation of GHB receptors. The same effect could be observed in NCB-20 cells in culture, maintained 24 h in contact with 1 mM GHB. The washout of GHB restores the initial B_{max} of GHB receptors (Ratomponirina *et al.* 1998).

Thus, it can be suggested that the hyperstimulation of brain GHB receptors *in vivo* by high concentrations of exogenous GHB led to a desensitization of GHB receptors after a time lapse of 10–20 min. This phenomenon eliminates the tonic inhibitory control of GHB

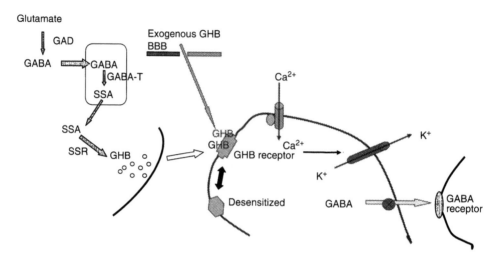

Figure 14.2 Proposed balance of GHB receptors between a sensitized/desensitized state, controlling the release of GABA by a GABAergic synapse. Under physiological conditions, GHB receptors stimulated by GHB released at GHBergic synapses induce a membrane hyperpolarization of the GABA neurons through modulation of calcium and potassium ion movements. These lead to a decrease in GABA release. If GHB receptors are desensitized by hyperstimulation (after the considerable increase in extracellular GHB concentration following GHB peripheral administration for therapeutic or recreative purposes), then the tonic inhibitory control of GHB receptor upon GABAergic neuron is lost and an increase in GABA release is registered. This could explain the neuropharmacological effects of exogenous GHB (sedation, anxiolysis, anesthesia, sleep modulation, absence epilepsy in rodents).

receptors over GABA and dopamine release via presynaptic GHB receptors. As a consequence, the release of both GABA and dopamine increases in various regions of the brain (Fig. 14.2). These increases could represent the basic phenomenon which explains both the therapeutic uses of high doses of GHB (sedation, anxiolysis, sleep induction) and the psychological benefits (euphoria, well-being state, induction of a Porsolt-positive test, preference for GHB in discriminative tests) (Zerbib *et al.* 1992; Colombo *et al.* 1995) observed after ingestion of several grams of GHB, but also the induction of absence epilepsy in rodents which is thought to be due to the potentiation of some $GABA_B$ pathways (Snead 1992).

Towards an explanation of the role of the GHB system in brain

Several pieces of evidence favor the regulated production in some region of the brain of the reductive catabolite of GABA metabolism whose levels fluctuate in the cytosol of some neurons. This fluctuation could be due to several phenomena, including the regulated transport of SSA outside the mitochondria, the intensity of SSADH and/or SSR and/or GABA-T activities. But the bioavailability of GABA, the precursor of GHB, could also influence the biosynthetic level of GHB. GABA is synthesized in the neuron via GAD activity but is also extracted from neighboring synapses by a family of transporters. The hyperactivity of these synapses induces the accumulation of transported GABA in neighboring cells and thus the increased accumulation of GABA catabolism. Then, GHB levels could be a direct index of the intensity of GABAergic mechanisms in closely related neurons.

In summary, the hyperactivity of GABA synapses could lead to an increased GHB level and release by specialized neurons in contact with GABA neurons. The increase in extracellular GHB will stimulate GHB receptors located presynaptically on the GABA neurons, inducing the hyperpolarization of this terminal and the reduction of calcium entry via calcium-dependent channels. This would in turn limit the release of GABA at the GABAergic synapse. Under these conditions, the activity of GABA synapses will induce a direct feed-back inhibitory control of its own activity via the synthesis and release of controlled amounts of GHB by neurons in the immediate vicinity (Fig. 14.3). This mechanism implicates a tonic inhibitory control of GHB synapses upon GABA release. This was confirmed by *in vivo* microdialysis in awake rats, provided the doses of GHB administered locally or peripherally remain low, close to the physiological amounts of GHB in brain.

If the extracellular amounts of GHB are greatly increased by exogenous administration of GHB (or in some pathological state like SSADH deficiency, see Gibson *et al.* 1984), then GHB receptors are desensitized within 10–20 min and are down-regulated thereafter. This desensitization would induce the disappearance of the tonic inhibitory control of GHB receptors on the GABA presynaptic element. As a consequence, both GABA release and GABAergic tone increase, leading to sedation, anesthesia and sometimes absence seizures. The equilibrium of GHB receptors between sensitization and desensitization followed by resensitization could be a mode for GABA release regulation in some regions of the brain. However, the precise events governing this phenomenon remain to be demonstrated, including the possible similar regulation of GABA$_B$ receptors whose sensitivity could be influenced by chronic GABA increase. But unlike most of G-protein-coupled receptor, GABA$_B$ R1a and R1b appear not to desensitize in heterologous expression systems.

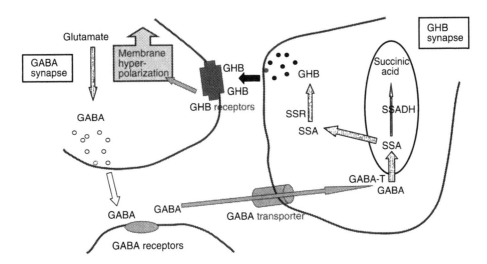

Figure 14.3 Sustained activity of a GABAergic synapse could be retroactively inhibited by a presynaptic input due to the stimulation of GHB receptors. The high concentrations of extracellular GABA due to the hyperactive GABA synapse lead to a high level of GABA uptake by local GHBergic neurons. Thus, GHB synthesis is probably potentiated by the regulated activity of GABA and/or SSA transport or by the regulated activity of GABA-T and/or SSR in the GHB neuron. As a consequence, the release of GHB is most probably potentiated, leading to the hyperpolarization of GABA neurons through GHB receptors and a decrease of GABA release.

References

Andriamampandry C., Siffert J. C., Schmitt M., Garnier J. M., Staub A., Muller C., Gobaille S., Mark J. and Maitre M. (1998) Cloning of a rat brain succinic semialdehyde reductase involved in the synthesis of the neuromodulator γ-hydroxybutyrate. *Biochem. J.* 334, 43–50.

Banerjee P. K. and Snead O. C. (1995) Presynaptic gamma-hydroxybutyric acid (GHB) and gamma-aminobutyric acid$_B$ (GABA$_B$) receptor-mediated release of GABA and glutamate (GLU) in rat thalamic ventrobasal nucleus (VB): a possible mechanism for the generation of absence-like seizures induced by GHB. *J. Pharmacol. Exper. Ther.* 273, 1534–43.

Benavides J., Rumigny J. F., Bourguignon J. J., Cash C., Wermuth C. G., Mandel P., Vincendon G. and Maitre M.(1982a) High affinity binding site for γ-hydroxybutyric acid in rat brain. *Life Sci.* 30, 953–61.

Benavides J., Rumigny J. F., Bourguignon J. J., Wermuth C. G., Mandel P. and Maitre M. (1982b) A high-affinity, Na$^+$-dependent uptake system for γ-hydroxybutyrate in membrane vesicles prepared from rat brain. *J. Neurochem.* 38, 1570–75.

Bernasconi R., Lauber J., Marescaux C., Vergnes M., Martin P., Rubio V., Leonhardt T., Reymann N. and Bittiger H. (1992) Experimental absence seizures: potential role of gamma-hydroxybutyric acid and GABA$_B$ receptors. *J. Neural Transm.* 35, 155–77.

Biggio G., Cibin M., Diana M., Fadda F., Ferrara S. D., Gallimberti L., Gessa G. L., Mereu G. P., Rossetti Z. L. and Serra M. (1992) Suppression of voluntary alcohol intake in rats and alcoholics by gamma-hydroxybutyric acid: a non-GABAergic mechanism. *Adv. Biochem. Psychopharmacol.* 47, 281–88.

Cash C. D., Maitre M. and Mandel P. (1979) Purification from human brain and some properties of two NADPH-linked aldehyde reductases which reduce succinic semialdehyde to 4-hydroxybutyrate. *J. Neurochem.* 33, 1169–75.

Cash C. D., Gobaille S., Kemmel V., Andriamampandry C. and Maitre M. (1999) γ-Hydroxybutyrate receptor function studied by the modulation of nitric oxide synthase activity in rat frontal cortex punches. *Biochem. Pharmacol.* 58, 1815–19.

Colombo G., Agabio R., Bourguignon J., Fadda F., Lobina C., Maitre M., Reali R., Schmitt M. and Gessa G. L. (1995) Blockade of the discriminative stimulus effects of γ-hydroxybutyric acid (GHB) by the GHB receptor antagonist NCS-382. *Physiol. Behav.* 58, 587–90.

Doherty J. D., Stout R. W. and Roth R. H. (1975) Metabolism of [1-^{14}C]γ-hydroxybutyric acid by rat brain after intraventricular injection. *Biochem. Pharmacol.* 24, 469–74.

Doherty J. D., Hattox S. E., Snead O. C. and Roth R. H. (1978) Identification of endogenous γ-hydroxybutyrate in human and bovine brain and its regional distribution in human, guinea pig and rhesus monkey brain. *J. Pharmacol. Exp. Ther.* 207, 130–39.

Engberg G. and Nissbrandt H. (1993) Gamma-hydroxybutyric acid (GHBA) induces pacemaker activity and inhibition of substantia nigra dopamine neurons by activating GABA$_B$-receptors. *Naunyn-Schmied. Arch. Pharmacol.* 348, 491–97.

Ferrara S. D., Zotti S., Tedeschi L., Frison G., Castagna F., Gallimberti L., Gessa G. L. and Palatini P. (1992) Pharmacokinetics of gamma-hydroxybutyric acid in alcohol dependent patients after single and repeated oral doses. *Br. J. Clin. Pharmacol.* 34, 231–35.

Gibson K. M., Sweetman L., Nyhan W. L. and Rating D. (1984) Succinic semialdehyde dehydrogenase deficiency. *J. Neurogenet.* 1, 213–18.

Gobaille S., Hechler V., Andriamampandry C., Kemmel V. and Maitre M. (1999) γ-Hydroxybutyrate modulates synthesis and extracellular concentration of γ-aminobutyric acid in discrete rat brain regions *in vivo. J. Pharmacol. Exp. Ther.* 290, 303–09.

Godbout R., Jelenic P., Labrie C., Schmitt M. and Bourguignon J. J. (1995) Effect of gamma-hydroxy-butyrate and its antagonist NSC-382 on spontaneous cell firing in the prefrontal cortex of the rat. *Brain Res.* 673, 157–60.

Gold B. I. and Roth R. H. (1977) Kinetics of *in vivo* conversion of γ-[^{3}H]aminobutyric acid to γ-[^{3}H]hydroxybutyric acid by rat brain. *J. Neurochem.* 28, 1069–73.

Harris N. C., Webb C. and Greenfield S. A. (1989) The effects of gamma-hydroxybutyrate on the membrane properties of guinea-pig pars compacta neurons in the substantia nigra *in vitro*. *Neuroscience* 31, 363–70.

Hechler V., Schmitt M., Bourguignon J. J. and Maitre M. (1990) Trans-gamma-hydroxycrotonic acid binding sites in brain: evidence for a subpopulation of gamma-hydroxybutyrate sites. *Neurosci. Lett.* 110, 204–09.

Hechler V., Gobaille S., Bourguignon J. J. and Maitre M. (1991) Extracellular events induced by gamma-hydroxybutyrate in striatum: a microdialysis study. *J. Neurochem.* 56, 938–44.

Hechler V., Gobaille S. and Maitre M. (1992) Selective distribution pattern of γ-hydroxybutyrate receptors in the rat forebrain and midbrain as revealed by quantitative autoradiography. *Brain Res.* 572, 345–48.

Hechler V., Ratomponirina C. and Maitre M. (1997) γ-Hydroxybutyrate conversion into GABA induces displacement of GABA$_B$ binding that is blocked by valproate and ethosuccimide. *J. pharmacol. Exp. Ther.* 281, 753–60.

Ishige K., Aizawa M., Ito Y. and Fukuda H. (1996) γ-Butyrolactone-induced absence-like seizures increase nuclear CRE- and AP-1 DNA-binding activities in mouse brain. *Neuropharmacology* 35, 45–55.

Ito Y., Ishige K., Zaitsu E., Anzai K. and Fukuda H. (1995) γ-Hydroxybutyric acid increases intracellular Ca^{2+}-concentration and nuclear cyclic AMP-responsive element- and activator protein 1 DNA-binding activities through GABA$_B$ receptor in cultured cerebellar granule cells. *J. Neurochem.* 65, 75–83.

Kam P. C. and Yoong F. F. (1998) Gamma-hydroxybutyric acid: an emerging recreational drug. *Anaesthesia* 53, 1195–98.

Kemmel V., Taleb O., Perard A., Andriamampandry C., Siffert J. C., Mark J. and Maitre M. (1998) Neurochemical and electrophysiological evidence for the existence of a functional γ-hydroxybutyrate system in NCB-20 neurons. *Neuroscience* 86, 989–1000.

Kozhechkin S. X. (1980) Microiontophoretic study of the mechanism of action of gamma-hydroxybutyric acid. *Bull. Exp. Biol. Med.* 88, 1293–96.

Laborit H. (1964) Sodium 4-hydroxybutyrate. *Int. J. Neuropharmacol.* 3, 433–52.

Laborit H. (1973) Gamma-hydroxybutyrate, succinic semialdehyde and sleep. *Prog. Neurobiol.* 1, 257–74.

Lettieri J. T. and Fung H. L. (1979) Dose-dependent pharmacokinetics and hypnotic effects of sodium γ-hydroxybutyrate in the rat. *J. Pharmacol. Exp. Ther.* 208, 7–11.

Lingenhoehl K., Brom R., Heid J., Beck P., Froestl W., Kaupmann K., Bettler B. and Mosbacher J. (1999) γ-Hydroxybutyrate is a weak agonist at recombinant GABA$_B$ receptors. *Neuropharmacology* 38, 1667–73.

McCormick S. J. and Tunnicliff G. (1998) Inhibitors of synaptosomal gamma-hydroxybutyrate transport. *Pharmacology* 57, 124–31.

Maitre M., Cash C., Weissmann-Nanopoulos D. and Mandel P. (1983) Depolarization-evoked release of γ-hydroxybutyrate from rat brain slices. *J. Neurochem.* 41, 287–90.

Maitre M., Hechler V., Vayer P., Gobaille S., Cash C. D., Schmitt M. and Bourguignon J. J. (1990) A specific gamma-hydroxybutyrate receptor ligand possesses both antagonistic and anticonvulsant properties. *J. Pharmacol. Exp. Ther.* 255, 657–63.

Maitre M., Ratomponirina C., Gobaille S., Hodé Y. and Hechler V. (1994) Displacement of [^{3}H]γ-hydroxybutyrate binding by benzamide neuroleptics and prochlorperazine but not other antipsychotics. *Eur. J. Pharmacol.* 256, 211–14.

Mamelak M., Scharf M. B. and Woods M. (1986) Treatment of narcolepsy with gamma-hydroxybutyrate. A review of clinical and sleep laboratory findings. *Sleep* 9, 285–89.

Mathivet P., Bernasconi R., De Barry J., Marescaux C. and Bittiger H. (1997) Binding characteristics of gamma-hydroxybutyric acid as a weak but selective GABA$_B$ receptor agonist. *Eur. J. Pharmacol.* 321, 67–75.

Nelson T., Kaufman E., Kline J. and Sokoloff L. (1981) The extraneural distribution of γ-hydroxybutyrate. *J. Neurochem.* 37, 1345–48.

Olpe H. R. and Koella W. P. (1979) Inhibition of nigral and neocortical cells by γ-hydroxybutyrate: a microiontophoretic investigation. *Eur. J. Pharmacol.* 53, 359–64.

Osorio I. and Davidoff R. A. (1979) γ-Hydroxybutyric acid is not a GABA-mimetic agent in the spinal cord. *Ann. Neurol.* 6, 111–16.

Ostrovskaya R. U. (1982) Pharmacological and electrophysiological differences between γ-hydroxybutyric acid and GABA-mimetic drugs. *Ann. Ist. Super Sanita* 18, 91–94.

Ratomponirina C., Hodé Y., Hechler V. and Maitre M. (1995) γ-Hydroxybutyrate receptor binding in rat brain is inhibited by guanyl nucleotides and pertussis toxin. *Neurosci. Lett.* 189, 51–53.

Ratomponirina C., Gobaille S., Hodé Y., Kemmel V. and Maitre M. (1998) Sulpiride, but not haloperidol, up-regulates γ-hydroxybutyrate receptors *in vivo* and in cultured cells. *Eur. J. Pharmacol.* 346, 331–37.

Roth R. H. (1970) Formation and regional distribution of γ-hydroxybutyric acid in mammalian brain. *Biochem. Pharmacol.* 19, 3013–19.

Roth R. H. and Giarman J. (1970) Natural occurrence of gamma-hydroxybutyrate in mammalian brain. *Biochem. Pharmacol.* 19, 1087–93.

Rumigny J. F., Maitre M., Cash C. and Mandel P. (1980) Specific and non-specific succinic semialdehyde reductases from rat brain: isolation and properties. *FEBS Lett.* 117, 111–16.

Rumigny J. F., Cash C., Mandel P., Vincendon G. and Maitre M. (1981) Evidence that a specific succinic semialdehyde reductase is responsible for γ-hydroxybutyrate synthesis in brain tissue slices. *FEBS Lett.* 134, 96–98.

Scharf M. B., Fletcher K. A. and Jennings S. W. (1988) Current pharmacologic management of narcolepsy. *Am. Fam. Physician* 38, 143–48.

Shumate J. S. and Snead O. C. (1979) Plasma and central nervous system kinetics of gamma-hydroxybutyrate. *Res. Commun. Chem. Pathol. Pharmacol.* 25, 241–56.

Snead O. C. (1987) γ-Hydroxybutyric acid in subcellular fractions of rat brain. *J. Neurochem.* 48, 196–201.

Snead O. C. (1992) Evidence for GABA_B-mediated mechanisms in experimental generalized absence seizures. *Eur. J. Pharmacol.* 213, 343–49.

Snead O. C. (1994) The ontogeny of [^{3}H] γ-hydroxybutyrate and [^{3}H]GABA_B binding sites: relation to the development of experimental absence seizures. *Brain Res.* 659, 147–56.

Snead O. C. and Liu C. C. (1984) Gamma-hydroxybutyric acid binding sites in rat and human brain synaptosomal membranes. *Biochem. Pharmacol.* 33, 2587–90.

Vayer P. and Maitre M. (1988) Regional differences in depolarization-induced release of γ-hydroxybutyrate from rat brain slices. *Neurosci. Lett.* 87, 99–103.

Vayer P., Mandel P. and Maitre M. (1985a) Conversion of γ-hydroxybutyrate to γ-aminobutyrate *in vitro*. *J. Neurochem.* 45, 810–14.

Vayer P., Schmitt M., Bourguignon J. J., Mandel P. and Maitre M. (1985b) Evidence for a role of high Km aldehyde reductase in the degradation of endogenous γ-hydroxybutyrate from rat brain. *FEBS Lett.* 190, 55–60.

Vayer P., Gobaille S., Mandel P. and Maitre M. (1987a) 3′–5′ Cyclic-guanosine monophosphate increase in rat brain hippocampus after gamma-hydroxybutyrate administration. Prevention by valproate and naloxone. *Life Sci.* 41, 605–10.

Vayer P., Mandel P. and Maitre M. (1987b) Gamma-hydroxybutyrate, a possible neurotransmitter. *Life Sci.* 41, 1547–57.

Vayer P., Ehrhardt J. D., Gobaille S., Mandel P. and Maitre M. (1988) Gamma hydroxybutyrate distribution and turnover rates in discrete brain regions of the rat. *Neurochem. Int.* 12, 53–59.

Waldmeier P. C. (1991) The GABA-B antagonist, CGP 35348, antagonizes the effects of baclofen, γ-butyrolactone and HA 966 on rat striatal dopamine synthesis. *Naunyn-Schmied. Arch. Pharmacol.* 343, 173–78.

Waldmeier P. C. and Baumann P. A. (1990) GABA_B receptors and transmitter release. In *GABA_B Receptors in Mammalian Function* (Bowery N.G., Bittiger H. and Olpe H.R., eds), pp. 63–80. John Wiley, UK.

Xie X. and Smart T. G. (1992) Gamma-hydroxybutyrate hyperpolarizes hippocampal neurones by activating GABA$_B$ receptors. *Eur. J. Pharmacol.* 212, 291–94.

Zerbib R., Pierrefiche G., Ferran C. and Laborit H. (1992) Potential antidepressant activity of gamma-hydroxybutyrate in the mouse 'behavioral despair' test: correlation with the central dopaminergic system. *Res. Commun. Psychol. Psychiat. Behav.* 17, 109–22.

Index